AF576042

New Insights into Molecular Mechanisms Underlying Neurodegenerative Disorders

New Insights into Molecular Mechanisms Underlying Neurodegenerative Disorders

Editors

Chiara Villa
Yam Nath Paudel
Christina Piperi

MDPI • Basel • Beijing • Wuhan • Barcelona • Belgrade • Manchester • Tokyo • Cluj • Tianjin

Editors

Chiara Villa
School of Medicine
and Surgery
University of Milano-Bicocca
Monza
Italy

Yam Nath Paudel
Department of Neurology
Mayo Clinic
Rochester
United States

Christina Piperi
Department of Biological
Chemistry
University of Athens
Medical School
Athens
Greece

Editorial Office
MDPI
St. Alban-Anlage 66
4052 Basel, Switzerland

This is a reprint of articles from the Special Issue published online in the open access journal *Brain Sciences* (ISSN 2076-3425) (available at: www.mdpi.com/journal/brainsci/special_issues/Neurodegenerative_Mechanisms).

For citation purposes, cite each article independently as indicated on the article page online and as indicated below:

LastName, A.A.; LastName, B.B.; LastName, C.C. Article Title. *Journal Name* **Year**, *Volume Number*, Page Range.

ISBN 978-3-0365-5486-0 (Hbk)
ISBN 978-3-0365-5485-3 (PDF)

Cover image courtesy of Chiara Villa

© 2022 by the authors. Articles in this book are Open Access and distributed under the Creative Commons Attribution (CC BY) license, which allows users to download, copy and build upon published articles, as long as the author and publisher are properly credited, which ensures maximum dissemination and a wider impact of our publications.
The book as a whole is distributed by MDPI under the terms and conditions of the Creative Commons license CC BY-NC-ND.

Contents

About the Editors

Chiara Villa

Dr. Chiara Villa is currently Assistant Professor in Pathology at University of Milano-Bicocca. She started her research activity in 2006 at Dept. of Neurological Sciences, University of Milan, Fondazione Cà Granda IRCCS Ospedale Maggiore Policlinico, where she obtained her PhD in Molecular Medicine in 2009. Her research activity was mainly focused on the study of novel biomarkers (e.g., microRNAs, pro-inflammatory cytokines) and genetic risk factors in key genes for early detection and/or progression of two neurodegenerative diseases: Alzheimer's disease and frontotemporal lobar degeneration. In 2014, she moved to the University of Milano-Bicocca and joined a research group working in the study of molecular bases of autism spectrum disorders and sleep disorders, including restless legs syndrome and nocturnal frontal lobe epilepsy. Her work has been now carried out by applying her previous research experience in this field through a molecular approach.

At the moment, she is co-author of 71 research articles published in peer-reviewed journals with a personal H-index of 23.

Yam Nath Paudel

Dr. Yam Nath Paudel is currently a Postdoctoral research fellow at Department of Neurology, Mayo Clinic, Rochester, USA. Dr. Paudel graduated with a PhD in Neuroscience from Monash University in June 2021, where his research work was mainly focused on investigating the role of neuroinflammatory protein HMGB1 against chronic epilepsy and related cognitive decline. Dr Paudel has an immense interest in understanding the underlying mechanism of neurodegenerative diseases and developing effective disease modifying strategies against the same.

Dr. Paudel has published around 65 scholarly publications to date in journals of high repute and his work has gained a citation of around 1125 with H-index of 16 and i10 index of 31.

Christina Piperi

Dr. Christina Piperi is currently holding a position of Professor in Medical Biochemistry at the National and Kapodistrian University of Athens. She started her research activity in 1995 at the Sir William Dunn School of Pathology of the University of Oxford, where she obtained her PhD in Molecular and Cellular Immunology in 1999. Her research project was focused on structural and functional studies of the CD22 molecule in B cell leukemia. In 2000, she carried on her studies on novel biomarkers in lung cancer at the Institute of Biomedical Research and Biotechnology in Athens, and in 2004, she obtained a lecturer position at the Department of Biological Chemistry of the University of Athens. Her research was mainly focused on the study of novel biomarkers (e.g., advanced glycation end products, pro-inflammatory cytokines, transcription factors) and genetic risk factors for early detection and/or progression of metabolic diseases and neurodegenerative disorders. In 2007, she started to work on signal transduction mechanisms, and epigenetic alterations involved in the pathogenesis of neurodegenerative diseases and human cell carcinogenesis (gliomas, melanomas, colon cancer). Currently, she is the principal investigator of molecular oncology group at the Department of Biological Chemistry. Her research work is focused on chromatin remodeling alterations and histone modifications implicated in the pathology of pediatric and adult brain tumors.

At present, she has co-authored 232 articles published in peer-reviewed journals with a personal H-index of 50.

MDPI

Editorial

New Insights into Molecular Mechanisms Underlying Neurodegenerative Disorders

Chiara Villa [1,*], Yam Nath Paudel [2,*] and Christina Piperi [3,*]

1 School of Medicine and Surgery, University of Milano-Bicocca, 20900 Monza, Italy
2 Neuropharmacology Research Laboratory, Jeffrey Cheah School of Medicine and Health Sciences, Monash University Malaysia, Bandar Sunway 47500, Malaysia
3 Department of Biological Chemistry, Medical School, National and Kapodistrian University of Athens, 11527 Athens, Greece
* Correspondence: chiara.villa@unimib.it (C.V.); yam.paudel@monash.edu (Y.N.P.); cpiperi@med.uoa.gr (C.P.)

Citation: Villa, C.; Paudel, Y.N.; Piperi, C. New Insights into Molecular Mechanisms Underlying Neurodegenerative Disorders. *Brain Sci.* **2022**, *12*, 1190. https://doi.org/10.3390/brainsci12091190

Received: 30 July 2022
Accepted: 4 August 2022
Published: 3 September 2022

Publisher's Note: MDPI stays neutral with regard to jurisdictional claims in published maps and institutional affiliations.

Copyright: © 2022 by the authors. Licensee MDPI, Basel, Switzerland. This article is an open access article distributed under the terms and conditions of the Creative Commons Attribution (CC BY) license (https://creativecommons.org/licenses/by/4.0/).

Neurodegenerative disorders remain a major burden for our society, affecting millions of people worldwide. Due to the symptomatic nature of current treatments without affecting the underlying cause of the disease, these disorders present an ongoing clinical challenge. Extensive research efforts demonstrate the important impact of molecular mechanisms in driving the primary pathological aspects of neurodegenerative diseases and the need for g reater understanding to facilitate targeted drug development.

This Special Issue collects six reviews, five original research articles and one communication in order to provide new insights in the molecular mechanisms that underlie neurodegenerative disorders.

The first entry is an elaborate review of signal transduction mediated by Toll-like receptors (TLRs) on the resident macrophages of the central nervous system (CNS) and neurons bridge immune system response to the pathogenesis of neurodegenerative disorders. A critical evaluation of several studies from animal models and humans highlights the important role of TLR2, TLR3, TLR4, TLR7, and TLR9 in Parkinson's disease (PD) and Alzheimer's disease (AD) [1].

In turn, clinical research studies indicate the emerging role of the anti-inflammatory cytokine IL-37 in common CNS diseases. Although IL-37 has been previously associated with the pathogenesis of autoimmune diseases and cancer, the present review by Li et al. focuses on the mechanism of action of IL-37 in CNS and the inhibition of inflammatory cytokines, such as IL-1β, IL-6, and tumor necrosis factor-α (TNF-α), indicating therapeutic potential [2].

In search of the epigenetic mechanisms associated with the pathogenesis of PD, Angelopoulou et al. explored the impact of environmental exposures in altering gene expression [3]. Smoking, coffee consumption, pesticide exposure, and heavy metals (manganese, arsenic, lead, etc.) have been revealed as potential epigenetic modifiers underlying PD development, stimulating the prospects for future research in this direction.

Novel applications of the graph theory and the contribution of electrophysiological techniques in neurodegenerative disorders are further described in the article of Miraglia et al. [4]. They particularly explore the graph theory as an emerging method for the study of functional connectivity in electrophysiological recordings, providing a simple representation of a complex system applied in AD and PD. The benefits of electrophysiological techniques, including their low cost, broad availability, and non-invasive nature, make them potential tools for large population screening.

The potential of the human monoclonal antibody, Aducanumab, as the first approved disease-modifying treatment for AD has been addressed in the article of Gunawardena et al. A critical overview of the promises and controversies associated with Aducanumab in low- and middle-income countries is given, along with contradicting evidence from two clinical trials [5].

As emerging evidence supports the notion that infections by human herpesviruses (HHVs) have been involved in the AD pathogenesis, other authors highlight the current knowledge about the potential molecular interplay between HHVs and AD. In particular, they focus on the main pathological processes of AD, including amyloid beta deposition, tau protein hyperphosphorylation, oxidative stress, autophagy, and neuroinflammation. Therefore, a deeper understanding of these links may help the identification of novel therapeutic targets to prevent or halt the neurodegenerative processes in AD [6].

Moving towards novel research articles, changes in the activity of the rate-limiting enzyme of dopamine synthesis, tyrosine hydroxylase, in the nigrostriatal system of mice has been associated with neurodegeneration and neuroplasticity in an acute model of PD. Detecting differences in the regulation of dopamine synthesis between DA-neuron bodies and their axons can be further considered for the development of symptomatic pharmacotherapy aimed at increasing tyrosine hydroxylase activity [7].

An Italian longitudinal study investigating gut microbiota alterations in fecal samples of PD patients over the period of a year demonstrated stability in microbiota findings. Any differences in the microbiota composition between PD patients and healthy controls also remained stable, without the detection of any worsening in the disease staging or motor impairment in PD patients, paving the way to more extensive longitudinal evaluations [8].

An elegant two-sample Mendelian randomization study with summary statistics from large-scale genome-wide association studies (GWAS) detected a shared genetic background between PD and schizophrenia. Kim et al. evaluated whether genetic variants which increase PD risk influence the risk of developing schizophrenia, and vice versa and detected increased risk of schizophrenia per one-standard deviation (SD) increase in the genetically predicted PD risk. This evidence supports the intrinsic nature of the psychotic symptoms in PD and points out that future studies are needed to investigate possible comorbidities and shared genetic structure between the two diseases [9].

Another intriguing research study employed homology modelling, molecular docking, and molecular dynamics simulation of Calcium homeostasis modulator 1 (CALHM1) in order to test secondary metabolites of *Bauhinia variegata* for AD treatment. Among various flavonoids and alkaloids from *Bauhinia variegata*, quercetin was revealed as a good inhibitor for treating AD, requiring future in vitro and in vivo analyses in order to confirm its effectiveness [10].

A transcriptome sequencing study was performed to screen differentially ex-pressed circular RNAs (DEcircRNAs) in the brains of a rat model of levodopa-induced dyskinesia (LID), a common complication after chronic dopamine-replacement therapy in the treatment of PD. Among a set of 99 DEcircRNAs in the striatum of LID rats, the authors identified high levels of rno-Rsf1_0012 which can competitively bind rno-mir-298-5p, thus abolishing its inhibitory effect on the expression of the target genes, *PCP4* and *TBP*, already associated with other movement disorders. Although these promising results, further investigations are needed to clarify the specific roles of rno-Rsf1_0012 in LID occurrence [11].

Lastly, an important communication proposes a theoretical framework to explain the stochastic processes, at the protein, DNA and RNA levels, which are involved in the development of adult sporadic neurodegenerative disorders [12]. This model of interacting degenerative proteins helps us to elucidate the existence of multiple misfolded proteinopathies in adult sporadic neurodegenerative disorders and may prove highly valuable in the future.

In conclusion, neurodegenerative diseases need integrative understanding of the underlying molecular mechanisms at both the theoretical and practical levels, in order to enable successful clinical management. This Special Issue provides up-to-date information on several aspects of the molecular pathology that underlies some major neurodegenerative disorders, in order to reinforce current thinking and therapeutic approaches.

Author Contributions: C.V., Y.N.P. and C.P. conceptualized and wrote the manuscript. All authors have read and agreed to the published version of the manuscript.

Funding: This research received no external funding.

Institutional Review Board Statement: Not applicable.

Data Availability Statement: Not applicable.

Conflicts of Interest: The authors declare no conflict of interest.

References

1. Adhikarla, S.V.; Jha, N.K.; Goswami, V.K.; Sharma, A.; Bhardwaj, A.; Dey, A.; Villa, C.; Kumar, Y.; Jha, S.K. TLR-Mediated Signal Transduction and Neurodegenerative Disorders. *Brain Sci.* **2021**, *11*, 1373. [CrossRef] [PubMed]
2. Li, X.; Yan, B.; Du, J.; Xu, S.; Liu, L.; Pan, C.; Kang, X.; Zhu, S. Recent Advances in Progresses and Prospects of IL-37 in Central Nervous System Diseases. *Brain Sci.* **2022**, *12*, 723. [CrossRef] [PubMed]
3. Angelopoulou, E.; Paudel, Y.N.; Papageorgiou, S.G.; Piperi, C. Environmental Impact on the Epigenetic Mechanisms Underlying Parkinson's Disease Pathogenesis: A Narrative Review. *Brain Sci.* **2022**, *12*, 175. [CrossRef]
4. Miraglia, F.; Vecchio, F.; Pappalettera, C.; Nucci, L.; Cotelli, M.; Judica, E.; Ferreri, F.; Rossini, P.M. Brain Connectivity and Graph Theory Analysis in Alzheimer's and Parkinson's Disease: The Contribution of Electrophysiological Techniques. *Brain Sci.* **2022**, *12*, 402. [CrossRef]
5. Gunawardena, I.P.C.; Retinasamy, T.; Shaikh, M.F. Is Aducanumab for LMICs? Promises and Challenges. *Brain Sci.* **2021**, *11*, 1547. [CrossRef] [PubMed]
6. Athanasiou, E.; Gargalionis, A.N.; Anastassopoulou, C.; Tsakris, A.; Boufidou, F. New Insights into the Molecular Interplay between Human Herpesviruses and Alzheimer's Disease—A Narrative Review. *Brain Sci.* **2022**, *12*, 1010. [CrossRef]
7. Kolacheva, A.; Alekperova, L.; Pavlova, E.; Bannikova, A.; Ugrumov, M.V. Changes in Tyrosine Hydroxylase Activity and Dopamine Synthesis in the Nigrostriatal System of Mice in an Acute Model of Parkinson's Disease as a Manifestation of Neurodegeneration and Neuroplasticity. *Brain Sci.* **2022**, *12*, 779. [CrossRef] [PubMed]
8. Cerroni, R.; Pietrucci, D.; Teofani, A.; Chillemi, G.; Liguori, C.; Pierantozzi, M.; Unida, V.; Selmani, S.; Mercuri, N.B.; Stefani, A. Not just a Snapshot: An Italian Longitudinal Evaluation of Stability of Gut Microbiota Findings in Parkinson's Disease. *Brain Sci.* **2022**, *12*, 739. [CrossRef] [PubMed]
9. Kim, K.; Kim, S.; Myung, W.; Shim, I.; Lee, H.; Kim, B.; Cho, S.K.; Yoon, J.; Kim, D.K.; Won, H.-H. Shared Genetic Background between Parkinson's Disease and Schizophrenia: A Two-Sample Mendelian Randomization Study. *Brain Sci.* **2021**, *11*, 1042. [CrossRef] [PubMed]
10. Khare, N.; Maheshwari, S.K.; Rizvi, S.M.D.; Albadrani, H.M.; Alsagaby, S.A.; Alturaiki, W.; Iqbal, D.; Zia, Q.; Villa, C.; Jha, S.K.; et al. Homology Modelling, Molecular Docking and Molecular Dynamics Simulation Studies of CALMH1 against Secondary Metabolites of Bauhinia variegata to Treat Alzheimer's Disease. *Brain Sci.* **2022**, *12*, 770. [CrossRef] [PubMed]
11. Han, C.-L.; Wang, Q.; Liu, C.; Li, Z.-B.; Du, T.-T.; Sui, Y.-P.; Zhang, X.; Zhang, J.-G.; Xiao, Y.-L.; Cai, G.-E.; Meng, F.-G. Transcriptome Sequencing Reveal That Rno-Rsf1_0012 Participates in Levodopa-Induced Dyskinesia in Parkinson's Disease Rats via Binding to Rno-mir-298-5p. *Brain Sci.* **2022**, *12*, 1206. [CrossRef]
12. Panegyres, P.K. Stochasticity, Entropy and Neurodegeneration. *Brain Sci.* **2022**, *12*, 226. [CrossRef] [PubMed]

MDPI

Review

TLR-Mediated Signal Transduction and Neurodegenerative Disorders

Shashank Vishwanath Adhikarla [1], Niraj Kumar Jha [2], Vineet Kumar Goswami [3,4], Ankur Sharma [4,5], Anuradha Bhardwaj [2], Abhijit Dey [6], Chiara Villa [7], Yatender Kumar [1,*] and Saurabh Kumar Jha [2,4,*]

1 Department of Biological Sciences and Engineering, Netaji Subhas University of Technology (Formerly NSIT, University of Delhi), New Delhi 110078, India; adhikarlav.bt.16@nsit.net.in
2 Department of Biotechnology, School of Engineering & Technology (SET), Sharda University, Greater Noida 201310, India; niraj.jha@sharda.ac.in (N.K.J.); anu.consensus@gmail.com (A.B.)
3 Department of Biotechnology, Delhi Technological University, Delhi 110042, India; vineetgoswami@gmail.com
4 Department of Science and Engineering, Novel Global Community Educational Foundation, Hebersham 2770, Australia; ankur.sharma7@sharda.ac.in
5 Department of Life Science, School of Basic Science & Research (SBSR), Sharda University, Greater Noida 201310, India
6 Department of Life Sciences, Presidency University, 86/1 College Street, Kolkata 700073, India; abhijit.dbs@presiuniv.ac.in
7 School of Medicine and Surgery, University of Milano-Bicocca, 20900 Monza, Italy; chiara.villa@unimib.it
* Correspondence: yatender.kumar@nsut.ac.in (Y.K.); saurabh.jha@sharda.ac.in (S.K.J.)

Abstract: A special class of proteins called Toll-like receptors (TLRs) are an essential part of the innate immune system, connecting it to the adaptive immune system. There are 10 different Toll-Like Receptors that have been identified in human beings. TLRs are part of the central nervous system (CNS), showing that the CNS is capable of the immune response, breaking the long-held belief of the brain's "immune privilege" owing to the blood–brain barrier (BBB). These Toll-Like Receptors are present not just on the resident macrophages of the central nervous system but are also expressed by the neurons to allow them for the production of proinflammatory agents such as interferons, cytokines, and chemokines; the activation and recruitment of glial cells; and their participation in neuronal cell death by apoptosis. This study is focused on the potential roles of various TLRs in various neurodegenerative diseases such as Parkinson's disease (PD) and Alzheimer's disease (AD), namely TLR2, TLR3, TLR4, TLR7, and TLR9 in AD and PD in human beings and a mouse model.

Keywords: TLRs; neuroinflammation; neuroprotection; neurodegeneration Parkinson's disease; Alzheimer's disease

Citation: Adhikarla, S.V.; Jha, N.K.; Goswami, V.K.; Sharma, A.; Bhardwaj, A.; Dey, A.; Villa, C.; Kumar, Y.; Jha, S.K. TLR-Mediated Signal Transduction and Neurodegenerative Disorders. *Brain Sci.* **2021**, *11*, 1373. https://doi.org/10.3390/brainsci11111373

Academic Editor: David Brown

Received: 22 September 2021
Accepted: 16 October 2021
Published: 20 October 2021

Publisher's Note: MDPI stays neutral with regard to jurisdictional claims in published maps and institutional affiliations.

Copyright: © 2021 by the authors. Licensee MDPI, Basel, Switzerland. This article is an open access article distributed under the terms and conditions of the Creative Commons Attribution (CC BY) license (https://creativecommons.org/licenses/by/4.0/).

1. Introduction

Toll-Like Receptors (TLRs) are a family of evolutionarily conserved transmembrane proteins and are classified as membrane-spanning pattern recognition receptor (PRR) proteins [1]. They are mammalian orthologs of *Drosophila melanogaster*'s Toll receptors [1]. TLRs are a part of the innate immune system and are pattern recognition receptors that recognize small molecular motifs from microorganisms called pathogen-associated molecular patterns or PAMPs, and molecules produced endogenously by tissue during inflammation called damage-associated molecular patterns, or DAMPs [2]. TLRs can initiate an acute immune response, and hence stimulate, coordinate, and tune the quality of the adaptive immune response [3].

To date, 19 TLRs have been observed and reported in humans, though many different TLRs are also present in other mammals in different combinations with different functions [4]. The different TLRs are denoted with a different number suffix added to "TLR". In humans and mice together, we have observed 13 different TLRs: TLR1 to TLR13

(Table 1). Different TLRs are expressed in different species. For example, mice have a gene for TLR10, which does not seem to be expressed due to damage caused by a retrovirus during the course of their evolution [5]. On the other hand, humans do not seem to have genes for TLR11, TLR12, and TLR13, which are present and expressed in mice. Other non-mammalian species also have TLRs which are distinct from those of mammals, such as TLR14 expressed in *Takifugu* pufferfish, which recognizes cell-wall components [6].

Table 1. The different TLRs found in humans and mice, and the ligands that activate those TLRs (TLR10 is not functional in mice, while TLR11 is not detected in humans).

TLR	CD	Ligand	Pathogen Recognised	Localisation
TLR1	CD281	Tri-acyl lipopeptides	Gram-positive bacteria	Extracellular
TLR2	CD282	Di- and tri-acyl Lipopeptides	Gram-positive bacteria; plasmodium	Extracellular
TLR3	CD283	dsRNA	dsRNA viruses	Endosomal
TLR4	CD284	Lipopolysaccharide	Gram-negative bacteria	Extracellular
TLR5	CD285	Flagellin	Motile bacteria	Extracellular
TLR6	CD286	Di-acyl lipopeptides	Gram-positive bacteria	Extracellular
TLR7	CD287	ssRNA	ssRNA viruses	Endosomal
TLR8	CD288	ssRNA; GC rich RNA	ssRNA viruses	Endosomal
TLR9	CD289	CpG DNA	DNA viruses; viral and bacterial DNA	Endosomal
TLR10	CD290	Unknown	-	Extracellular

Even though TLRs are transmembrane proteins, it is a common misconception that they are all expressed on the cell surface. Some TLRs are expressed inside the cell and are localized within the endosomal compartment of cells, or are present on the membrane of the endosome [7]. TLRs which recognize bacterial and yeast cell products such as peptidoglycans and lipopolysaccharides are present on the cell surface, while oligonucleotide-sensing TLRs are present inside the cell. TLR1, TLR2, TLR4, TLR5, and TLR6 are present on the plasma membrane of mammalian cells [8,9], while TLR3, TLR7, TLR8, and TLR9 are localized within the cells in the endosome [9,10].

TLRs are Type-1 transmembrane proteins containing three structural domains: an extracellular leucine-rich repeat (LRR) motif, a transmembrane domain, and an intracellular cytoplasmic Toll/interleukin-1 receptor (TIR) domain. The pattern recognition of TLRs is brought about by the LRR domain, while the TIR domain interacts with various downstream adaptors and initiates signal transduction [11].

TLRs make use of a variety of adaptors, often in combination with a diverse range of signaling. The most commonly used adaptor is the myeloid differentiation primary response 88 factor (MyD88) [12]. MyD88 is a pivotal molecule in the modulation of the innate immune response because it is the adaptor known to transduce the signal from TLRs by activation of IL-1 receptor associated kinases (IRAKs) via homotypic protein–protein interaction, which eventually leads to the activation of nuclear factor-kappa B (NFκB), mitogen-activated protein kinases (MAP kinases), and activator protein 1 (AP1), making MyD88 a central node for inflammation pathways [13]. MyD88 has two functional domains: the C-terminal TIR domain which allows it to interact with other TIR-containing receptors and adaptors, and an N-terminal death domain (DD), which is involved in the interaction with IRAKs. The DD of MyD88 can independently activate NFκB and c-Jun N-terminal kinase (JNK). The TLR1–TLR2 heterodimer and the TLR2–TLR6 heterodimer make use of TIRAP and MyD88 as adaptors for transduction, while TLR5, TLR7, TLR8, and TLR9 solely make use of MyD88 [14,15].

TLR3 is unique in the sense that it makes use of MyD88-independent signaling pathways, using TRIF as an adaptor [16]. TLR4 is only one of its kind, as it utilizes both

MyD88-dependent and MyD88-independent pathways [17]. However, there are a variety of other adaptors which TLRs can make use of, apart from their primary adapters (MyD88 and TRIF), such as adapter protein 3 (AP3) used by TLR9 [15]. Various TLRs and their respective adaptors are represented in Figure 1.

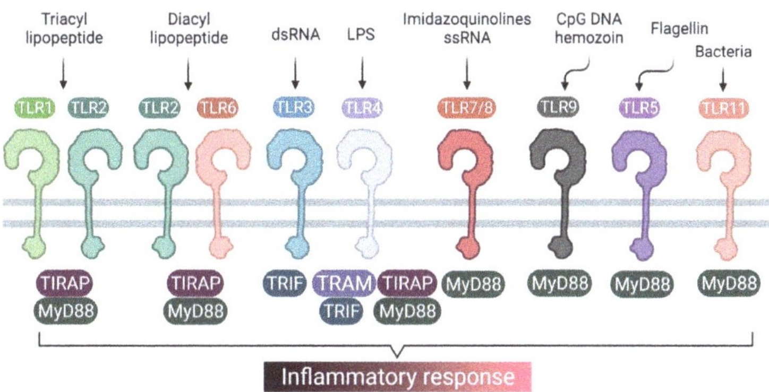

Figure 1. TLRs and their different adaptors in signal transduction pathways.

The MyD88 pathway, which makes use of MyD88 and TIRAP (Figure 2), activates IRAKs, which, in turn, phosphorylate IκBα protein, which is an inhibitor protein that binds to NFκB, masking its NLS signal and keeping it within the cytoplasm in the inactivated form [18]. Phosphorylation of IκBα results in the release of free NFκB. NFκB is a transcription factor that promotes the transcription of and results in the production of pro-inflammatory cytokines such as TNFα, IL1β, and IL6. The MyD88-independent pathway, which makes use of TRIF as an adaptor, eventually leads to the activation of IRF3, which leads to the production of interferon alpha and beta and other interferon-induced genes. IRF3 plays a crucial role in the production of the antiviral Type 1 interferon [19].

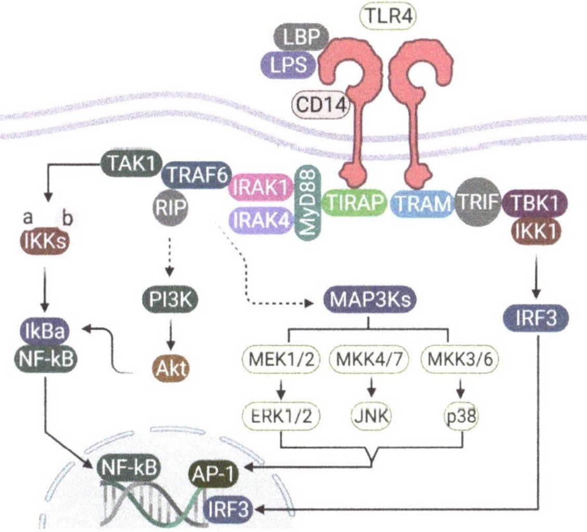

Figure 2. Schematic representation of the various elements involved in MyD88-dependent and MyD88-independent signal transduction, using TLR4 as a model molecule.

2. TLRs in the Nervous System

For quite a long time, the brain was considered to be immunologically privileged, as there seems to be no or very limited immune response within the brain. It made sense, as the brain is a sensitive organ having tissues with poor regenerative capacity. The selective nature of the blood–brain barrier made the belief even stronger. However, over the years, it has been realized that the immunological privilege of the brain is not absolute, and is rather complex, compartmentalized, and region-specific. The "privilege" comes not from the absence of immunological components but a very elaborate and dynamic regulation process. This regulation process is indispensable for damage limitation and mitigation in this sensitive organ [20]. It has also been realized that the cells of the central nervous system (CNS) are quite capable of mounting a dynamic immune response to a variety of stimuli [21].

TLRs play an essential role in this complex regulation within the central and peripheral nervous systems. The expression of TLRs on cells of both the central and peripheral nervous system have been observed and reported [22–28]. Various cell lines and animal models have been used to study the expression of TLRs in the cells of the nervous system. The expression of TLRs within the nervous systems of mice and humans has been found to be very different, though many of those differences arise from the lack of in vivo studies in humans, with the expression of many TLRs having not yet been studied in the human body. Cell lines, though very helpful in studying TLRs, are not a very accurate representation of in vivo cells owing to the chromosomal aberrations, changes in gene expression, and physiological differences between cancerous cell lines and actual neurons and glia.

In the central nervous system, neurons express TLRs 1 to 9 in both mice and humans. Microglia cells also express TLRs 1 to 9 in both mice and humans. Astrocytes, though known to express TLRs 1 to 9 in mice, only show significant levels of TLRs 1, 3, 4, 5, and 9, while TLRs 2, 6, 7, and 8 have not been reported yet. Oligodendrocytes have also been reported to express only TLR2 and TLR3 in humans, while the other TLRs are yet to be detected. In the peripheral nervous system, the neurons and the resident macrophages express TLRs 1 to 9 in both mice and humans. While the Schwann cells in mice are known to express TLRs 1 to 9, only TLR2 has been very well studied and characterized in human Schwann cells [21,29,30].

In the enteric nervous system (ENS), which is part of the peripheral nervous system of human beings, glial cells are known to express TLRs 2 to 9, while TLR1 seems to be absent in glia. Neurons in the ENS express TLRs 1 to 9 [29]. It is important to note that TLR signaling in neurons does not necessarily use IKKs and NFκB, and may involve the glycogen synthase kinase 3β (GSK3β), Jun N-terminal kinase (JNK), and phosphatidylinositol 3-kinase/protein kinase B (PI3K/AKT) pathways as well.

3. TLRs and Alzheimer's Disease

3.1. TLR2

TLR2 recognizes bacterial lipopeptides and peptidoglycans. TLR2 forms heterodimers with TLR1 and TLR6, mediating the response against Gram-positive bacteria and yeast. TLR2 works via the MyD88-dependent pathway, using TIRAP-MyD88 as adaptors and stimulating the production of proinflammatory cytokines using NFκB. TLR2 was found to be upregulated in the microglia surrounding amyloid β (Aβ) plaques, both in human post mortem brains as well as Alzheimer's Disease (AD)mouse models. Aβ cannot induce an inflammatory response in microglia deficient of TLR2 or in the cortex of TLR2-deficient mice, which acts as further proof that TLR2 has a role in AD [31]. TLR2-deficient AD mouse models show more pronounced cognitive impairment because of greater levels of Aβ proteins and greater white-matter damage [32,33]. The microglial phagocytic response to Aβ is also TLR2-dependent, as Aβ acts as an agonist for microglial TLR2 [34]. Neuronal viability in AD also seems to be affected by microglial TLR2 activation, since neuronal death was shown to be conferred by the release of pro-inflammatory signals by activated microglia [35]. Table 2 summarizes the various possible effects that different TLRs can have on AD.

Neuronal TLR2 can also be implicated in the inflammatory response against Aβ in AD, since TLR2 in neurons is upregulated when neurons are exposed to AD-specific metabolites such as 4-hydroxynonenal (HNE; an AD-related lipid peroxidation product). HNE exposure to neurons also leads to an increase in phosphorylated JNK and cleaved caspase 3, which could indicate AD-related metabolites causing apoptosis in neurons via TLR2 activation [36].

3.2. TLR4

TLR4 received attention when an Italian study found that the TLR4 polymorphism is responsible for late-onset of Alzheimer's disease in the Italian population [37]. An Asp299Gly mutation in TLR4 could confer neuroprotection, since it renders the TLR less responsive to lipopolysaccharides by bringing about structural changes [38]. Another study of post mortem human brains showed TLR4 to be upregulated in the glia surrounding Aβ plaques [39]. Knocking out TLR4 in AD mouse models showed reduced expression of TNFα and the chemokine macrophage inflammatory protein 1β in the cortex [40], though knocking TLR4 out of AD mice led to an increase in the activated microglia, astrocytes, and Aβ protein in the brain [41]. Data from in vitro cultures of microglia revealed TLR4-mediated activation of the microglia in the degeneration of neurons. Microglia seem to play a pro-inflammatory role in AD and have a phagocytic response to Aβ through TLR4, resulting in neuronal cell death. Much like the case of TLR2, neurons respond to Aβ and AD-related metabolites through TLR4, leading to their apoptosis [36]. TLR4 seems to induce a pro-inflammatory response against Aβ aiming to stimulate the activated microglia to clear out the Aβ through microglial uptake. A leading hypothesis regarding the role of TLR4 in AD is that insufficient removal of Aβ leads to its accumulation in the intercellular space, which subsequently activates the microglia and astrocytes which induce apoptosis in affected neurons, causing neuronal cell death.

3.3. TLR7

TLR7 is unique, since this TLR induces neuronal cell death without the activation of glia. More interestingly, the TLR itself is activated by endogenous overexpression of a microRNA, making TLR7′s case very different from that of TLR2 and TLR4. Let-7 is a family of microRNAs (miRNA) which are highly abundant in the brain. The family is known to have a specific GUUGUGU motif in the core sequence of the miRNA. The GUUGUGU motif just happens to match the sequence ssRNA40, a motif in the HIV ssRNA that TLR7 recognizes. It has been found that Let-7; specifically Let-7b, is overexpressed in the neurons of patients with Alzheimer's disease. This overexpression seems to cause TLR7 to become activated, triggering the production of cytokines such as TNFα, eventually leading to the production of cleaved caspases that cause neurons to undergo apoptosis [10].

Both in vitro and in vivo studies have shown that neither microglia nor astrocytes have a role in neurodegeneration via TLR7 activation. It was also shown that extracellular Let-7 is capable of activating TLR7, essentially showing that neurons that undergo apoptosis release Let-7, which stimulates the TLR7 of neighboring neurons and causes them to undergo apoptosis as well. Beside Let-7, overexpression of any miRNAs with a seed sequence having the GUUGUGU motif such as miR-599 could induce the production of TNFα and negatively affect neuronal viability both in vitro and in vivo [42].

Even in humans, cerebrospinal fluid (CSF) collected from AD patients showed larger amounts of Let-7b as compared with the control group, supporting the hypothesis that Let-7 from dying neurons stimulated the surrounding neurons to undergo apoptosis [10]. RNAs containing the ssRNA40 motif have also been shown to cause neurodegeneration via TLR7 induced by microglia [43]. Apart from TLR7 inducing autophagy, it has also been documented to help with the clearance of Aβ proteins from the system [44]. Knockout models specific for microglia would be needed for deeper studies and better understanding.

Table 2. The various possible effects of different TLRs on Alzheimer's disease (AD).

TLR	Effects on Alzheimer's Disease	Reference
TLR2	Activation of microglial, reduces neuronal viability.	[34]
	Is activated in AD patients because of metabolites associated with AD such as HNE, which, in turn, pushes the cell towards apoptosis.	[36]
TLR4	Upregulated in glia surrounding AB plaques.	[39]
	Activation of microglia, reduces neuronal viability.	[41]
TLR7	Activation of TLR7 on neurons can lead to neuronal cell death; does not require activation of glia.	[10]
	Certain RNAs which are overexpressed in AD can trigger TLR7 activation of other surrounding neurons and hence trigger apoptosis in the neurons.	[10]

4. TLRs and Parkinson's Disease

4.1. TLR2

Clinical studies have shown that TLR2 is overexpressed in the microglia of patients with Parkinson's disease (PD), especially in the substantia nigra and the hippocampus region of the brain in the early stages of the disease [33]. In the late stages, TLR2 is upregulated in the striatum [45]. This indicates that TLR2 expression is time-dependent and brain-region-specific. Another set of evidence linking TLR2 with PD is the fact that TLR2 polymorphism is associated with an increased risk of PD. The polymorphisms often lead to changes in the TLR2 promoter, which leads to reduced TLR2 expression [46].

The α-Synuclein has been shown to activate microglia in in vitro cultures via TLR2 [47]. The hypothesis about the role of TLR2 is that it helps microglia clear out excess α-synuclein, but activation of the microglia via the TLR2 pathway induces neurotoxicity. α-synuclein seems to induce a positive feedback loop, which activates the microglia via TLR2 and eventually leads to neurodegeneration, though it is not yet clear why TLR2 shows region- and time-specificity in expression. Table 3 summarizes the effects of TLRs on Parkinson's disease.

4.2. TLR3

TLR3 recognizes double-stranded RNA associated with viruses. Stimulation of TLR3 leads to the production of antiviral interferons such as IFNα and IFNβ via the MyD88-independent TRIF-mediated pathway. It also stimulates natural killer (NK) cells and macrophages to elicit an antiviral response [48]. It is well known that stimulation of TLR3 during the process of neurogenesis causes the process to stop and causes neurodegeneration to occur. The expression of TLR3 in embryonic stem cells stops when neurogenesis begins [49]. Even in adult neurons, stimulation of TLR3 with agonists like poly (I:C) results in growth inhibition of neurons and neurodegeneration. In vivo, injecting postnatal mice with poly (I:C) leads to sensory-motor deficits and fewer axons in the spinal cord [50]. This observation of TLR3 causing neurons to undergo apoptosis led to the hypothesis that TLR3 is involved in virus-associated PD. Parkinson-like symptoms have been observed in several patients of diseases caused by viruses such as HIV [51], Epstein–Barr virus [52], and hepatitis C virus [53]. Even the influenza virus has been reported to cause neurodegeneration [54]. It is possible that activation of TLR3 upon recognition of viral RNA could signal for an immune response, leading to neuronal cell death, or could trigger the neuron to undergo apoptosis, either way leading to neurodegeneration upon activation of TLR3 mediated by viral RNA.

4.3. TLR4

TLR4 seems to have both neuroprotective and neurodegenerative roles in the context of PD. TLR4 is upregulated in the post mortem brains of PD patients, suggesting that TLR4 could potentially have a role in neurodegeneration. However, in mouse models of PD, it has been observed that TLR4-deficient mice are more vulnerable to dopaminergic neuronal loss and motor impairment due to α-synuclein overexpression than mice that express TLR4 [55]. At the same time, TLR4-deficient mice are less likely to develop PD symptoms in PD mouse models induced by 1-methyl-4-phenyl-1,2,3,6-tetrahydropyridine (MPTP) [56]. These observations suggest that TLR4 might be neuroprotective in the context of PD, as TLR4 could help in the clearance of toxic aggregates but could cause neurodegeneration in the context of toxin-induced PD, for example in patients who are in contact with agents such as MPTP or rotenone.

Microglial TLR4 has been found to play role in α-synuclein-dependent activation of microglia [57], and microglia activated by α-synuclein tend to downregulate TLR4. It leads to the disabling of a neuroinflammatory positive feedback loop but also reduces the ability of microglia to take up α-synuclein from their environment [47]. Hence, it can be proposed that the balance between the neuroinflammation caused by microglia TLR activation and the endocytosis of α-synuclein would eventually determine if TLR4 plays a neuroprotective role or facilitates neurodegeneration in the case of PD.

4.4. TLR9

TLR9 has been observed to be overexpressed in human brain regions such as the substantia nigra and putamen in PD patients and in the brain stem in PD mouse models [58]. Some studies have demonstrated that the activation of TLR9 signaling exacerbates neurodegeneration by inducing oxidative stress and inflammation. Microglia are activated by CpG DNA and induce TNFα and nitric oxide [59]. In a co-culture containing microglia and neurons, the activation of microglia cells by CpG-DNA via TLR9 induced neuronal toxicity, mediated partly through TNF-α [60]. Intracerebroventricular infusions of CpG-DNA caused impairment in spatial memory, microglia activation, and acute axonal damage [61]. Furthermore, intrathecal injection of CpG oligodeoxynucleotide (ODN) induced loss of neurons, axonal injury in the cerebral cortex, and pronounced microglia activation [62].

Table 3. Effects of different TLRs on Parkinson's disease.

TLR	Effects on Parkinson's Disease (PD)	Reference
TLR2	Upregulated in PD patients, especially in the substantia nigra and in late stages in the striatum.	[45]
	α-Synuclein activates microglia via TLR2.	[48]
	Activation of microglial TLR2 reduces neuronal viability.	
TLR4	Activation of TLR4 can be neurotoxic or neuroprotective, depending on context.	[47]
	It is shown to help clear α-synuclein aggregates in PD.	[47]
	In oxidative stress-induced PD models, TLR4 activation is associated with cell death.	[56]
TLR9	Overexpressed in PD models in regions such as the substantia nigra and the putamen.	[58]
	Its activation leads to an enhancement of oxidative stress for neuronal cells, which exacerbates cell death in PD.	[59]
	Activation of microglia TLR9 reduces neuronal viability.	[59,61]

5. Conclusions

However, the current knowledge on the effects and pathways modulated by TLRs in microglia is still modest and further studies are necessary to establish their exact roles in neuropathological events. As evident from the literature, TLRs are involved in a variety of physiological pathways and pathological conditions, and studies that specifically delete these receptors in microglia using models of neurodegeneration could contribute to further clarifying their roles in neuropathological conditions.

It is also evident from the literature that the activation of both the endosomal and plasma membrane receptors control microglial activity and may alter phenotypes, which could control the evolution of neurodegenerative processes. Thus, TLRs could represent potential pharmaco-pathological targets for the development of neuroprotective drugs.

Author Contributions: Conceptualization, S.V.A., Y.K. and S.K.J.; validation, S.K.J., N.K.J., V.K.G. and C.V.; resources, S.K.J., N.K.J., A.B. and C.V.; writing—original draft preparation, S.V.A., Y.K., A.D., A.S. and S.K.J.; writing—review and editing, S.V.A., Y.K., S.K.J., N.K.J., V.K.G., A.S. and C.V.; funding acquisition, C.V. All authors have read and agreed to the published version of the manuscript.

Funding: This research received no external funding.

Institutional Review Board Statement: Not applicable.

Informed Consent Statement: Not applicable.

Data Availability Statement: Not applicable.

Conflicts of Interest: The authors declare no conflict of interest.

References

1. El-Zayat, S.R.; Sibaii, H.; Mannaa, F.A. Toll-like Receptors Activation, Signaling, and Targeting: An Overview. *Bull. Natl. Res. Cent.* **2019**, *43*, 187–199. [CrossRef]
2. Chaturvedi, A.; Pierce, S.K. How Location Governs Toll-like Receptor Signaling. *Traffic* **2009**, *10*, 621–628. [CrossRef]
3. Manicassamy, S.; Pulendran, B. Modulation of adaptive immunity with Toll-like receptors. *Semin. Immunol.* **2009**, *21*, 185–193. [CrossRef]
4. Mahla, R.S.; Reddy, M.C.; Prasad, D.V.; Kumar, H. Sweeten PAMPs: Role of Sugar Complexed PAMPs in Innate Immunity and Vaccine Biology. *Front. Immunol.* **2013**, *4*, 248. [CrossRef]
5. Basith, S.; Manavalan, B.; Lee, G.; Kim, S.G.; Choi, S. Toll-like receptor modulators: A patent review (2006–2010). *Expert. Opin. Ther. Pat.* **2011**, *21*, 927–944. [CrossRef]
6. Li, N.; Shi-Yu, C.; Jian-Zhong, S.; Jiong, C. Toll-Like Receptors, Associated Biological Roles, and Signaling Networks in Non-Mammals. *Front. Immunol.* **2018**, *9*, 1523. [CrossRef]
7. Kawasaki, T.; Kawai, T. Toll-like receptor signaling pathways. *Front. Immunol.* **2014**, *5*, 461. [CrossRef]
8. Triantafilou, M.; Gamper, F.G.; Haston, R.M.; Mouratis, M.A.; Morath, S.; Hartung, T.; Triantafilou, K. Membrane sorting of toll-like receptor (TLR)-2/6 and TLR2/1 heterodimers at the cell surface determines heterotypic associations with CD36 and intracellular targeting. *J. Biol. Chem.* **2006**, *281*, 31002–31011. [CrossRef]
9. Hu, Y.; Zhang, Y.; Jiang, L.; Wang, S.; Lei, C.; Sun, M.; Shu, H.; Liu, Y. WDFY1 mediates TLR3/4 signaling by recruiting *Trif. Embo Rep.* **2015**, *16*, 447–455. [CrossRef]
10. Lehmann, S.M.; Krüger, C.; Park, B.; Derkow, K.; Rosenberger, K.; Baumgart, J.; Trimbuch, T.; Eom, G.; Hinz, M.; Kaul, D.; et al. An unconventional role for miRNA: Let-7 activates Toll-like receptor 7 and causes neurodegeneration. *Nat. Neurosci.* **2012**, *15*, 827–835. [CrossRef]
11. Takeda, K.; Kaisho, T.; Akira, S. Toll-like receptors. *Annu. Rev. Immunol.* **2003**, *21*, 335–376. [CrossRef]
12. Falck-Hansen, M.; Kassiteridi, C.; Monaco, C. Toll-like receptors in atherosclerosis. *Int. J. Mol. Sci.* **2013**, *14*, 14008–14023. [CrossRef]
13. Deguine, J.; Barton, G.M. MyD88: A central player in innate immune signaling. *F1000Prime Rep.* **2014**, *6*, 97. [CrossRef]
14. Tahoun, A.; Jensen, K.; Corripio-Miyar, Y.; McAteer, S.; Smith, D.G.E.; McNeilly, T.N.; Gally, D.L.; Glass, E.J. Host species adaptation of TLR5 signalling and flagellin recognition. *Sci. Rep.* **2017**, *7*, 17677. [CrossRef]
15. Sasai, M.; Linehan, M.M.; Iwasaki, A. Bifurcation of Toll-like receptor 9 signaling by adaptor protein 3. *Science* **2010**, *329*, 1530–1534. [CrossRef]
16. Yamamoto, M.; Sato, S.; Hemmi, H.; Hoshino, K.; Kaisho, T.; Sanjo, H.; Takeuchi, O.; Sugiyama, M.; Okabe, M.; Takeda, K.; et al. Role of adaptor TRIF in the MyD88-independent toll like receptor signaling pathway. *Science* **2003**, *301*, 640–643. [CrossRef]
17. Fang, W.; Bi, D.; Zheng, R.; Cai, N.; Xu, H.; Zhou, R.; Lu, J.; Wan, M.; Xu, X. Identification and activation of TLR4-mediated signaling pathways by alginate-derived guluronate oligosaccharide in RAW264.7 macrophages. *Sci. Rep.* **2017**, *7*, 1663. [CrossRef]
18. Jacobs, M.D.; Harrison, S.C. Structure of an IkappaBalpha/NF-kappaB complex. *Cell* **1998**, *95*, 749–758. [CrossRef]
19. Collins, S.E.; Noyce, R.S.; Mossman, K.L. Innate cellular response to virus particle entry requires IRF3 but not virus replication. *J. Virol.* **2004**, *78*, 1706–1717. [CrossRef]
20. Galea, I.; Bechmann, I.; Perry, V.H. What is immune privilege (not)? *Trends Immunol.* **2007**, *28*, 12–18. [CrossRef]
21. McKimmie, C.S.; Fazakerley, J.K. In response to pathogens, glial cells dynamically and differentially regulate Toll-like receptor gene expression. *J. Neuroimmunol.* **2005**, *169*, 116–125. [CrossRef]

22. Kurt-Jones, E.A.; Chan, M.; Zhou, S.; Wang, J.; Reed, G.; Bronson, R.; Arnold, M.M.; Knipe, D.M.; Finberg, R.W. Herpes simplex virus 1 interaction with Toll-like receptor 2 contributes to lethal encephalitis. *Proc. Natl. Acad. Sci. USA* **2004**, *101*, 1315–1320. [CrossRef]
23. Prehaud, C.; Megret, F.; Lafage, M.; Lafon, M. Virus infection switches TLR-3-positive human neurons to become strong producers of beta interferon. *J. Virol.* **2005**, *79*, 12893–12904. [CrossRef]
24. Wadachi, R.; Hargreaves, K.M. Trigeminal nociceptors express TLR-4 and CD14: A mechanism for pain due to infection. *J. Dent. Res.* **2006**, *85*, 49–53. [CrossRef]
25. Lafon, M.; Megret, F.; Lafage, M.; Prehaud, C. The innate immune facet of brain: Human neurons express TLR-3 and sense viral dsRNA. *J. Mol. Neurosci.* **2006**, *29*, 185–194. [CrossRef]
26. Jackson, A.C.; Rossiter, J.P.; Lafon, M. Expression of Toll-like receptor 3 in the human cerebellar cortex in rabies, herpes simplex encephalitis, and other neurological diseases. *J. Neurovirol.* **2006**, *12*, 229–234. [CrossRef]
27. Flo, T.H.; Halaas, O.; Torp, S.; Ryan, L.; Lien, E.; Dybdahl, B.; Sundan, A.; Espevik, T. Differential expression of Toll-like receptor 2 in human cells. *J. Leuk. Biol.* **2001**, *69*, 474–481.
28. Goethals, S.; Ydens, E.; Timmerman, V.; Janssens, S. Toll-like receptor expression in the peripheral nerve. *Glia* **2010**, *58*, 1701–1709. [CrossRef]
29. Barajon, I.; Serrao, G.; Arnaboldi, F.; Opizzi, E.; Ripamonti, G.; Balsari, A.; Rumio, C. Toll-like receptors 3, 4, and 7 are expressed in the enteric nervous system and dorsal root ganglia. *J. Histochem. Cytochem.* **2009**, *57*, 1013–1023. [CrossRef] [PubMed]
30. Rietdijk, C.D.; Wezel, R.J.A.; Garssen, J.; Kraneveld, A.D. Neuronal toll-like receptors and neuro-immunity in Parkinson's disease, Alzheimer's disease and stroke. *Neuroimmunol. Neuroinflamm.* **2016**, *3*, 27–37. [CrossRef]
31. Jana, M.; Palencia, C.A.; Pahan, K. Fibrillar amyloid-beta peptides activate microglia via TLR2: Implications for Alzheimer's disease. *J. Immunol.* **2008**, *181*, 7254–7262. [CrossRef] [PubMed]
32. Richard, K.L.; Filali, M.; Prefontaine, P.; Rivest, S. Toll-like receptor 2 acts as a natural innate immune receptor to clear amyloid beta 1–42 and delay the cognitive decline in a mouse model of Alzheimer's disease. *J. Neurosci.* **2008**, *28*, 5784–5793. [CrossRef] [PubMed]
33. Zhou, C.; Sun, X.; Hu, Y.; Song, J.; Dong, S.; Kong, D.; Wang, Y.; Hua, X.; Han, J.; Zhou, Y.; et al. Genomic deletion of TLR2 induces aggravated white matter damage and deteriorated neurobehavioral functions in mouse models of Alzheimer's disease. *Aging (Albany NY)* **2019**, *11*, 7257–7273. [CrossRef] [PubMed]
34. Schutze, S.; Loleit, T.; Zeretzke, M.; Bunkowski, S.; Brück, W.; Ribes, S.; Nau, R. Additive Microglia-Mediated Neuronal Injury Caused by Amyloid-beta and Bacterial TLR Agonists in Murine Neuron-Microglia Co-Cultures Quantified by an Automated Image Analysis using Cognition Network Technology. *J. Alzheimers Dis.* **2012**, *31*, 651–657. [CrossRef]
35. Lotz, M.; Ebert, S.; Esselmann, H.; Iliev, A.I.; Prinz, M.; Wiazewicz, N.; Wiltfang, J.; Gerber, J.; Nau, R. Amyloid beta peptide 1-40 enhances the action of Toll-like receptor-2 and -4 agonists but antagonizes Toll-like receptor-9-induced inflammation in primary mouse microglial cell cultures. *J. Neurochem.* **2005**, *94*, 289–298. [CrossRef]
36. Tang, S.C.; Lathia, J.D.; Selvaraj, P.K.; Hoffmann, O.; Röhr, C.; Prinz, V.; König, J.; Lehrach, H.; Nietfeld, W.; Trendelenburg, G. Toll-like receptor-4 mediates neuronal apoptosis induced by amyloid beta-peptide and the membrane lipid peroxidation product 4-hydroxynonenal. *Exp. Neurol.* **2008**, *213*, 114–121. [CrossRef]
37. Minoretti, P.; Gazzaruso, C.; Vito, C.D.; Emanuele, E.; Bianchi, M.; Coen, E.; Reino, M.; Geroldi, D. Effect of the functional toll-like receptor 4 Asp299Gly polymorphism on susceptibility to late-onset Alzheimer's disease. *Neurosci. Lett.* **2006**, *391*, 147–149. [CrossRef]
38. Poltorak, A.; He, X.; Smirnova, I.; Liu, M.-Y.; Van Huffel, C.; Du, X.; Birdwell, D.; Alejos, E.; Silva, M.; Galanos, C.; et al. Defective LPS signaling in C3H/HeJ and C57BL/10ScCr mice: Mutations in Tlr4 gene. *Science* **1998**, *282*, 2085–2088. [CrossRef]
39. Walter, S.; Letiembre, M.; Liu, Y.; Heine, H.; Penke, B.; Hao, W.; Bode, B.; Manietta, N.; Walter, J.; Schulz-Schuffer, W.; et al. Role of the toll-like receptor 4 in neuroinflammation in Alzheimer's disease. *Cell Physiol. Biochem.* **2007**, *20*, 947–956. [CrossRef]
40. Jin, J.-J.; Kim, H.-D.; Maxwell, J.A.; Li, L.; Fukuchi, K. Toll-like receptor 4-dependent upregulation of cytokines in a transgenic mouse model of Alzheimer's disease. *J. Neuroinflamm.* **2008**, *5*, 3006–3019. [CrossRef]
41. Calvo-Rodriguez, M.; García-Rodríguez, C.; Villalobos, C.; Núñez, L. Role of Toll Like Receptor 4 in Alzheimer's Disease. *Front. Immunol.* **2020**, *11*, 1588. [CrossRef]
42. Zhang, T.; Ma, G.; Zhang, Y.; Huo, H.; Zhao, Y. miR-599 inhibits proliferation and invasion of glioma by targeting periostin. *Biotechnol. Lett.* **2017**, *39*, 1325–1333. [CrossRef]
43. Lehmann, S.M.; Rosenberger, K.; Krüger, C.; Habbel, P.; Derkow, K.; Kaul, D.; Rybak-Wolf, A.; Brandt, C.; Schott, E.; Wulczyn, F.G.; et al. Extracellularly delivered single-stranded viral RNA causes neurodegeneration dependent on TLR7. *J. Immunol.* **2012**, *189*, 1448–1458. [CrossRef]
44. Gambuzza, M.E.; Sofo, V.; Salmeri, F.M.; Soraci, L.; Marino, S.; Bramanti, P. Toll-like receptors in Alzheimer's disease: A therapeutic perspective. *CNS Neurol. Disord.-Drug Targets* **2014**, *13*, 1542–1558. [CrossRef]
45. Doorn, K.J.; Moors, T.; Drukarch, B.; van de Berg, W.D.; Lucassen, P.J.; van Dam, A.M. Microglial phenotypes and toll-like receptor 2 in the substantia nigra and hippocampus of incidental Lewy body disease cases and Parkinson's disease patients. *Acta Neuropathol. Commun.* **2014**, *2*, 90–91.
46. Kalinderi, K.; Bostantjopoulou, S.; Katsarou, Z.; Fidani, L. TLR9 -1237 T/C and TLR2 -194 to 174 del polymorphisms and the risk of Parkinson's disease in the Greek population: A pilot study. *Neurol. Sci.* **2013**, *34*, 679–682. [CrossRef] [PubMed]

47. Beraud, D.; Twomey, M.; Bloom, B.; Mittereder, A.; Ton, V.; Neitzke, K.; Chasovskikh, S.; Mhyre, T.R.; Maguire-Zeiss, K.A. Alpha-Synuclein Alters Toll-Like Receptor Expression. *Front. Neurosci.* **2011**, *5*, 80. [CrossRef] [PubMed]
48. Lien, E.; Ingalls, R.R. Toll-like receptors. *Crit. Care Med.* **2002**, *30*, S1–S11. [CrossRef]
49. Lathia, J.D.; Okun, E.; Tang, S.C.; Griffioen, K.; Cheng, A.; Mughal, M.R.; Laryea, G.; Selvaraj, P.K.; ffrench-Constant, C.; Magnus, T.; et al. Toll-like receptor 3 is a negative regulator of embryonic neural progenitor cell proliferation. *J. Neurosci.* **2008**, *28*, 13978–13984. [CrossRef] [PubMed]
50. Cameron, J.S.; Alexopoulou, L.; Sloane, J.A.; DiBernardo, A.B.; Ma, Y.; Kosaras, B.; Flavell, R.; Strittmatter, S.M.; Volpe, J.; Sidman, R.; et al. Toll-like receptor 3 is a potent negative regulator of axonal growth in mammals. *J. Neurosci.* **2007**, *27*, 13033–13041. [CrossRef]
51. Moulignier, A.; Gueguen, A.; Lescure, F.X.; Ziegler, M.; Girard, P.M.; Cardon, B.; Pialoux, G.; Molina, J.M.; Brandel, J.P.; Lamirel, C. Does HIV infection alter parkinson's disease? *J. Acquir. Immune. Defic. Syndr.* **2015**, *70*, 129–136. [CrossRef]
52. Woulfe, J.M.; Gray, M.T.; Gray, D.A.; Munoz, D.G.; Middeldorp, J.M. Hypothesis: A role for EBV induced molecular mimicry in Parkinson's disease. *Parkinsonism Relat. Disord.* **2014**, *20*, 685–694. [CrossRef]
53. Wu, W.Y.; Kang, K.H.; Chen, S.L.; Chiu, S.Y.; Yen, A.M.; Fann, J.C.; Su, C.W.; Liu, H.C.; Lee, C.Z.; Fu, W.M.; et al. Hepatitis C virus infection: A risk factor for Parkinson's disease. *J. Viral. Hepat.* **2015**, *22*, 784–791. [CrossRef] [PubMed]
54. Jang, H.; Boltz, D.; Sturm-Ramirez, K. Highly pathogenic H5N1 influenza virus can enter the central nervous system and induce neuroinflammation and neurodegeneration. Highly pathogenic H5N1 influenza virus can enter the central nervous system and induce neuroinflammation and neurodegeneration. *Proc. Natl. Acad. Sci. USA* **2009**, *106*, 14063–14068. [CrossRef] [PubMed]
55. Stefanova, N.; Fellner, L.; Reindl, M.; Masliah, E.; Poewe, W.; Wenning, G.K. Toll-like receptor 4 promotes alpha-synuclein clearance and survival of nigral dopaminergic neurons. *Am. J. Pathol.* **2011**, *179*, 954–963. [CrossRef]
56. Noelker, C.; Morel, L.; Lescot, T.; Osterloh, A.; Alvarez-Fischer, D.; Breloer, M.; Henze, C.; Depboylu, C.; Skrzydelski, D.; Michel, P.P.; et al. Toll like receptor 4 mediates cell death in a mouse MPTP model of Parkinson disease. *Sci. Rep.* **2013**, *3*, 1393. [CrossRef] [PubMed]
57. Fellner, L.; Irschick, R.B.; Schanda, K.; Reindl, M.; Klimaschewski, L.; Poewe, W.; Wenning, G.; Stefanova, N. Toll-like receptor 4 is required for α-synuclein dependent activation of microglia and astroglia. *Glia* **2013**, *61*, 349–360. [CrossRef] [PubMed]
58. Ros-Bernal, F.; Hunot, S.; Herrero, M.T.; Parnadeau, S.; Corvol, J.-C.; Lu, L.; Alvarez-Fischer, D.; Sauvage, M.A.C.-D.; Saurini, F.; Coussieu, C.; et al. Microglial glucocorticoid receptors play a pivotal role in regulating dopaminergic neurodegeneration in parkinsonism. *Proc. Natl. Acad. Sci. USA* **2011**, *108*, 6632–6637. [CrossRef]
59. Dalpke, A.H.; Schäfer, M.K.-H.; Frey, M.; Zimmermann, S.; Tebbe, J.; Weihe, E.; Heeg, K. Immunostimulatory CpG-DNA activates murine microglia. *J. Immunol.* **2002**, *168*, 4854–4863. [CrossRef]
60. Iliev, A.I.; Stringaris, A.K.; Nau, R.; Neumann, H. Neuronal injury mediated via stimulation of microglial toll-like receptor-9 (TLR9). *FASEB J.* **2004**, *18*, 412–414. [CrossRef]
61. Tauber, S.C.; Ebert, S.; Weishaupt, J.H.; Reich, A.; Nau, R.; Gerber, J. Stimulation of toll-like receptor 9 by chronic intraventricular unmethylated cytosine-guanine DNA infusion causes neuroinflammation and impaired spatial memory. *J. Neuropathol. Exp. Neurol.* **2009**, *68*, 1116–1124. [CrossRef] [PubMed]
62. Rosenberger, K.; Derkow, K.; Dembny, P.; Krüger, C.; Schott, E.; Lehnardt, S. The impact of single and pairwise Toll-like receptor activation on neuroinflammation and neurodegeneration. *J. Neuroinflamm.* **2014**, *11*, 166. [CrossRef] [PubMed]

MDPI

Review

Recent Advances in Progresses and Prospects of IL-37 in Central Nervous System Diseases

Xinrui Li [1], Bing Yan [2], Jin Du [1,3], Shanshan Xu [1], Lu Liu [1], Caifei Pan [1], Xianhui Kang [1,*] and Shengmei Zhu [1,*]

1 Department of Anesthesiology, The First Affiliated Hospital, Zhejiang University School of Medicine, Hangzhou 310003, China; 22118644@zju.edu.cn (X.L.); du4532380@163.com (J.D.); shanshan_xu2012@163.com (S.X.); liulu124@zju.edu.cn (L.L.); again_16@sina.com (C.P.)
2 Department of Anesthesiology, Haining People's Hospital, Haining 314499, China; hnphyb@163.com
3 China Coast Guard Hospital of the People's Armed Police Force, Jiaxing 314000, China
* Correspondence: kxhui66@zju.edu.cn (X.K.); smzhu20088@zju.edu.cn (S.Z.)

Abstract: Interleukin-37 (IL-37) is an effective anti-inflammatory factor and acts through intracellular and extracellular pathways, inhibiting the effects of other inflammatory cytokines, such as IL-1β, IL-6, and tumor necrosis factor-α (TNF-α), thereby exerting powerful anti-inflammatory effects. In numerous recent studies, the anti-inflammatory effects of IL-37 have been described in many autoimmune diseases, colitis, and tumors. However, the current research on IL-37 in the field of the central nervous system (CNS) is not only less, but mainly for clinical research and little discussion of the mechanism. In this review, the role of IL-37 and its associated inflammatory factors in common CNS diseases are summarized, and their therapeutic potential in CNS diseases identified.

Keywords: Interleukin-37; central nervous system diseases; inflammatory; anti-inflammatory; cytokines

Citation: Li, X.; Yan, B.; Du, J.; Xu, S.; Liu, L.; Pan, C.; Kang, X.; Zhu, S. Recent Advances in Progresses and Prospects of IL-37 in Central Nervous System Diseases. *Brain Sci.* **2022**, *12*, 723. https://doi.org/10.3390/brainsci12060723

Academic Editors: Chiara Villa, Christina Piperi, Yam Nath Paudel and Andrew Clarkson

Received: 30 March 2022
Accepted: 25 May 2022
Published: 31 May 2022

Publisher's Note: MDPI stays neutral with regard to jurisdictional claims in published maps and institutional affiliations.

Copyright: © 2022 by the authors. Licensee MDPI, Basel, Switzerland. This article is an open access article distributed under the terms and conditions of the Creative Commons Attribution (CC BY) license (https://creativecommons.org/licenses/by/4.0/).

1. Introduction

Interleukin-37 (IL-37), a novel cytokine which was once considered as a member of IL-1 family, was reported to be a natural innate immune inhibitor [1]. Similar to other IL-1 family cytokines, IL-37 is encoded by chromosome 2 [2]. All members of this family share a common b-trefoil structure that includes 12 β-chains [3]. IL-37 is encoded by six exons. At present, five subtypes (IL-37a–e) formed by alternative splicing have been found, among which IL-37b has the largest molecular weight and is the most researched [4]. IL-37b contains five exons except exon 3, and IL-37c and IL-37e are predicted to be non-functional proteins because they lack one or more exons [5]. Based on the fact that most of the current research is on the IL-37b subtype, the IL-37 mentioned below in this article refers to IL-37b. In addition, IL-37a is the only form found in the brain. An unstable mRNA motif exists in exon 5. IL-37a is generally considered to be the functional subtype, but little research has been concentrated on this subtype alone [4]. Although IL-37 is widely expressed in many cell tissues in the human body, the concentration in the blood is extremely low (100 pg/mL) [6]. In some diseases involving inflammation, IL-37 is elevated due to inflammatory stimulation [7].

Previous research has shown the function of IL-37 in autoimmune diseases, including systemic lupus erythematosus [8], colitis [9], sepsis [10], asthma [1,11], and cancer [1,10,12]. However, the function and potential mechanisms of IL-37 in central nervous system (CNS) diseases have been investigated in only a few reviews. In the present review, current progress regarding IL-37 in CNS diseases is summarized and its therapeutic potential for CNS diseases identified.

2. About IL-37

2.1. Function

IL-37 was found in 2000 by several groups independently [11]; a homologue has not yet been identified in a mouse [4,7]. IL-37, originally named IL-1F7 [13], is mainly produced by Toll-like receptor (TLR)-activated macrophages [14]. The precursor of IL-37 (pre-IL-37) is cleaved by caspase-1 into mature IL-37, of which approximately 20% enters the nucleus, with the remaining released out of the cell with pre-IL-37 [15]. Many cells in the human body, including epithelial cells, keratinocytes, renal tubular epithelial cells, monocytes, activated B cells, plasma cells, dendritic cells (DCs), macrophages, and CD4+ Tregs, express IL-37 [5,7,16–19]. Reportedly, IL-37 expression is low in unstimulated peripheral blood mononuclear cells (PBMCs) and M1 macrophages, and IL-37 expression significantly increases after being activated by lipopolysaccharide (LPS) [20,21]. Human IL-37 precursor undergoes alternative splicing to form five different subtypes, and the IL-37b subtype has been the focus in most of the current studies [4]. IL-37 protein exists at low levels in human PBMCs and can be upregulated by inflammatory stimuli and cytokines, such as IL-1, IL-18, tumor necrosis factor (TNF), interferons (IFNs), and transforming growth factor (TGF) [7]. In addition, IL-37 is downregulated by IL-4, IL-12, IL-32, and granulocyte macrophage colony-stimulating factor [6,12].

IL-37 is constitutively expressed at low levels in various tissues, including lymph nodes, thymus, bone marrow, brain, intestines, airways, adipose, thymus, placenta, uterus, testis, heart, kidney, bone marrow, prostate, and breast [5,7,12,17–19]. IL-37 has a protective effect in a variety of diseases. Hui-min Chen et al. reported that compared with wild-type (WT) mice, IL-37 expression in DCs attenuates the ability of DCs to initiate contact hypersensitivity (CHS) responses in mice transgenic with human IL-37 (IL-37tg), demonstrating that IL-37 may be an immune tolerance factor [16]. Dov B. Ballak et al. revealed that IL-37 is expressed in human adipose tissue, and IL-37 can reduce diet-induced obesity in IL-37tg mice. In addition, IL-37 ameliorated diet-induced insulin resistance and improved insulin sensitivity in IL-37tg mice compared with WT mice, indicating potential as a treatment for obesity and type 2 diabetes [22]. Jilin Li et al. reported that IL-37 expression in an old endotoxemic mouse model suppressed myocardial inflammation-associated endotoxemia and improved left ventricle (LV) function, indicating its protective function in septic myocarditis [23]. Tianheng Hou et al. proved that IL-37 could reduce Der p1-induced thymic stromal lymphopoietin (TSLP) overexpression in HaCa T cells, and decreased TSLP receptors and basophil activation marker CD203c in vitro. In vivo experiments in an atopic dermatitis mouse model showed alternative depletion of basophils rescued atopic dermatitis symptoms and significantly lowered the helper T cell 2 (Th2) and eosinophil populations in the ear and spleen [24]. In other studies, therapeutic effects of IL-37 on allergic diseases, autoimmune diseases, and other immune system diseases have been reported [6,25–28]. In addition, IL-37 reportedly exerts tumor-inhibiting effects in a variety of cancers, such as breast, cervical, melanoma, and non-small cell lung cancer [12,29]; refer to the relevant review for details.

2.2. Pathway

IL-37 exerts anti-inflammatory effects through intracellular and extracellular pathways [6]. In the intracellular pathway, the precursor IL-37 (pro-IL-37) is cleaved by caspase-1 to produce mature IL-37 after activation by lipopolysaccharides(LPS) [30,31]. A possible cleavage site for caspase-1 is located in exon 1 between the D20 and E21 residues of IL-37 [32]. However, Ana-Maria Bulau et al. previously demonstrated that caspase-1 inhibitors only partially inhibit the processing of IL-37, indicating that caspase-1 is not the only enzyme responsible for the processing of IL-37 [30]. Human embryonic kidney 293 (HEK 293) or Chinese hamster ovary (CHO) cells transfected with IL-37 precursor release IL-37 from amino acid V46, indicating there is a second cleavage site in the sequence encoded in exon 2 [33]. In addition to caspase-1, caspase-4 was shown to cleave pro-IL-37 to a certain extent [32]. IL-37 binds with drosophila mothers against decapentaplegic protein

3 (Smad3) to form a complex in the cytoplasm; then, the complex enters the nucleus to regulate the transcription process, such as nuclear factor-κB (NF-κB) and mitogen-activated protein kinase (MAPK) pathways, thereby inhibiting the transcription process of some inflammatory cytokines [6,12,34].

In the extracellular pathway, IL-37 binds to IL-18 receptor α chain (IL-18Rα) to exert an anti-inflammatory effect [35]. IL-37 binds to IL-18Rα to recruit IL-1 receptor 8 (IL-1R8, also named single Ig IL-1R-related molecule, SIGIRR), to form a trimeric complex (IL-37/IL-18Rα/IL-1R8) [36]. When adenosine 5′-monophosphate-activated protein kinase (AMPK) is increased, signal transducer and activator of transcription 3 (STAT3), STAT6, phosphatase and tensin homologs, and other factors inhibit the inflammatory response induced by IL-18 and downregulate the expression of IFN-γ and transcription factor NF-κB [37]. SIGIRR is the only receptor containing a TLR domain with a single immunoglobulin domain. Although SIGIRR has an immunoglobulin domain, it cannot bind to IL-1 or enhance IL-1-dependent signaling. SIGIRR is a negative regulator of the inflammatory response and inhibits the inflammation process of IL-1 and IL-18 [38]. Both pro-IL-37 and mature IL-37 can bind to IL18Rα but the binding of the mature form is approximately 5–10-fold stronger than the immature form [32]. IL-IR8 is necessary for activation of the anti-inflammatory signal transducer and transcriptional activator STAT3 in splenic DCs and macrophages [36]. However, IL-37 is not a receptor antagonist of IL-18Rα [39]. SIGIRR was shown unstable in response to IL-37, and IL-37 can mediate the ubiquitination and degradation of SIGIRR through glycogen synthesis kinase 3β (GSK3β) [40]. See Figure 1 for a summary of the IL-37 pathway.

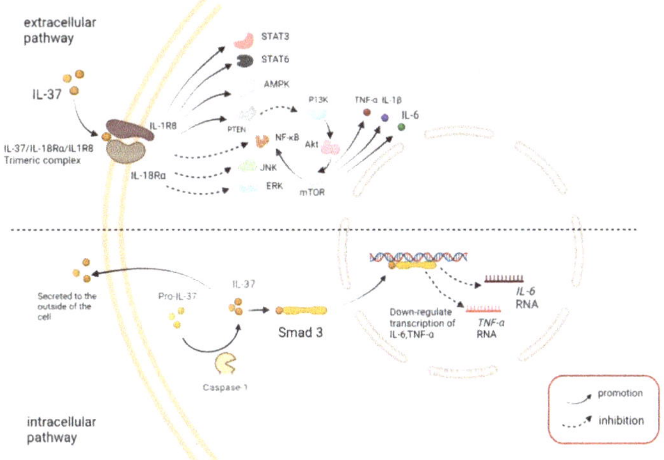

Figure 1. Pathway through which Interleukin-37 (IL-37) exerts anti-inflammatory effects. In extracellular pathway, IL-37 binds with IL-18 receptor α chain (IL-18Rα) and IL-1 receptor 8 (IL-1R8) to form a trimeric complex. The trimeric complex activates adenosine 5′-monophosphate-activated protein kinase (AMPK), signal transducer and activator of transcription 3 (STAT3), STAT6 and phosphate and tension homology deleted on chromosome ten (PTEN), and inhibits the pathway of c-Jun N-terminal kinase (JNK), extracellular regulated protein kinases (ERK) and nuclear factor-κB (NF-κB). At the same time, PTEN inhibits the PI3K/AKT/mTOR pathway, thereby inhibiting the production of NF-κB and pro-inflammatory cytokines, including IL-6, tumor necrosis factor-α (TNF-α) and IL-1β. In intracellular pathway, pro-IL-37 is cleaved by caspase-1 to become IL-37. A part of IL-37 is released outside the cell, and a part binds to drosophila mothers against decapentaplegic 3 (Smad3) in the cytoplasm. After entering the nucleus, it inhibits the transcription of other inflammatory cytokines, such as IL-6 and TNF-α.

3. IL-37 in CNS Diseases

3.1. Acute Spinal Cord Injury (ASCI)

Acute spinal cord injury (ASCI) is common in serious trauma caused by transport accidents, low-height falls, and other accidents [41]. The quality of life of ASCI patients is reduced, the prognosis very poor, and the mortality rate is high [42]. According to survey results, the mortality rate of individuals over 65 years of age suffering from SCI is 36.5% [43]. Therefore, improving the prognosis and survival rate of ASCI patients is important.

Marina Coll-Miró et al. transferred the human IL-37 gene into mice to produce hIL-37tg mice. The WT and hIL-37tg mice were subjected to spinal cord contusion injury. The authors found that compared with WT mice, hIL-37tg mice had more myelin and neurons preserved, and maintained a lower level of cytokines and chemokines (e.g., an 80% reduction in IL-6). The authors infused recombinant human IL-37 (rIL-37) into the lesion site via a glass capillary 5 min after the contusion injury, and found the mice injected with rIL-37 had a greater extensive movement of the ankle restored and increased their speed on a treadmill by 50% [44]. An experiment conducted on 148 patients showed that serum IL-37 levels were significantly higher within 24 h of ASCI compared with the control group ($p < 0.05$). Serum IL-37 concentration in patients with SCI is negatively correlated with American Spinal Cord Injury Association (ASIA) exercise score ($p < 0.05$) [45]. The above results indicate IL-37 may be a potential therapeutic target and a biomarker after ASCI. IL-37 may inhibit the inflammatory response after ASCI to produce neurological protection and recovery. Jesus Amo-Aparicio et al. used a hIL-37D20ATg transgenic mouse model lacking the IL-37 intracellular pathway to prove that the neuroprotection role of IL-37 after SCI does not rely on the intracellular pathway rather than the extracellular pathway. Their study demonstrated IL-37 exerts an anti-inflammatory effect by binding IL-1R8 [46].

3.2. Demyelinating Disease

Multiple sclerosis (MS) is a chronic, predominantly immune-mediated disease of the brain and spinal cord, and a common cause of neurological disability in young adults, affecting more than 2.5 million individuals globally [47,48]. Alba Sánchez-Fernández et al. used the experimental autoimmune encephalomyelitis (EAE), a murine model of MS, hypothesizing that IL-37 reduces inflammation and protects against neurological deficits and myelin loss in EAE mice by combining with IL-1R5/IL-1R8. The authors found that transgenic expression of IL-37 reduces neurological deficits and inflammation in the spinal cord of EAE mice. However, in the transgenic homozygote of human IL-37 (hIL-37tg) mice lacking IL-1R8, the beneficial effects of IL-37 were completely absent in demyelinating disease of the CNS, indicating IL-37 acts with IL-1R8 [49].

In another study, IL-37 level in a cluster of differentiating CD4+ T cells from MS patients was decreased in vitro compared with healthy controls based on in silico analysis [50]. IL-37 expression was observed in PBMCs from MS patients during the exacerbation of the disease [50]. In addition, higher IL-37 levels showed an inhibitory effect on MS recurrence; however, obvious effects were not observed with IL-1R8 and IL-18R1. Higher IL-37 levels were associated with younger age and lower Multiple Sclerosis Severity Score (MSSS) [50]. After testing serum IL-37 levels in 84 MS patients and 75 healthy controls, Ebrahim Kouchaki et al. found IL-37 levels in MS patients were higher than in the control group ($p < 0.001$) [51]. The research by Farrokhi. M et al. of 122 MS patients and 49 healthy subjects showed the IL-37 levels were higher in the MS patients than in the controls [52].

Guillain-Barré syndrome (GBS) is the most common and severe acute paralytic neuropathy. Globally, approximately 100,000 people develop this disorder annually [53]. GBS is considered an immune-mediated disease, possibly triggered by a recent infection, and driven by an immune attack targeting the peripheral nervous system [54]. Approximately 25% of patients have respiratory insufficiency and many patients show signs of autonomic dysfunction [55]. GBS is the most common cause of acute flaccid paralysis, which leads to disability and high risk of mortality. The specific pathogenic mechanism of GBS remains unclear. Some patients have been reported to have an infectious disease before the onset of the disease; thus, the disease is hypothetically an immune-mediated disease [54]. Cong Li et al. measured the IL-37 levels in the cerebrospinal fluid (CF) and plasma of 25 GBS patients and 20 healthy controls and found the IL-37 levels in the CF and plasma of GBS patients were significantly higher than in the healthy controls ($p = 0.0002$ and $p < 0.0001$) [56]. This result indicates that during the pathogenesis of GBS, pro-inflammatory cytokines may promote the expression of anti-inflammatory IL-37, thereby downregulating the excessive inflammatory response, similar to the results of several previous studies [9,57,58].

3.3. *Alzheimer's Disease (AD)*

Alzheimer's disease (AD) is a neurodegenerative disease highly correlated with age [59]. From 1999 to 2018, the number of deaths from AD in the United States increased [60]. Currently, an estimated 6.2 million Americans 65 years of age and older have AD, and if no effective treatment is found, this number will continue to increase [61]. In many current studies, the relationship between inflammation and AD is being investigated [62,63]. Aging tissue cells secrete inflammation and immune-related cytokines, such as interleukins, chemokines, growth factors, and proteases, which constitute senescence-associated secretory phenotype (SASP) [64]. Astrocytes and microglia surround neuropathic plaques composed of amyloid β-protein (Aβ) and neurofibrillary tangles [65]. Microglia release cytokines and cause neuroinflammation [66]. Among them, M1 type microglia, which are activated by LPS, IFN-γ or TNF-α, secrete classic inflammatory cytokines, such as IL-1β, TNF-α, STAT3, IL-6, IL-12, and IL-23, and free radicals such as reactive oxygen species (ROS) [67]. Another M2 anti-inflammatory phenotype promotes tissue remodeling by releasing high levels of anti-inflammatory cytokines, such as IL-10, IL-4, IL-13 and transforming growth factor-β, and low levels of pro-inflammatory cytokines/repair and angiogenesis [67].

In previous studies, a high-fat diet was shown to induce insulin resistance, reduce the transport of glucose into the brain, and ultimately lead to neuronal stress (elevated neuronal corticosterone) [68]. The intake of fructose promotes the synthesis of triglycerides, gluconeogenesis and insulin resistance, and ultimately accelerates the progression of AD [69]. A rat model of AD was established by Mohamed, R.A. et al. with a synergistic high fat/high fructose diet (HFFH) and LPS injection. The authors found the hippocampal AD marker Aβ1-42 and the inflammatory marker IL-1β in the mouse model were increased 3.2-fold and 5.6-fold, respectively, compared with the controls [70]. The combined use of palonosetron and methyllycaconitine (MLA) restored the activity of caspase-1 protein in AD rats and reduced the reactivity of astrocytes [70].

The role of many cytokines in AD has been elucidated [71]. Tau and Aβ modified by advanced glycation end products stimulate human neurons to produce IL-6 [72]. IL-6 can also activate janus kinase (JAK)/STATs, N-methyl-D-aspartate (NMDA) receptor, and mitogen-activated protein kinase (MAPK)-p38, which are involved in the hyperphosphorylation of tau [72]. However, in an in vivo study, IL-6 overexpression induced the reduction in neuroinflammation at the Aβ level rather than aggravating the pathology of Aβ plaques [73]. Irina Belaya et al. reported regular exercise can modulate iron homeostasis in WT mice and in a mouse model of AD via the IL-6/STAT3/JAK1 pathway [74], indicating that M2 type microglia are mainly activated, which enhances the phagocytic function of Aβ [73]. The combination of IL-18 and its receptor complex can activate c-Jun N-terminal kinase (JNK) and MAPK-p38, thereby activating endogenous and exogenous pro-apoptotic signaling pathways [75]. This effect may be achieved by inducing the expression of p53 and Fas ligand, indicating IL-18 can promote the progression of AD [75,76]. However, research on the role of IL-37 in AD is limited. Exploring the roles of IL-37 in AD can be a promising direction for future research because caspase-1 is also an enzyme that cleaves the precursor of IL-37 and renders it active [31]. In addition, IL-37 has been shown to inhibit the inflammatory effects of other inflammatory cytokines, such as IL-6 and IL-1β in many other diseases [77–82]. In a temporomandibular joint study, IL-37 could convert M1 type macrophages to M2 type, thus alleviating inflammation [83]. Therefore, future research should focus on the protective effect of IL-37 on the progression of AD.

3.4. Stroke

Stroke is the second leading cause of death and disability in the world, a non-communicable disease that seriously endangers the health of Chinese people [84]. In an epidemiological survey of 10,926 participants, 602 cases were diagnosed as ischemic stroke (IS), 151 cases as hemorrhagic stroke, and 22 cases as both hemorrhage and IS. The crude prevalence rates of total stroke, IS, and hemorrhagic stroke were 6690.5/100,000, 5509.8/100,000, and 1382.0/100,000, respectively, and the standardized rates were 4903.8/100,000, 4041.7/100,000, and 990.9/100,000, respectively [85].

Feng Zhang et al. measured the serum IL-37 levels using enzyme-linked immunosorbent assay in 152 patients admitted to the hospital due to acute IS, and in 45 healthy controls. The authors found serum IL-37 levels in IS patients were significantly higher than in controls (182.26 vs. 97.89 pg/mL, $p < 0.001$) and associated with the National Institutes of Health Stroke Scale (NIHSS) scores (r = 0.521, $p < 0.0001$) and lesion volume (r = 0.442, $p < 0.0001$). Notably, elevated plasma IL-37 levels were independently associated with unfavorable 3-month outcomes (adjusted odds ratio = 1.033, $p = 0.001$, 95% confidence interval, CI:1.015–1.056) [86]. In another study, the serum IL-37 levels were measured in 310 IS patients who were followed up for 3 months to determine the relationship between serum IL-37 and recurrence of IS. The study results showed the median IL-37 serum level in the IS patients was 344.1 pg/mL (interquartile range, IQR, 284.4–405.3) and 122.3 pg/mL (IQR, 104.4–1444.0) in the controls, which was significantly lower. The size of the lesion area observed on magnetic resonance imaging (MRI) positively correlated with serum IL-37 levels. In that study, 36 patients experienced relapse of IS within 3 months, and their serum IL-37 levels were higher than in patients who did not relapse (417.0 pg/mL; IQR, 359.3–436.1 vs. 333.3 pg/mL; IQR, 279.0–391.0), indicating IL-37 levels are associated with the recurrence of IS. Based on receiver operating characteristic (ROC) analysis, the authors determined the IL-37 level cut-off value to diagnose IS was 193.0 pg/mL, the cut-off value to diagnose moderate-to-high clinical severity (NIHSS score > 5) was 374.0 pg/mL, and the cut-off value to predict recurrence was 406.8 pg/mL [87].

To date, research on the relationship between IL-37 and hemorrhagic stroke is scarce. However, an increase in the white blood cell/lymphocyte ratio in patients with IS was positively correlated with the probability of hemorrhagic transformation [88], indicating hemorrhagic transformation, a serious complication of severe IS, may be associated with inflammation possibly related to the destruction of the blood–brain barrier by neutrophils [89,90].

4. Conclusions

IL-37 is a potent endogenous anti-inflammatory factor of the IL-1 family. IL-37 suppresses other inflammatory cytokines, thus inhibiting the progression of disease [3]. In an in vitro experiment, siRNA to IL-37 (siIL-37) was transfected into human PBMCs stimulated by LPS. After IL-37 expression was specifically silenced, the production of IL-6 and other cytokines increased in a dose-dependent manner [82], indicating IL-37 can inhibit the inflammatory effect of IL-6 and play an anti-inflammatory role. Irene Tsilioni et al. found that neuropeptide NT can stimulate human microglia to secrete IL-1β, CXCL8, and other cytokines that can be inhibited by IL-37 [91]; however, the specific mechanism underlying this inhibition remains unclear. In 293T cells, overexpression of IL-37 restored the viability of cells damaged by homocysteine and reduced the release of lactate dehydrogenase, pro-inflammatory cytokines IL-1β, IL-6, and TNF-α [92]. In liver cancer cells, IL-37 inhibited IL-6 expression by hindering the STAT3 pathway, thereby inhibiting the inflammatory response of IL-6 [10]. IL-37 inhibited the expression and phosphorylation of STAT3, thus hindering the inflammatory effects of TNF-γ and IL-1β mediated by STAT3 [93].

IL-18, a member of the IL-1 family, was first discovered in 1989 and is an inflammatory factor [35]. IL-18 is the main inducer of IFN-γ, playing an important role in promoting the activation of inflammatory helper T cell 1 (Th1) and natural killer (NK) cells [94]. The IL-37/IL-18Rα complex combines with IL-1R8 to promote anti-inflammatory effects by activating STAT3 and transmitting inhibitory signals [7,32,36,39]. IL-37 is an endogenous factor that inhibits IL-18 effects. IL-37 has high homology with IL-18 and IL-18BP binds to IL-37. IL-18BP is a structural secreted protein with a high affinity for IL-18 [37,95], which when combined with IL-37 can enhance the ability of IL-18BP to inhibit IL-18-stimulated ITF-γ induction and inflammation [7,35,37,39]. However, Suzhao Li et al. reported that at micromolar concentrations, IL-37 binds to IL-18Rα and recruits IL-1R8, which may result in anti-inflammatory effects. At picomolar concentrations, the IL-37/IL-18Rα complex may recruit IL-18Rβ and the corresponding IL-18 signal, which may be associated with the inflammation process [21].

Based on the above studies, IL-37 has significant therapeutic potential through suppressing inflammation by inhibiting the transcription and expression of other inflammatory factors, including IL-6, IL-1β, IL-18 [6,12,34]. In tumors and some autoimmune diseases, IL-37 has exhibited its powerful anti-inflammatory ability [12]. Based on that, many CNS diseases are closely related to inflammation, and future research on CNS diseases may focus on the anti-inflammation function in some diseases, as well as expound and reveal its anti-inflammatory effect and mechanism. Due to the lack of research on the a subtype, and as IL-37a is the only subtype expressed in the brain, future research may focus on the protective effect of this subtype on the brain and explore the therapeutic and protective effects.

Author Contributions: X.L. wrote the manuscript; X.L. and B.Y. drew the figure; S.X., J.D. and C.P. provided information and references; L.L., X.K. and S.Z. reviewed this manuscript. All authors have read and agreed to the published version of the manuscript.

Funding: This study was supported by the National Science Foundation of China under Grant No. 81971008.

Conflicts of Interest: The authors declare no conflict of interest.

References

1. Bello, R.O.; Chin, V.K.; Abd Rachman Isnadi, M.F.; Abd Majid, R.; Atmadini Abdullah, M.; Lee, T.Y.; Amiruddin Zakaria, Z.; Hussain, M.K.; Basir, R. The Role, Involvement and Function(s) of Interleukin-35 and Interleukin-37 in Disease Pathogenesis. *Int. J. Mol. Sci.* **2018**, *19*, 1149. [CrossRef] [PubMed]
2. Kumar, S.; McDonnell, P.C.; Lehr, R.; Tierney, L.; Tzimas, M.N.; Griswold, D.E.; Capper, E.A.; Tal-Singer, R.; Wells, G.I.; Doyle, M.L.; et al. Identification and initial characterization of four novel members of the interleukin-1 family. *J. Biol. Chem.* **2000**, *275*, 10308–10314. [CrossRef] [PubMed]
3. Boraschi, D.; Lucchesi, D.; Hainzl, S.; Leitner, M.; Maier, E.; Mangelberger, D.; Oostingh, G.J.; Pfaller, T.; Pixner, C.; Posselt, G.; et al. IL-37: A new anti-inflammatory cytokine of the IL-1 family. *Eur. Cytokine Netw.* **2011**, *22*, 127–147. [CrossRef] [PubMed]
4. Taylor, S.L.; Renshaw, B.R.; Garka, K.E.; Smith, D.E.; Sims, J.E. Genomic organization of the interleukin-1 locus. *Genomics* **2002**, *79*, 726–733. [CrossRef]
5. Dinarello, C.A.; Nold-Petry, C.; Nold, M.; Fujita, M.; Li, S.; Kim, S.; Bufler, P. Suppression of innate inflammation and immunity by interleukin-37. *Eur. J. Immunol.* **2016**, *46*, 1067–1081. [CrossRef]
6. Bai, J.; Li, Y.; Li, M.; Tan, S.; Wu, D. IL-37 As a Potential Biotherapeutics of Inflammatory Diseases. *Curr. Drug Targets* **2020**, *21*, 855–863. [CrossRef]
7. Nold, M.F.; Li, Y.; Li, M.; Tan, S.; Wu, D. IL-37 is a fundamental inhibitor of innate immunity. *Nat. Immunol.* **2010**, *11*, 1014–1022. [CrossRef]
8. Wu, C.; Ma, J.; Yang, H.; Zhang, J.; Sun, C.; Lei, Y.; Liu, M.; Cao, J. Interleukin-37 as a biomarker of mortality risk in patients with sepsis. *J. Infect.* **2021**, *82*, 346–354. [CrossRef]
9. McNamee, E.N.; Masterson, J.C.; Jedlicka, P.; McManus, M.; Grenz, A.; Collins, C.B.; Nold, M.F.; Nold-Petry, C.; Bufler, P.; Dinarello, C.A.; et al. Interleukin 37 expression protects mice from colitis. *Proc. Natl. Acad. Sci. USA* **2011**, *108*, 16711–16716. [CrossRef]
10. Pu, X.Y.; Zheng, D.F.; Shen, A.; Gu, H.T.; Wei, X.F.; Mou, T.; Zhang, J.B.; Liu, R. IL-37b suppresses epithelial mesenchymal transition in hepatocellular carcinoma by inhibiting IL-6/STAT3 signaling. *Hepatobiliary Pancreat. Dis. Int.* **2018**, *17*, 408–415. [CrossRef]

11. Smith, D.E.; Renshaw, B.R.; Ketchem R.R.; Kubin, M.; Garka K.E.; Sims J.E. Four new members expand the interleukin-1 superfamily. *J. Biol. Chem.* **2000**, *275*, 1169–1175. [CrossRef] [PubMed]
12. Abulkhir, A.; Samarani, S.; Amre, D.; Duval, M.; Haddad, E.; Sinnett, D.; Leclerc, J.M.; Diorio, C.; Ahmad, A. A protective role of IL-37 in cancer: A new hope for cancer patients. *J. Leukoc. Biol.* **2017**, *101*, 395–406. [CrossRef]
13. Dinarello, C.; Arend, W.; Sims, J.; Smith, D.; Blumberg, H.; O'Neill, L.; Goldbach-Mansky, R.; Pizarro, T.; Hoffman, H.; Bufler, P.; et al. IL-1 family nomenclature. *Nat. Immunol.* **2010**, *11*, 973. [CrossRef] [PubMed]
14. Garlanda, C.; Riva, F.; Polentarutti, N.; Buracchi, C.; Sironi, M.; De Bortoli, M.; Muzio, M.; Bergottini, R.; Scanziani, E.; Vecchi, A.; et al. Intestinal inflammation in mice deficient in Tir8, an inhibitory member of the IL-1 receptor family. *Proc. Natl. Acad. Sci. USA* **2004**, *101*, 3522–3526. [CrossRef]
15. Li, S.; Amo-Aparicio, J.; Neff, C.P.; Tengesdal, I.W.; Azam, T.; Palmer, B.E.; López-Vales, R.; Bufler, P.; Dinarello, C.A. Role for nuclear interleukin-37 in the suppression of innate immunity. *Proc. Natl. Acad. Sci. USA* **2019**, *116*, 4456–4461. [CrossRef] [PubMed]
16. Chen, H.M.; Fujita, M. IL-37: A new player in immune tolerance. *Cytokine* **2015**, *72*, 113–114. [CrossRef]
17. Dinarello, C.A.; Bufler, P. Interleukin-37. *Semin. Immunol.* **2013**, *25*, 466–468. [CrossRef] [PubMed]
18. Fonseca-Camarillo, G.; Furuzawa-Carballeda, J.; Yamamoto-Furusho, J.K. Interleukin 35 (IL-35) and IL-37: Intestinal and peripheral expression by T and B regulatory cells in patients with Inflammatory Bowel Disease. *Cytokine* **2015**, *75*, 389–402. [CrossRef]
19. Shuai, X.; Wei-min, L.; Tong, Y.L.; Dong, N.; Sheng, Z.Y.; Yao, Y.M. Expression of IL-37 contributes to the immunosuppressive property of human CD4+CD25+ regulatory T cells. *Sci. Rep.* **2015**, *5*, 14478. [CrossRef]
20. Bufler, P.; Gamboni-Robertson, F.; Azam, T.; Kim, S.H.; Dinarello, C.A. Interleukin-1 homologues IL-1F7b and IL-18 contain functional mRNA instability elements within the coding region responsive to lipopolysaccharide. *Biochem. J.* **2004**, *381 Pt 2*, 503–510. [CrossRef]
21. Li, S.; Neff, C.P.; Barber, K.; Hong, J.; Luo, Y.; Azam, T.; Palmer, B.E.; Fujita, M.; Garlanda, C.; Mantovani, A.; et al. Extracellular forms of IL-37 inhibit innate inflammation in vitro and in vivo but require the IL-1 family decoy receptor IL-1R8. *Proc. Natl. Acad. Sci. USA* **2015**, *112*, 2497–2502. [CrossRef] [PubMed]
22. Ballak, D.B.; van Diepen, J.A.; Moschen, A.R.; Jansen, H.J.; Hijmans, A.; Groenhof, G.J.; Leenders, F.; Bufler, P.; Boekschoten, M.V.; Müller, M.; et al. IL-37 protects against obesity-induced inflammation and insulin resistance. *Nat. Commun.* **2014**, *5*, 4711. [CrossRef] [PubMed]
23. Li, J.; zhai, Y.; Ao, L.; Hui, H.; Fullerton, D.A.; Dinarello, C.A.; Meng, X. Interleukin-37 suppresses the inflammatory response to protect cardiac function in old endotoxemic mice. *Cytokine* **2017**, *95*, 55–63. [CrossRef] [PubMed]
24. Hou, T.; Tsang, M.S.; Kan, L.L.; Li, P.; Chu, I.M.; Lam, C.W.; Wong, C.K. IL-37 Targets TSLP-Primed Basophils to Alleviate Atopic Dermatitis. *Int. J. Mol. Sci.* **2021**, *22*, 7393. [CrossRef] [PubMed]
25. Wang, J.; Shen, Y.; Li, C.; Liu, C.; Wang, Z.H.; Li, Y.S.; Ke, X.; Hu, G.H. IL-37 attenuates allergic process via STAT6/STAT3 pathways in murine allergic rhinitis. *Int. Immunopharmacol.* **2019**, *69*, 27–33. [CrossRef] [PubMed]
26. Wang, L.; Wang, Y.; Xia, L.; Shen, H.; Lu, J. Elevated frequency of IL-37- and IL-18Rα-positive T cells in the peripheral blood of rheumatoid arthritis patients. *Cytokine* **2018**, *110*, 291–297. [CrossRef]
27. Ragab, D.; Mobasher, S.; Shabaan, E. Elevated levels of IL-37 correlate with T cell activation status in rheumatoid arthritis patients. *Cytokine* **2019**, *113*, 305–310. [CrossRef]
28. Meng, P.; Chen, Z.G.; Zhang, T.T.; Liang, Z.Z.; Zou, X.L.; Yang, H.L.; Li, H.T. IL-37 alleviates house dust mite-induced chronic allergic asthma by targeting TSLP through the NF-κB and ERK1/2 signaling pathways. *Immunol. Cell Biol.* **2019**, *97*, 403–415. [CrossRef]
29. Osborne, D.G.; Domenico, J.; Luo, Y.; Reid, A.L.; Zhai, Z.; Gao, D.; Ziman, M.; Dinarello, C.A.; Robinson, A.W.; Fujita, M.; et al. Interleukin-37 is highly expressed in regulatory T cells of melanoma patients and enhanced by melanoma cell secretome. *Mol. Carcinog.* **2019**, *58*, 1670–1679. [CrossRef]
30. Sharma, S.; Kulk, N.; Nold, M.F.; Gräf, R.; Kim, S.H.; Reinhardt, D.; Dinarello, C.A.; Bufler, P. The IL-1 family member 7b translocates to the nucleus and down-regulates proinflammatory cytokines. *J. Immunol.* **2008**, *180*, 5477–5482. [CrossRef]
31. Bulau, A.M.; Nold, M.F.; Li, S.; Nold-Petry, C.A.; Fink, M.; Mansell, A.; Schwerd, T.; Hong, J.; Rubartelli, A.; Dinarello, C.A.; et al. Role of caspase-1 in nuclear translocation of IL-37, release of the cytokine, and IL-37 inhibition of innate immune responses. *Proc. Natl. Acad. Sci. USA* **2014**, *111*, 2650–2655. [CrossRef] [PubMed]
32. Kumar, S.; Hanning, C.R.; Brigham-Burke, M.R.; Rieman, D.J.; Lehr, R.; Khandekar, S.; Kirkpatrick, R.B.; Scott, G.F.; Lee, J.C.; Lynch, F.J.; et al. Interleukin-1F7B (IL-1H4/IL-1F7) is processed by caspase-1 and mature IL-1F7B binds to the IL-18 receptor but does not induce IFN-gamma production. *Cytokine* **2002**, *18*, 61–71. [CrossRef] [PubMed]
33. Pan, G.; Risser, P.; Mao, W.; Baldwin, D.T.; Zhong, A.W.; Filvaroff, E.; Yansura, D.; Lewis, L.; Eigenbrot, C.; Henzel, W.J.; et al. IL-1H, an interleukin 1-related protein that binds IL-18 receptor/IL-1Rrp. *Cytokine* **2001**, *13*, 1–7. [CrossRef] [PubMed]
34. Grimsby, S.; Jaensson, H.; Dubrovska, A.; Lomnytska, M.; Hellman, U.; Souchelnytskyi, S. Proteomics-based identification of proteins interacting with Smad3: SREBP-2 forms a complex with Smad3 and inhibits its transcriptional activity. *FEBS Lett.* **2004**, *577*, 93–100. [CrossRef]
35. Yasuda, K.; Nakanishi, K.; Tsutsui, H. Interleukin-18 in Health and Disease. *Int. J. Mol. Sci.* **2019**, *20*, 649. [CrossRef]

36. Nold-Petry, C.A.; Lo, C.Y.; Rudloff, L.; Elgass, K.D.; Li, S.; Gantier, M.P.; Lotz-Havla, A.S.; Gersting, S.W.; Cho, S.X.; Lao, J.C.; et al. IL-37 requires the receptors IL-18Rα and IL-1R8 (SIGIRR) to carry out its multifaceted anti-inflammatory program upon innate signal transduction. *Nat. Immunol.* **2015**, *16*, 354–365. [CrossRef]
37. Novick, D.; Kim, S.; Kaplanski, G.; Dinarello, C.A. Interleukin-18, more than a Th1 cytokine. *Semin. Immunol.* **2013**, *25*, 439–448. [CrossRef]
38. Wald, D.; Qin, J.; Zhao, Z.; Qian, Y.; Naramura, M.; Tian, L.; Towne, J.; Sims, J.E.; Stark, G.R.; Li, X. SIGIRR, a negative regulator of Toll-like receptor-interleukin 1 receptor signaling. *Nat. Immunol.* **2003**, *4*, 920–927. [CrossRef]
39. Bufler, P.; Azam, T.; Gamboni-Robertson, F.; Reznikov, L.L.; Kumar, S.; Dinarello, C.A.; Kim, S.H. A complex of the IL-1 homologue IL-1F7b and IL-18-binding protein reduces IL-18 activity. *Proc. Natl. Acad. Sci. USA* **2002**, *99*, 13723–13728. [CrossRef]
40. Li, L.; Wei, J.; Suber, T.L.; Ye, Q.; Miao, J.; Li, S.; Taleb, S.J.; Tran, K.C.; Tamaskar, A.S.; Zhao, J.; et al. IL-37-induced activation of glycogen synthase kinase 3β promotes IL-1R8/Sigirr phosphorylation, internalization, and degradation in lung epithelial cells. *J. Cell. Physiol.* **2021**, *236*, 5676–5685. [CrossRef]
41. Beck, B.; Cameron, P.A.; Braaf, S.; Nunn, A.; Fitzgerald, M.C.; Judson, R.T.; Teague, W.J.; Lennox, A.; Middleton, J.W.; Harrison, J.E. Traumatic spinal cord injury in Victoria 2007–2016. *Med. J. Aust.* **2019**, *210*, 360–366. [CrossRef] [PubMed]
42. Inglis, T.; Banaszek, D.; Rivers, C.S.; Kurban, D.; Ewaniew, N.; Fallah, N.; Waheed, Z.; Christie, S.; Fox, R.; Thiong, J.M.; et al. In-Hospital Mortality for the Elderly with Acute Traumatic Spinal Cord Injury. *J. Neurotrauma* **2020**, *37*, 2332–2342. [CrossRef] [PubMed]
43. Zeitouni, D.; Catalino, M.; Kessler, B.; Pate, V.; Stürmer, T.; Quinsey, C.; Bhowmick, D.A. 1-Year Mortality and Surgery Incidence in Older US Adults with Cervical Spine Fracture. *World Neurosurg.* **2020**, *141*, e858–e863. [CrossRef] [PubMed]
44. Coll-Miró, M.; Francos-Quijorna, I.; Santos-Nogueira, E.; Torres-Espin, A.; Bufler, P.; Dinarello, C.A.; López-Vales, R. Beneficial effects of IL-37 after spinal cord injury in mice. *Proc. Natl. Acad. Sci. USA* **2016**, *113*, 1411–1416. [CrossRef] [PubMed]
45. Chen, Y.; Wang, D.; Cao, S.; Hou, G.; Ma, G.; Shi, B. Association between Serum IL-37 and Spinal Cord Injury: A Prospective Observational Study. *BioMed Res. Int.* **2020**, *2020*, 6664313. [CrossRef] [PubMed]
46. Amo-Aparicio, J.; Sanchez-Fernandez, A.; Li, S.; Eisenmesser, E.Z.; Garlanda, C.; Dinarello, C.A.; Lopez-Vales, R. Extracellular and nuclear roles of IL-37 after spinal cord injury. *Brain Behav. Immun.* **2021**, *91*, 194–201. [CrossRef]
47. Wekerle, H. Lessons from multiple sclerosis: Models, concepts, observations. *Ann. Rheum. Dis.* **2008**, *67* (Suppl. 3), iii56–iii60. [CrossRef]
48. Oh, J.; Vidal-Jordana, A.; Montalban, X. Multiple sclerosis: Clinical aspects. *Curr. Opin. Neurol.* **2018**, *31*, 752–759. [CrossRef]
49. Sánchez-Fernández, A.; Zandee, S.; Amo-Aparicio, J.; Charabati, M.; Prat, A.; Garlanda, C.; Eisenmesser, E.Z.; Dinarello, C.A.; López-Vales, R. IL-37 exerts therapeutic effects in experimental autoimmune encephalomyelitis through the receptor complex IL-1R5/IL-1R8. *Theranostics* **2021**, *11*, 1–13. [CrossRef]
50. Cavalli, E.; Mazzon, E.; Basile, M.S.; Mammana, S.; Pennisi, M.; Fagone, P.; Kalfin, R.; Martinovic, V.; Ivanovic, J.; Andabaka, M.; et al. In Silico and In Vivo Analysis of IL37 in Multiple Sclerosis Reveals Its Probable Homeostatic Role on the Clinical Activity, Disability, and Treatment with Fingolimod. *Molecules* **2019**, *25*, 20. [CrossRef]
51. Kouchaki, E.; Tamtaji, O.R.; Dadgostar, E.; Karami, M.; Nikoueinejad, H.; Akbari, H. Correlation of Serum Levels of IL-33, IL-37, Soluble Form of Vascular Endothelial Growth Factor Receptor 2 (VEGFR2), and Circulatory Frequency of VEGFR2-expressing Cells with Multiple Sclerosis Severity. *Iran. J. Allergy Asthma Immunol.* **2017**, *16*, 329–337. [PubMed]
52. Farrokhi, M.; Rezaei, A.; Amani-Beni, A.; Etemadifar, M.; Kouchaki, E.; Zahedi, A. Increased serum level of IL-37 in patients with multiple sclerosis and neuromyelitis optica. *Acta Neurol. Belg.* **2015**, *115*, 609–614. [CrossRef] [PubMed]
53. Willison, H.J.; Jacobs, B.C.; van Doorn, P.A. Guillain-Barré syndrome. *Lancet* **2016**, *388*, 717–727. [CrossRef]
54. Malek, E.; Salameh, J. Guillain-Barre Syndrome. *Semin. Neurol.* **2019**, *39*, 589–595. [CrossRef]
55. van den Berg, B.; Walgaard, C.; Drenthen, J.; Fokke, C.; Jacobs, B.C.; van Doorn, P.A. Guillain-Barré syndrome: Pathogenesis, diagnosis, treatment and prognosis. *Nat. Rev. Neurol.* **2014**, *10*, 469–482. [CrossRef]
56. Li, C.; Zhao, P.; Sun, X.; Che, Y.; Jiang, Y. Elevated levels of cerebrospinal fluid and plasma interleukin-37 in patients with Guillain-Barré syndrome. *Mediat. Inflamm.* **2013**, *2013*, 639712. [CrossRef]
57. Sakai, N.; Van Sweringen, H.L.; Belizaire, R.M.; Quillin, R.C.; Schuster, R.; Blanchard, J.; Burns, J.M.; Tevar, A.D.; Edwards, M.J.; Lentsch, A.B. Interleukin-37 reduces liver inflammatory injury via effects on hepatocytes and non-parenchymal cells. *J. Gastroenterol. Hepatol.* **2012**, *27*, 1609–1616. [CrossRef]
58. Bulau, A.M.; Fink, M.; Maucksch, C.; Kappler, R.; Mayr, D.; Wagner, K.; Bufler, P. In vivo expression of interleukin-37 reduces local and systemic inflammation in concanavalin A-induced hepatitis. *Sci. World J.* **2011**, *11*, 2480–2490. [CrossRef]
59. Avila, J.; Perry, G. A Multilevel View of the Development of Alzheimer's Disease. *Neuroscience* **2021**, *457*, 283–293. [CrossRef]
60. Zhao, X.; Li, C.; Ding, G.; Heng, Y.; Li, A.; Wang, W.; Hou, H.; Wen, J.; Zhang, Y. The Burden of Alzheimer's Disease Mortality in the United States 1999–2018. *J. Alzheimers Dis.* **2021**, *82*, 803–813. [CrossRef]
61. 2021 Alzheimer's disease facts and figures. *Alzheimers Dement.* **2021**, *17*, 327–406. [CrossRef]
62. Guerrero, A.; de Strooper, B.; Arancibia-Cárcamo, I.L. Cellular senescence at the crossroads of inflammation and Alzheimer's disease. *Trends Neurosci.* **2021**, *44*, 714–727. [CrossRef] [PubMed]
63. Franceschi, C.; Campisi, J. Chronic inflammation (inflammaging) and its potential contribution to age-associated diseases. *J. Gerontol. A Biol. Sci. Med. Sci.* **2014**, *69* (Suppl. 1), S4–S9. [CrossRef] [PubMed]

64. Coppé, J.P.; Patil, C.K.; Rodier, F.; Sun, Y.; Muñoz, D.P.; Goldstein, J.; Nelson, P.S.; Desprez, P.Y. Senescence-associated secretory phenotypes reveal cell-nonautonomous functions of oncogenic RAS and the p53 tumor suppressor. *PLoS Biol.* **2008**, *6*, 2853–2868. [CrossRef] [PubMed]
65. Serrano-Pozo, A.; Mielke, M.L.; Gómez-Isla, T.; Betensky, R.A.; Growdon, J.H.; Frosch, M.P.; Hyman, B.T. Reactive glia not only associates with plaques but also parallels tangles in Alzheimer's disease. *Am. J. Pathol.* **2011**, *179*, 1373–1384. [CrossRef]
66. Wang, W.Y.; Tan, M.S.; Yu, J.T.; Tan, L. Role of pro-inflammatory cytokines released from microglia in Alzheimer's disease. *Ann. Transl. Med.* **2015**, *3*, 136.
67. Czeh, M.; Gressens, P.; Kaindl, A.M. The yin and yang of microglia. *Dev. Neurosci.* **2011**, *33*, 199–209. [CrossRef]
68. Pratchayasakul, W.; Kerdphoo, S.; Petsophonsakul, P.; Pongchaidecha, A.; Chattipakorn, N.; Chattipakorn, S.C. Effects of high-fat diet on insulin receptor function in rat hippocampus and the level of neuronal corticosterone. *Life Sci.* **2011**, *88*, 619–627. [CrossRef]
69. Chou, L.M.; Lin, C.I.; Chen, Y.H.; Liao, H.; Lin, S.H. A diet containing grape powder ameliorates the cognitive decline in aged rats with a long-term high-fructose-high-fat dietary pattern. *J. Nutr. Biochem.* **2016**, *34*, 52–60. [CrossRef]
70. Mohamed, R.A.; Mohamed, R.A.; Abdallah, D.M.; El-Brairy, A.I.; Ahmed, K.A.; El-Abhar, H.S. Palonosetron/Methyllycaconitine Deactivate Hippocampal Microglia 1, Inflammasome Assembly and Pyroptosis to Enhance Cognition in a Novel Model of Neuroinflammation. *Molecules* **2021**, *26*, 5068. [CrossRef]
71. Ogunmokun, G.; Dewanjee, S.; Chakraborty, P.; Valupadas, C.; Chaudhary, A.; Kolli, V.; Anand, U.; Vallamkondu, J.; Goel, P.; Paluru, H.P.R. The Potential Role of Cytokines and Growth Factors in the Pathogenesis of Alzheimer's Disease. *Cells* **2021**, *10*, 2790. [CrossRef] [PubMed]
72. Spooren, A.; Kolmus, K.; Laureys, G.; Clinckers, R.; De Keyser, J.; Haegeman, G.; Gerlo, S. Interleukin-6, a mental cytokine. *Brain Res. Rev.* **2011**, *67*, 157–183. [CrossRef] [PubMed]
73. Chakrabarty, P.; Jansen-West, K.; Beccard, A.; Ceballos-Diaz, C.; Levites, Y.; Verbeeck, C.; Zubair, A.C.; Dickson, D.; Golde, T.E.; Das, P. Massive gliosis induced by interleukin-6 suppresses Abeta deposition in vivo: Evidence against inflammation as a driving force for amyloid deposition. *FASEB J.* **2010**, *24*, 548–559. [CrossRef] [PubMed]
74. Belaya, I.; Kucháriková, N.; Górová, V.; Kysenius, K.; Hare, D.J.; Crouch, P.J.; Malm, T.; Atalay, M.; White, A.R.; Liddell, J.R.; et al. Regular Physical Exercise Modulates Iron Homeostasis in the 5xFAD Mouse Model of Alzheimer's Disease. *Int. J. Mol. Sci.* **2021**, *22*, 8715. [CrossRef]
75. Sutinen, E.M.; Pirttilä, T.; Anderson, G.; Salminen, A.; Ojala, J.O. The Role of Interleukin-18, Oxidative Stress and Metabolic Syndrome in Alzheimer's Disease. *J. Neuroinflamm.* **2012**, *9*, 199. [CrossRef]
76. Yu, J.T.; Chang, R.C.; Tan, L. Calcium dysregulation in Alzheimer's disease: From mechanisms to therapeutic opportunities. *Prog. Neurobiol.* **2009**, *89*, 240–255. [CrossRef]
77. Charrad, R.; Berraïes, A.; Hamdi, B.; Ammar, J.; Hamzaoui, K.; Hamzaoui, A. Anti-inflammatory activity of IL-37 in asthmatic children: Correlation with inflammatory cytokines TNF-α, IL-β, IL-6 and IL-17A. *Immunobiology* **2016**, *221*, 182–187. [CrossRef]
78. Zeng, M.; Dang, W.; Chen, B.; Qing, Y.; Xie, W.; Zhao, M.; Zhou, J. IL-37 inhibits the production of pro-inflammatory cytokines in MSU crystal-induced inflammatory response. *Clin. Rheumatol.* **2016**, *35*, 2251–2258. [CrossRef]
79. van Geffen, E.W.; van Caam, A.P.; van Beuningen, H.M.; Vitters, E.L.; Schreurs, W.; van de Loo, F.A.; van Lent, P.L.; Koenders, M.I.; Blaney Davidson, E.N.; van der Kraan, P.M. IL37 dampens the IL1β-induced catabolic status of human OA chondrocytes. *Rheumatology* **2017**, *56*, 351–361. [CrossRef]
80. Jiang, M.; Wang, Y.; Zhang, H.; Ji, Y.; Zhao, P.; Sun, R.; Zhang, C. IL-37 inhibits invasion and metastasis in non-small cell lung cancer by suppressing the IL-6/STAT3 signaling pathway. *Thorac. Cancer* **2018**, *9*, 621–629. [CrossRef]
81. Yan, P.; Zhang, Y.; Wang, C.; Lv, F.; Song, L. Interleukin-37 (IL-37) Suppresses Pertussis Toxin-Induced Inflammatory Myopathy in a Rat Model. *Med. Sci. Monit.* **2018**, *24*, 9187–9195. [CrossRef] [PubMed]
82. Cavalli, G.; Dinarello, C.A. Suppression of inflammation and acquired immunity by IL-37. *Immunol. Rev.* **2018**, *281*, 179–190. [CrossRef]
83. Luo, P.; Peng, S.; Yan, Y.; Ji, P.; Xu, J. IL-37 inhibits M1-like macrophage activation to ameliorate temporomandibular joint inflammation through the NLRP3 pathway. *Rheumatology* **2020**, *59*, 3070–3080. [CrossRef]
84. Global, regional, and national burden of neurological disorders during 1990–2015: A systematic analysis for the Global Burden of Disease Study 2015. *Lancet Neurol.* **2017**, *16*, 877–897. [CrossRef]
85. Xing, L.; Jing, L.; Tian, Y.; Liu, S.; Lin, M.; Du, Z.; Ren, G.; Sun, Q.; Shi, L.; Dai, D.; et al. High prevalence of stroke and uncontrolled associated risk factors are major public health challenges in rural northeast China: A population-based study. *Int. J. Stroke* **2020**, *15*, 399–411. [CrossRef] [PubMed]
86. Zhang, F.; Zhu, T.; Li, H.; He, Y.; Zhang, Y.; Huang, N.; Zhang, G.; Li, Y.; Chang, D.; Li, X. Plasma Interleukin-37 is Elevated in Acute Ischemic Stroke Patients and Probably Associated With 3-month Functional Prognosis. *Clin. Interv. Aging* **2020**, *15*, 1285–1294. [CrossRef] [PubMed]
87. Zhang, Y.; Zhang, Y.; Xu, C.; Wang, H.; Nan, S. Serum Interleukin-37 Increases in Patients after Ischemic Stroke and Is Associated with Stroke Recurrence. *Oxid. Med. Cell. Longev.* **2021**, *2021*, 5546991. [CrossRef]
88. Zhang, W.B.; Zeng, Y.Y.; Wang, F.; Cheng, L.; Tang, W.J.; Wang, X.Q. A high neutrophil-to-lymphocyte ratio predicts hemorrhagic transformation of large atherosclerotic infarction in patients with acute ischemic stroke. *Aging* **2020**, *12*, 2428–2439. [CrossRef]
89. Perez-de-Puig, I.; Miró-Mur, F.; Ferrer-Ferrer, M.; Gelpi, E.; Pedragosa, J.; Justicia, C.; Urra, X.; Chamorro, A.; Planas, A.M. Neutrophil recruitment to the brain in mouse and human ischemic stroke. *Acta Neuropathol.* **2015**, *129*, 239–257. [CrossRef]

90. Kolaczkowska, E.; Kubes, P. Neutrophil recruitment and function in health and inflammation. *Nat. Rev. Immunol.* **2013**, *13*, 159–175. [CrossRef]
91. Tsilioni, I.; Patel, A.B.; Pantazopoulos, H.; Berretta, S.; Conti, P.; Leeman, S.E.; Theoharides, T.C. IL-37 is increased in brains of children with autism spectrum disorder and inhibits human microglia stimulated by neurotensin. *Proc. Natl. Acad. Sci. USA* **2019**, *116*, 21659–21665. [CrossRef] [PubMed]
92. Wang, S.; Huang, Z.; Li, W.; He, S.; Wu, H.; Zhu, J.; Li, R.; Liang, Z.; Chen, Z. IL-37 expression is decreased in patients with hyperhomocysteinemia and protects cells from inflammatory injury by homocysteine. *Mol. Med. Rep.* **2020**, *21*, 371–378. [CrossRef] [PubMed]
93. Wang, S.; An, W.; Yao, Y.; Chen, R.; Zheng, X.; Yang, W.; Zhao, Y.; Hu, X.; Jiang, E.; Bie, Y.; et al. Interleukin 37 Expression Inhibits STAT3 to Suppress the Proliferation and Invasion of Human Cervical Cancer Cells. *J. Cancer* **2015**, *6*, 962–969. [CrossRef] [PubMed]
94. Conti, P.; Stellin, L.; Caraffa, A.; Gallenga, C.E.; Ross, R.; Kritas, S.K.; Frydas, I.; Younes, A.; Di Emidio, P.; Ronconi, G. Advances in Mast Cell Activation by IL-1 and IL-33 in Sjögren's Syndrome: Promising Inhibitory Effect of IL-37. *Int. J. Mol. Sci.* **2020**, *21*, 4297. [CrossRef] [PubMed]
95. Zhou, T.; Zhou, T.; Damsky, W.; Weizman, O.E.; McGeary, M.K.; Hartmann, K.P.; Rosen, C.E.; Fischer, S.; Jackson, R.; Flavell, R.A.; et al. IL-18BP is a secreted immune checkpoint and barrier to IL-18 immunotherapy. *Nature* **2020**, *583*, 609–614. [CrossRef] [PubMed]

Review

Environmental Impact on the Epigenetic Mechanisms Underlying Parkinson's Disease Pathogenesis: A Narrative Review

Efthalia Angelopoulou [1,2], Yam Nath Paudel [3], Sokratis G. Papageorgiou [2] and Christina Piperi [1,*]

[1] Department of Biological Chemistry, Medical School, National and Kapodistrian University of Athens, 11527 Athens, Greece; angelthal@med.uoa.gr

[2] 1st Department of Neurology, Medical School, National and Kapodistrian University of Athens, Eginition University Hospital, 11527 Athens, Greece; sokpapa@med.uoa.gr

[3] Neuropharmacology Research Laboratory, Jeffrey Cheah School of Medicine and Health Sciences, Monash University Malaysia, Bandar Sunway 47500, Malaysia; yam.paudel@monash.edu

* Correspondence: cpiperi@med.uoa.gr; Tel.: +30-2107462610

Abstract: Parkinson's disease (PD) is the second most common neurodegenerative disorder with an unclear etiology and no disease-modifying treatment to date. PD is considered a multifactorial disease, since both genetic and environmental factors contribute to its pathogenesis, although the molecular mechanisms linking these two key disease modifiers remain obscure. In this context, epigenetic mechanisms that alter gene expression without affecting the DNA sequence through DNA methylation, histone post-transcriptional modifications, and non-coding RNAs may represent the key mediators of the genetic–environmental interactions underlying PD pathogenesis. Environmental exposures may cause chemical alterations in several cellular functions, including gene expression. Emerging evidence has highlighted that smoking, coffee consumption, pesticide exposure, and heavy metals (manganese, arsenic, lead, etc.) may potentially affect the risk of PD development at least partially via epigenetic modifications. Herein, we discuss recent accumulating pre-clinical and clinical evidence of the impact of lifestyle and environmental factors on the epigenetic mechanisms underlying PD development, aiming to shed more light on the pathogenesis and stimulate future research.

Keywords: Parkinson disease; epigenetics; DNA methylation; histone modifications; lcRNAs; environmental toxins; smoking; pesticides; coffee; diagnosis

Citation: Angelopoulou, E.; Paudel, Y.N.; Papageorgiou, S.G.; Piperi, C. Environmental Impact on the Epigenetic Mechanisms Underlying Parkinson's Disease Pathogenesis: A Narrative Review. *Brain Sci.* **2022**, *12*, 175. https://doi.org/10.3390/brainsci12020175

Academic Editor: Alicia M. Pickrell

Received: 31 December 2021
Accepted: 22 January 2022
Published: 28 January 2022

Publisher's Note: MDPI stays neutral with regard to jurisdictional claims in published maps and institutional affiliations.

Copyright: © 2022 by the authors. Licensee MDPI, Basel, Switzerland. This article is an open access article distributed under the terms and conditions of the Creative Commons Attribution (CC BY) license (https://creativecommons.org/licenses/by/4.0/).

1. Introduction

Parkinson's disease (PD) is the most common age-related neurodegenerative movement disorder, affecting approximately 1–2% of the population above the age of 60 years [1]. Neuropathologically, it is mainly characterized by the accumulation of Lewy bodies and Lewy neurites mainly constituting of α-synuclein and the progressive loss of dopaminergic neurons in the substantia nigra pars compacta (SNpc), resulting in nigrostriatal degeneration [2].

Although the exact pathogenic mechanisms remain obscure, several interconnecting pathophysiological processes have been demonstrated to be involved in PD development, including mitochondrial dysfunction, oxidative stress, dysregulation of apoptosis, autophagy impairment, proteosomal dysfunction, and excessive neuroinflammation [2]. Inflammation is rather a "double-edged sword", since it protects against pathogens and helps to clear out neurotoxins, but it can also induce cytotoxicity and degeneration [3,4]. An inflammatory imbalance, favoring excessive microglial activation (M1 phenotype, pro-inflammatory) against anti-inflammatory responses (M2 phenotype, anti-inflammatory) have been consistently shown to contribute to neurodegenerative disorders, including PD [3,5,6]. Over-activated microglia release pro-inflammatory cytokines, such as TNF-α, IL-6 and IL-1β, induce oxidative stress, α-synuclein accumulation, as well as autophagy and mitochondrial dysfunction, finally leading to a vicious cycle of neurodegeneration [3,7,8].

Viral infections have been also associated with PD progression, potentially acting as triggering factors promoting neuroinflammation [4]. Of note, recent evidence has highlighted the potential role of human endogenous retroviruses (HERVs) in several neurodegenerative diseases [5]. HERVs represent ancient retroviral elements accounting for about 8% of the human genome that can be re-activated by various environmental factors including viruses, resulting in the potential formation of viral particles [5,6]. Although their role in PD pathogenesis has not been elucidated, they could constitute another pathophysiological component of the disease that should be further explored. The peripheral gastrointestinal dysfunction and inflammation have been demonstrated to possibly contribute to dopaminergic neuronal cell death [4]. In addition, the spread and transmission of α-synuclein pathology in a prion-like manner represents another emerging concept in the field of PD pathophysiology that is still under investigation [7]. These diverse pathophysiological contexts highlight the complex but overlapping hypotheses considered to contribute to PD pathogenesis.

Patients with PD display both motor and non-motor manifestations. According to the Movement Disorder Society's (MDS) PD criteria, parkinsonism is the necessary core feature of PD, defined as bradykinesia plus rigidity or a rest tremor [8]. For the identification of PD as the cause of parkinsonism, there are also absolute exclusion criteria ruling out PD, including cerebellar abnormalities, downward vertical supranuclear gaze palsy, no response to high-dose levodopa treatment, and drug-induced parkinsonism. Red flags for PD diagnosis include rapid progression, early bulbar dysfunction or severe autonomic failure, bilateral symmetric parkinsonism at onset, and early recurrent falls among others. These features should be counterbalanced by additional supportive criteria for PD diagnosis, such as levodopa-induced dyskinesia, rest tremor of a limb, olfactory loss, and a dramatic response to dopaminergic therapy.

Currently, available treatment options mainly involve levodopa and dopamine receptor agonists, which do not halt disease progression, exert only partial symptomatic relief, and are often accompanied by severe adverse effects, such as motor complications (fluctuations and dyskinesias), orthostatic hypotension, and impulse control disorder [2].

Several genetic causes of PD have been identified, including mutations in the *SNCA*, *LRRK2*, *GBA*, *PINK1*, and *Parkin* genes following a Mendelian inheritance pattern, which exhibit variable penetrance and account for only 5–10% of all PD cases [9]. Except for these rare genetic forms, the etiology of most cases of PD remains obscure [10]. Several genes potentially associated with PD risk have also been identified by genome wide association studies (GWAS), including variants of causative genes of PD, such as *MAPT* H1 haplotype, the promoter region of *SNCA* [10], a common polymorphism of *UCHL1*, and a variant of *LRRK2* [11]. Key pathways related to PD susceptibility genes include dopamine transport and metabolism (*DRD2*, *MAO-B*, *DAT*, and *COMT*), oxidative stress (*SOD2* and *NOS*), and xenobiotics metabolism (*CYP2D6*, *NAT2*, and *GSTs*) [11,12]. However, many of these studies have provided conflicting results, attributed to ethnic genetic heterogeneities and potential interactions with environmental exposures [11].

Several studies have demonstrated that particular gene polymorphisms interact with exposure to cigarette smoking, pesticides, or coffee to differentially affect the risk for PD development, although with inconsistent results [13]. Advancing age is the strongest risk factor for PD, since its incidence increases exponentially after the age of 60 years [9]. 1-methyl-4-phenyl-1,2,3,6-tetrahydropyridine (MPTP) was first observed to cause parkinsonism in drug users as a contaminant in heroin. MPTP, which is currently widely used as a neurotoxin for mimicking PD in animal models, shares common chemical properties with paraquat, a herbicide [9]. Since then, accumulating epidemiological evidence has shown that other environmental factors, such as tobacco smoking and coffee consumption may protect against PD development, while exposure to pesticides or specific heavy metals including manganese or lead, may increase this risk [14].

However, neither the genetic nor the toxin-based animal models of PD accurately reflect human PD pathology [9]. Therefore, PD is proposed to be a rather multifactorial disorder and a complex interplay between genetic–environmental interactions may underlie its pathogenesis [9]. Pre-clinical evidence has revealed that the cellular effects of several PD-related genes and environmental factors share some common mechanisms, including mitochondrial and autophagy impairment, oxidative stress, and inflammation [9]. Despite extensive research efforts, the exact molecular mechanisms linking these two key disease modifiers are still obscure. In this context, emerging evidence has shown that epigenetics, referring to alterations in gene expression without affecting the DNA sequence, may play an important role in the pathophysiology of neurodegenerative disorders, including PD, potentially representing a mechanistic bridge between sporadic PD and exposure to environmental factors.

In addition, it has been suggested that differences in age of disease onset, severity, and penetrance can possibly be explained by epigenetic modifications [11]. However, till now, very few environmental factors have been demonstrated to cause epigenetic modifications in PD, including smoking, pesticides, and heavy metals [11].

In this narrative review, we discuss the accumulating evidence on the role of environmental and lifestyle factors in epigenetic modifications regulating PD pathogenesis, focusing on smoking, coffee consumption, exposure to pesticides, and specific heavy metals, whose role in PD has been extensively studied. Although the role of epigenetics in PD has already been addressed elsewhere [15–18], there is no recent review focusing specifically on the epigenetic mechanisms that underlie the pathogenesis of PD from the scope of specific environmental factors. Despite the fact that the field is still in its infancy, we have compiled the available preclinical and clinical results of relative studies, aiming to address the way that exposure to these factors may affect PD development through epigenetic regulation, in order to pave the way for future research on PD pathogenesis, diagnosis, and treatment. Based on the current evidence, potential novel pathways are suggested that need to be explored for better clarification of this interesting relationship.

2. Methods

A literature search was performed in the PubMed and Scopus Databases, aiming to identify studies exploring and discussing any concepts on the epigenetic mechanisms underlying the impact of environmental factors on the pathogenesis of PD. We focused on the most well studied environmental factors associated with PD, including tobacco smoking, coffee, pesticides, and heavy metals, and the following keywords were used in various combinations: "epigenetics", "epigenetic", "DNA methylation", "histone modifications", "histone acetylation", "miRNA", "microRNA", "non-coding RNA", "lncRNA", "Parkinson's disease", "environmental", "environment", "smoking", "tobacco", "coffee", "caffeine", "pesticides", "herbicides", "heavy metals", "environmental toxin", "gene polymorphism", "gene-environment interaction", and "genetic-environmental interaction". We searched for both original pre-clinical and clinical studies, as well as reviews, published in the English language, until December of 2021. Although our aim was not to perform a systematic review, we also screened each relevant study for additional results to identify possible further articles that could help to better explore the epigenetic mechanisms mediating the effects of the environmental factors on PD.

3. Epigenetic Modifications in PD

Epigenetic modifications refer to changes in gene expression without affecting the DNA sequence; they occur throughout a lifetime, depending on several environmental factors [10]. Epigenetic mechanisms constitute heritable but possibly reversible alterations, mainly including DNA methylation and post-transcriptional histone modifications, as well as microRNAs (miRNAs) and other non-coding RNAs (Figure 1) [10].

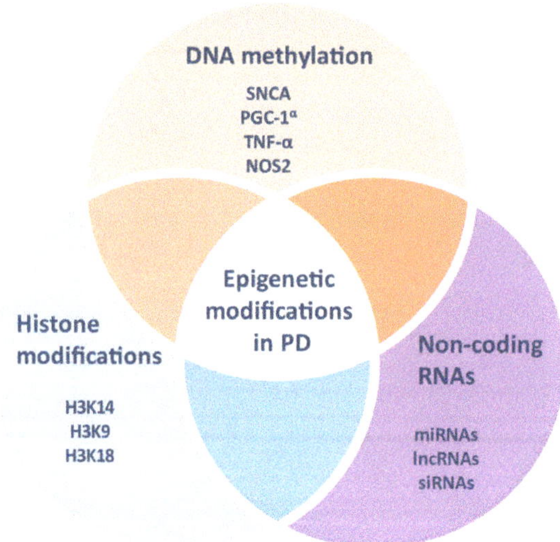

Figure 1. Epigenetic mechanisms regulating the pathogenesis of Parkinson's disease.

Epigenetics are majorly involved in several cellular functions of the central nervous system (CNS), such as neurogenesis, synaptic plasticity, and the learning process. Given the implication of epigenetic mechanisms in both physiological and pathophysiological processes in the CNS [19], it is not surprising that they also play a crucial role in the development of multifactorial disorders, including PD (Table 1).

Table 1. Studies revealing the main epigenetic modifications in Parkinson's disease.

Epigenetic Modifications	Reference
DNA methylation	
- DNMT1 is downregulated in the brain of PD patients	[20]
- The *SNCA* gene is hypomethylated in the brain and blood of PD patients	[17]
- Altered DNA methylation patterns on gene variants of PARK16/1q32, GPNMB/7p15, and STX1B/16p11 loci in post-mortem brain samples have also been identified between PD patients and controls	[21]
- The *DAT* gene is hypermethylated in PD patients	[22]
- The *TNF-α* gene is hypomethylated in the SN of PD patients	[23]
- The *CYP2E1* gene is hypomethylated in the brain of PD patients	[24]
- The *COMT*, *PRNP* and *DCTN1* genes are differentially methylated in PD	[25]
- The DNA methylation status is different in blood samples from PD patients compared to controls in mitochondria-related genes, including *LARS2*, *MIR1977*, and *DDAH2*	[26]
Histone modifications	
- The acetylation levels of histones H2A, H,3 and H4 are higher in the dopaminergic neurons from the midbrain of PD patients	[27]
- HDACs are reduced in MPP(+)-treated neuronal cells, and the brain of MPTP-treated mice and α-synuclein accumulation is associated with H3 hypoacetylation	[28]
- SIRT2, a HDAC, is implicated in α-synuclein aggregation, autophagy, oxidative stress, and neuroinflammation, although with conflicting results	[29]
- Under oxidative stress, nuclear α-synuclein can bind to the promoter of the *PGC-1α* gene, which leads to its hypoacetylation and the downregulation of its expression, finally resulting in mitochondrial impairment and neurotoxicity	[30]

Table 1. *Cont.*

Epigenetic Modifications	Reference
Non-coding RNAs	
- A total of 125 miRs are differentially expressed in the prefrontal cortex of PD patients compared to controls	[31]
- MiR-7, miR-203a-3p, and miR-153 bind to and downregulate the expression of the *SNCA* gene	[32]
- MiR-132 is downregulated in rat models of PD, accompanied by lower levels of Nurr1, its molecular target	[33]
- MiR-133b levels are reduced in the midbrain of PD patients	[34]
- MiR-205 can bind to the 3′ UTR of LRRK2 and downregulate its expression	[35]
- MiR-214 levels are reduced after the treatment of cells or mice with MPP+ or MPTP, respectively, accompanied by increased α-synuclein levels in dopaminergic cells	[36]
- Mir-124 modulates dopaminergic neuronal loss, mitochondrial function, autophagy, oxidative stress, and neuroinflammation in PD animal models via several signaling pathways	[33]
- Mir-26 is upregulated in the SN and CSF of PD patients and downregulated in the blood of PD patients	[37]
- MiR-30, miR-34, miR-99, miR-124, miR-125, miR-146, miR-219, and miR-222 are differentially expressed in PD	[32]
- MiR-144 and miR-15b-5p are associated with PD	[32]
- A total of 87 lncRNAs are differentially expressed in the SN of PD patients	[38]

3.1. DNA Methylation in PD

DNA methylation is the most well-studied epigenetic mechanism [39]. The expression levels of many genes depend on the degree of methylation of their promoters [11]. DNA methylation involves the transfer of a methyl group from S-adenosyl methionine (SAM) to the fifth carbon position of the cytosine most common in CpG dinucleotides (CpGs) via DNA methyltransferases (DNMTs), leading to the formation of 5-methylcytosine (5-mC) [40]. Concomitantly, SAM is also converted to S-adenosylhomocysteine (SAH) and finally to homocysteine. CpGs are located at the promoter region of several genes throughout the human genome, known as CpG islands [17]. There are three main types of DNMTs in humans: DNMT1, DNMT3a, and DNMT3b [41]. DNMT1 can bind to hemimethylated DNA, maintaining the patterns of methylation following the replication of DNA, whereas DNMT3a and DNMT3b can bind to both hemimethylated and unmethylated DNA, resulting in de novo methylation [16]. Although the expression of DNMTs is significantly decreased after cellular differentiation, post-mitotic neurons of the mature human brain continue to express DNMTs, suggesting that this epigenetic mechanism could be implicated in neuronal function [40]. DNA methylation may critically affect the interaction between DNA and histones, being functionally associated with gene silencing. This can be mediated directly via the inhibition of the interaction of the DNA machinery with chromatin, or indirectly by recruiting methyl CpG-binding domain (MBD) proteins (MBPs) [17]. On the contrary, unmethylated DNA generally results in gene activation [16]. Thereby, the altered DNA methylation status of gene promoters significantly affects gene expression, contributing to a plethora of pathological conditions.

The ageing process triggers a general reduction in DNA methylation, whereas hypermethylation occurs at certain gene promoters [10]. Disturbed DNA methylation patterns have been identified in the brain and blood of PD patients [42]. DNMT1 has been shown to be downregulated in the brain of PD patients, and reduced DNA methylation was associated with DNMT1 accumulation outside the cell nucleus [20], suggesting that DNMT1 dysregulation may underlie the impaired DNA methylation status in PD. Several genes—including those related to PD development—have been shown to be hyper- or hypomethylated, suggesting that the dysregulation of DNA methylation may be highly implicated in PD pathogenesis [42]. The promoter of the *SNCA* gene was significantly hypomethylated in the brain and blood samples of PD patients [17]. Altered DNA methylation patterns on gene variants of *PARK16/1q32*, *GPNMB/7p15*, and *STX1B/16p11* loci in post-mortem brain samples have also been identified between PD patients and controls in a genome-wide association (QWA) meta-analysis [21]. Another study demonstrated that the 5-UTR region of the *dopamine transporter* (*DAT*) gene is hypermethylated in PD patients compared to controls [22]. *DAT* is involved in the maintenance of dopaminergic neurons, and specific *DAT* gene variants have been demonstrated to increase PD risk, although not consistently [11]. Furthermore, the promoter of the *tumor necrosis factor-alpha* (*TNF-α*) gene was found to be significantly hypomethylated in the SN of PD patients compared to controls [23], suggesting that this mechanism could play a role in the excessive neuroinflammation observed in PD. The promoter of the *CYP2E1* gene has also been shown to be hypomethylated in the brain of PD patients [24], and a specific single nucleotide polymorphism (SNP) of this gene has been found to be a genetic risk factor for PD in a Swedish study [43]. Importantly, cytochrome P450 2E1, the protein encoded by *CYP2E1*, is critically involved in the formation of potentially toxic metabolites related to dopaminergic degeneration [16], suggesting that exposure to environmental toxins might contribute to PD pathogenesis via this mechanism. Other genes, including *COMT*—another genetic risk factor for PD—*PRNP*, and *DCTN1* have also shown different DNA methylation levels between PD patients and controls [25].

Collectively, causative genes and gene polymorphisms affecting the risk for PD have been demonstrated to be hypo- or hyper- methylated in PD patients. Some of them are implicated in inflammation and toxin-induced cytotoxicity, two mechanisms critically involved in environmental toxin-induced degeneration [44,45]. Recent evidence has also revealed that specific environmental factors could possibly affect DNA methylation patterns in PD [46]. Thus, our deeper understanding of the exact mechanisms bridging these concepts may open the way for the better elucidation of PD pathophysiology.

3.2. Histone Acetylation and Other Modifications in PD

Histone modifications constitute additional epigenetic mechanisms involved in PD. Histones are important nuclear proteins enriched in arginine and lysine residues and allow DNA winding and the formation of nucleosomes [47]. In this way, histones are highly implicated in DNA replication, the regulation of gene expression, and protection against DNA damage [48]. The core five histones H1/H5, H2A, H2B, H3, and H4 [47] are subjected to many types of post-transcriptional modifications at their N-terminal tails, including lysine acetylation and deacetylation, lysine or arginine methylation and demethylation, serine, threonine or tyrosine phosphorylation, SUMOylation, crotonylation, ubiquitination, hydroxylation, and proline isomerization [17,48]. Histone acetyltransferases (HATs) catalyze histone acetylation, thereby opening the chromatin, making the DNA more accessible to transcription factors and subsequently enhancing gene transcription, while histone deacetylases (HDACs) remove the acetyl groups from histones, leading to tighter chromatin packing and the inhibition of gene expression [47]. In this way, HATs and HDACs significantly affect the level of gene expression, as they are crucially implicated in both normal and pathological conditions.

Although histone post-transcriptional modifications have been demonstrated to be majorly involved in the maintenance and differentiation of dopaminergic neurons [49], their role and the factors affecting these molecular processes underlying PD development have not been clarified yet. The acetylation levels of histones H2A, H3, and H4 were shown to be significantly higher in dopaminergic neurons isolated from the midbrain of PD patients compared to controls [27]. In this study, HDACs were also reduced in 1-methyl-4-phenylpyridinium (MPP(+)-treated neuronal cells, and the brain of 1-methyl-4-phenyl-1,2,3,6-tetrahydropyridine (MPTP)-treated mice, implying that environmental toxins could alter histone acetylation levels in PD [27]. A-synuclein accumulation has been associated with H3 hypoacetylation, and HDAC inhibitors may protect against dopaminergic degeneration in pre-clinical studies [28]. Furthermore, sirtuin 2 (SIRT2), a HDAC, is highly implicated in the core pathophysiological mechanisms of PD, such as α-synuclein aggregation, autophagy, oxidative stress, and neuroinflammation, although with conflicting results; there is evidence on the neuroprotective but also detrimental role of SIRT2 in dopaminergic degeneration [29]. Under conditions of oxidative stress, nuclear α-synuclein can bind to the promoter of the *PGC-1α* gene, encoding a protein acting as a mitochondrial transcription factor [30]. This process leads to its hypoacetylation and downregulation of its expression, finally resulting in mitochondrial impairment and neurotoxicity [30]. In summary, emerging evidence highlights the impaired histone acetylation and HDAC levels in PD, which may be associated with oxidative stress, inflammation, and neurotoxin-induced neurodegeneration, suggesting their potential implication in environmental toxin-induced PD-related pathology.

3.3. MiRNAs and Other Non-Coding RNAs in PD

MiRNAs (miRs) are small endogenous single-stranded non-coding RNAs consisting of 20–25 nucleotides, which bind to specific RNAs at their 3′-untranslated region (3′UTR), resulting in the negative regulation of gene expression at a post-transcriptional level.

Accumulating evidence has highlighted the role of several miRNAs in PD pathogenesis; a post-mortem study has shown that 125 miRs were differentially expressed in the prefrontal cortex of PD patients compared to controls [31]. MiR-7, miR-203a-3p, and miR-153 are able to bind to and downregulate the expression of the *SNCA* gene [32]. Furthermore, miR-132 has been demonstrated to be downregulated in rat models of PD, accompanied by lower levels of the nuclear receptor related 1 protein (Nurr1), its molecular target [33]. Clinical evidence has shown that miR-133b levels are reduced in the midbrain of PD patients. MiR-133b significantly regulates the maintenance of dopaminergic neurons via its implication in a circuit that involves the paired-like homeodomain transcription factor Pitx3 [34]. Mir-124 has been shown to modulate dopaminergic neuronal loss, mitochondrial function, autophagy, oxidative stress, and neuroinflammation in PD animal models via several signaling pathways [33]. MiR-205 has been found to bind to the 3′ UTR of *LRRK2* and downregulate its expression, while brain samples from the frontal cortex of PD patients displayed increased levels of the LRRK2 protein and low levels of miR-205, supporting its crucial role in *LRRK2* gene suppression [35]. Furthermore, miR-214 levels have been demonstrated to be reduced after the treatment of cells or mice with MPP+ or MPTP, respectively, accompanied by increased α-synuclein levels in dopaminergic cells [36] while resveratrol reversed these effects. Hence, it could be speculated that other environmental toxins could trigger α-synuclein accumulation by altering miR-214 levels. Mir-26 is also upregulated in the SN and CSF of PD patients compared to controls and downregulated in the blood of PD patients [37]. Therefore, miRNAs can mediate at least some of the main pathophysiological processes of PD, such as inflammation, α-synuclein accumulation, mitochondrial impairment, and autophagy dysfunction via a variety of mechanisms. Since neurotoxins, tobacco smoking, and pesticides also affect the levels of miRNAs in several studies [50,51], their role as mediators of the effects of environmental factors on PD risk deserves further elucidation.

Long non-coding RNAs (lncRNAs) are RNA transcripts longer than 200 nucleotides, which play critical roles in neurogenesis and neuroplasticity, and contribute to the pathogenesis of neurodegenerative diseases including PD [38]. A total of 87 lncRNAs have been identified to be differentially expressed in the SN of PD patients, suggesting that they might be actively involved in the pathogenesis of PD [38].

4. Environmental Impact on Epigenetic Modifications in PD

Accumulating evidence has indicated the contribution of several environmental factors on epigenetic modifications implicated in the pathogenesis of PD. Among them, the most prominent are smoking, exposure to pesticides and insecticides, coffee consumption, and exposure to heavy metals (Table 2).

Table 2. Studies of the environmental impact on the epigenetic modifications implicated in Parkinson's disease.

Environmental Factors	Reference
Tobacco smoking	
- Smoking-induced DMRs may display diverse distribution patterns in both hypomethylated and hypermethylated regions between smokers and non-smokers - The differentially expressed genes are implicated in "immunosuppression" pathways	[52]
- The DNA hypermethylation of *CYP2D6*, which is observed more commonly in poor metabolizers, is associated with a lower risk of heavy smoking	[53]
- Tobacco smoking is associated with the reduced methylation of the promoter region of *LINE-1* retrotransposons in the blood mononuclear cells only in controls, but not in the PD cases - The inverse relationship between smoking and PD is stronger in low levels of *LINE-1* methylation and is less evident as *LINE-1* methylation levels are increased	[54]
- Methylation levels of the first intron of *SLC6A3* gene are potentially related to nicotine dependence, an increased tendency to start smoking, and an impaired ability to quit	[55]
- *ANKK1/DRD2* genetic region variants are associated with nicotine dependence in males	[56]
- MiR-124 and let-7a are differentially expressed between smokers and non-smokers	[50]
- Nicotine attenuates inflammation by upregulating miR-124	[57]
- MiR-26, miR-30, miR-34, miR-99, miR-124, miR-125, miR-146, miR-219, and miR-222 are among the most significantly downregulated miRs in the lungs of rats exposed to smoking	[58]
Pesticides exposure	
- Dieldrin increases H3 and H4 acetylation, leading to proteosomal dysfunction and the accumulation of the cAMP response element-binding protein in dopaminergic neurons - Treatment with the HAT inhibitor anacardic acid prevents against dieldrin-induced histone hyper-acetylation, DNA fragmentation, and dopaminergic degeneration	[59]
- Exposure to the herbicide paraquat induces H3 acetylation in dopaminergic cells, and is associated with reduced HDAC levels - Anacardic acid protects against these effects	[60]
- In paraquat-treated mice, α-synuclein is accumulated in the nucleus near acetylated H3, and α-synuclein can directly bind to H1 and form a 2:1 complex	[61]
- Rotenone promotes H3K9 acetylation by downregulating SIRT1 and upregulating p53, thus promoting neurodegeneration	[62]
- SIRT3, a HDAC, can deacetylate SOD2, resulting in protection against MPTP-induced ROS accumulation and dopaminergic neurodegeneration	[63]
- SIRT5, another HDAC, is associated with increased SOD2 levels and improved mitochondrial function in MPTP-treated mice, thereby preventing nigrostriatal degeneration	[64]
- The miR-380-3p/Sp3-mRNA pathway is involved in MPTP-induced neurodegeneration	[65]
- Rotenone is associated with increased miR-26a and miR-34a levels and reduced miR-7 and let7a levels in rat models of PD	[66]
- MiR-34a, miR-141, and miR-9 are differentially expressed in MPP+-treated PC12 cells	[67]
- MiR-384-5p, which targets and downregulates SIRT1 expression, is increased in rotenone-induced mice and SH-SY5Y cell models of PD	[68]
- Differential expression levels of MiR-34a-5p are detected in the plasma of PD patients	[69]
Coffee consumption	
- Association between the DNA methylation status of CpG sites and coffee consumption in some genes causing familial PD, such as *GBA*, *Parkin*, and *PINK1* in the blood of non-PD individuals	[70]
- Caffeine may increase the expression of DAT, P450 1A2, and the adenosine A2A receptor in the striatum of MPTP-treated mice	[71]
- Theacrine protects against dopaminergic degeneration in in vitro and in vivo models of PD by directly activating SIRT3, resulting in SOD2 deacetylation, the prevention of apoptosis, a reduction in ROS accumulation, and the restoration of mitochondrial dysfunction	[72]
- Mir-144 and miR-15b-5p are upregulated following treatment with coffee compounds	[73]
- Coffee has been demonstrated to upregulate miR-30	[74]

Table 2. *Cont.*

Environmental Factors	Reference
Exposure to heavy metals	
- Manganese chloride can inhibit H3 and H4 acetylation, increase HDAC3 and HDAC4 expression, and reduce HAT expression	[75]
- Manganese is associated with lower levels of histone acetylation and expression levels of GLT-1 and astrocytic GLAST, thereby promoting neurotoxicity	[76]
- *Parkin* and *PINK1* gene activities are affected by increased DNA methylation in vitro upon exposure to manganese	[77]
- Reduced methylation levels of the promoter of *LINE-1* are related to exposure to lead	[78]
- Arsenic alters the status of *LINE-1* methylation	[79]

4.1. Smoking

Many case-control and prospective cohort studies, as well as several meta-analyses have confirmed the inverse relationship between cigarette smoking and PD, reducing the risk by approximately 50% [9,14]. Regarding the underlying molecular mechanisms, it has been shown that nicotine may act neuroprotectively by interacting with nicotinic acetylcholine receptors [80]. Downstream pathways include the cleavage of poly (ADP-ribose) polymerase-1 (PARP-1) and caspase-3 by nicotine [81]. The epigenetic modifications induced by smoking in PD have not been well-studied; however, it has been demonstrated that smoking-induced differentially methylated regions (DMRs) may display diverse distribution patterns in both hypomethylated and hypermethylated regions between smokers and non-smokers [52]. Interestingly, the differentially expressed genes were shown to be implicated in "immunosuppression" pathways, suggesting that smoking-induced epigenetic modifications may involve immune-related mechanisms in PD. For instance, given the differentially methylated promoter of the *TNF-α* gene in PD [23], the role of cigarette smoking in this relationship should be further explored. Additionally, smoking seems to affect histone acetylation and de-acetylation, and affect the expression of miRNAs in non-nerve tissues [82–85]. Thus, DNA methylation and possibly other epigenetic modifications might mediate the effects of smoking in PD.

To investigate potential gene–environment interactions in terms of epigenetics in PD, it is important to identify shared relative pathways; in particular, we need firstly to explore the common mechanisms between susceptibility genes and epigenetic alterations induced by specific environmental toxins related to PD. In this context, it has been demonstrated that the combination of AA and GA genotypes of rs4680 of the *COMT* gene and non-smoking was associated with a higher risk of PD compared to the combination of AA genotype and a positive smoking history [86]. Given the diverse methylation levels of COMT in PD patients [25], this epigenetic mechanism could potentially underlie the effects of smoking in PD and should be further investigated.

Specific *CYP2D6* gene variants have been shown to interact with cigarette smoking to alter the risk for PD development [87]. However, compared to the extensive metabolizer CYP2D6 genotype, the poor metabolizer has been revealed to reduce the risk of heavy smoking [53]. In addition, DNA hypermethylation of *CYP2D6*, which is observed more commonly in poor metabolizers, is associated with a lower risk of heavy smoking [53]. Therefore, the different methylation status of *CYP2D6* may also be related to smoking behavior, which could in turn alter the PD risk.

Glutathione-S-transferase (GST) plays a major role in the metabolism of both smoke and pesticides, acting as an endogenous antioxidant [88]. An interaction between smoking and specific *GSTP1* polymorphisms has been identified to affect the risk for PD [89,90], although not in all studies [88]. In addition, increased levels of GSTP1 have been observed in the peripheral leucocytes of PD patients upon MPP+ exposure [91], highlighting its implication in the exposure to environmental toxins in PD development. The role of epigenetic modifications of *GSTP1* has not been extensively studied in PD. However, the differential hypermethylation in the *GSTP1* gene has been identified in other neurodegenerative diseases such as Alzheimer's disease [92]. Another interesting relationship has been identified between *GSTM1* gene polymorphisms (a type of GST) and smoking in regard to PD risk. In particular, a more prominent negative association between smoking and PD has been observed in individuals expressing GSTM1-1 [90], but there is also evidence not confirming this interaction [89]. Thus, we could speculate that an altered *GST* methylation status induced by smoking could at least partially mediate its protective effects in PD, but this hypothesis deserves further study.

Interestingly, a case-control study has indicated that tobacco smoking may be associated with reduced methylation of the promoter region of *long interspersed nucleotide element-1* (*LINE-1*) retrotransposons in blood mononuclear cells only in controls, but not in PD cases [93]. In addition, the inverse relationship between smoking and PD was stronger in low levels of *LINE-1* methylation and became less evident as *LINE-1* methylation levels were elevated [93]. *LINE-1* DNA sequences exist in many repeats in the whole genome, and those carrying an intact promoter can replicate themselves and integrate in other DNA regions, possibly altering gene expression [46]. Importantly, *LINE-1* sequences can be inserted into neural progenitor cells expressing tyrosine hydroxylase (TH), which can be differentiated into dopaminergic neuronal cells, thereby implicating dopaminergic cell survival and differentiation [46]. Collectively, these results suggest that PD patients and controls may respond differently to smoking and display diverse DNA methylation profiles in *LINE-1*, which might affect PD risk. *LINE-1* methylation correlates with genome-wide methylation [93], and DNA hypomethylation has been associated with oxidative stress [94], organic pollutants [95], and heavy metals [96] in some studies. Therefore, it is tempting to propose that tobacco smoking may act protectively against PD especially in the cases of additional environmental exposures, and smoking-induced alterations of LINE-1 methylation status could underlie these effects.

Furthermore, a network-based meta-analysis of four blood microarray studies has demonstrated that the *Polypyrimidine Tract Binding Protein 1* (*PTBP1*) gene, encoding a protein highly implicated in the mRNA translation and stabilization of insulin and previously related to diabetes, was the most significantly downregulated gene of PD patients compared to controls [97]. Longitudinal analyses have demonstrated that the relative abundance of *PTBP1* mRNA significantly decreased over a 3-year follow-up period. Insulin resistance has been associated with PD in several studies, and PTBP1 modulates the expression of glucagon-like peptide 1 (GLP-1) [98], which constitutes a pharmacological target in clinical trials for PD [99]. Interestingly, a study in patients with intracranial aneurysms and controls has demonstrated that long-term tobacco smoking was significantly associated with increased DNA methylation levels in the promoter of the *PTBP1* gene in blood, resulting in a reduction in gene expression [100]. Taken together, it could be speculated that smoking could play a potential role in the dynamic methylation and expression of *PTBP1* gene in PD, and this interesting association should be investigated.

Smoking behavior, including the tendency to start or the ability to quit smoking, has been also hypothesized to affect the relationship between smoking and PD. The *SLC6A3* gene is a *DAT* gene, a variant of which has been also suggested as a genetic risk factor for PD [101]. It was recently demonstrated that the number of tandem repeats and methylation levels of the first intron of *SLC6A3* gene might be related to nicotine dependence and a potentially increased tendency to start smoking and an impaired ability to quit [55]. Another recent study demonstrated that the DNA methylation rates of *DRD2* in peripheral leukocytes were reduced in PD patients [102]. *DRD2* is considered a risk factor for PD, and *ANKK1/DRD2* genetic region variants have been associated with nicotine dependence in males in a Chinese study [56]. Hence, a reduced nicotine dependence associated with altered *DAT* or *DRD2* methylation could reflect at least partially an additional endogenous feature of patients susceptible to PD, suggesting that non-smoking may not represent an absolute "actual cause" of the disease.

Regarding the role of smoking in miRs, a study of the plasma levels of 84 miRs among smokers and non-smokers indicated that miR-124 and let-7a were differentially expressed between these two groups [50]. In addition, nicotine has been demonstrated to attenuate inflammation by upregulating miR-124 [57]. Given the important role of miR-124 in PD as abovementioned [103], the potential effects of smoking on miR-124 levels in PD could also be explored. Moreover, a study that investigated the differential expression of miRs in the lungs of rats exposed to smoking showed that miR-26, miR-30, miR-34, miR-99, miR-124, miR-125, miR-146, miR-219, and miR-222 were among the most significantly downregulated miRs [58]. At the same time, these specific miRs have been associated with PD in various studies [32,37]. These miRs are implicated in stress responses, cell apoptosis, and proliferation [58]. Despite the innate differences between lung and brain tissue, investigating the interaction between smoking and these miRs will enable the elucidation of the molecular mechanisms underlying the protective role of smoking in PD (Figure 2).

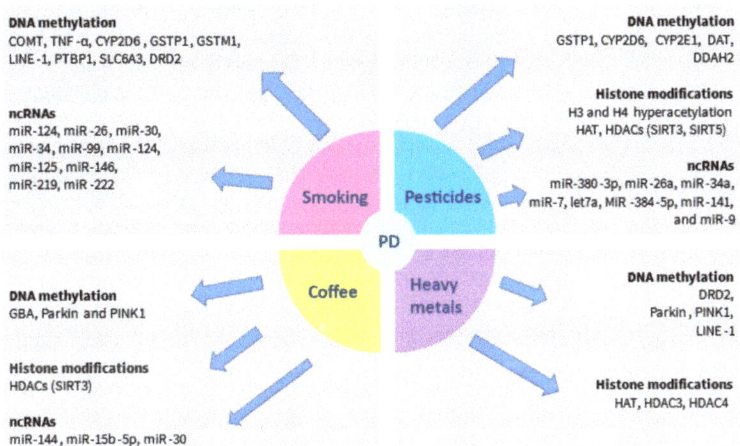

Figure 2. Schematic representation of the potential impact of environmental factors on epigenetic modifications in PD.

4.2. Exposure to Pesticides and Insecticides

Several epidemiological studies have demonstrated that exposure to pesticides may increase the risk of PD development [104]. Dieldrin, an organochlorine compound and a persistent organic pollutant widely used as an insecticide, has been associated with an increased PD risk [105]. Preclinical evidence has indicated that dieldrin may enhance reactive oxygen species (ROS) production and oxidative stress, mitochondrial damage, cytochrome c release, and the activation of caspase 3, leading to dopaminergic neuronal apoptosis [106]. Rotenone acts as a neurotoxin by suppressing the mitochondrial complex 1, elevating ROS levels, and inhibiting the production of ATP [107]. Given the effects of pesticides on epigenetic modifications [108], the specific pesticide-induced epigenetic mechanisms in PD have attracted increasing interest.

Several pesticides have been identified as regulators of gene expression at an epigenetic level, including DNA methylation, HDACs, and non-coding RNAs [109]. *Glutathione S-transferase pi* gene (*GSTP1*), which encodes GSTP1-1—a detoxification enzyme—has been shown to increase PD risk with exposure to pesticides [110]. The relationship between the age of PD onset in men and specific *GSTP1* polymorphisms is also affected by the occupational exposure to herbicides [111]. Interestingly, the expression of *GSTP1* is downregulated through the DNA hypermethylation of its promoter by the mutant form *G2019S* of *LRRK2* [112], a genetic cause of PD. In vivo evidence has also shown that the *G2019S LRRK2* mutation was associated with an increased paraquat-induced inflammatory response and an enhanced stress phenotype in transgenic mice [113]. Penetrance of

LRRK2 G2019S is highly variable, estimated at 24–100%, and it is supposed to be affected by ethnicity, gender, and the other genetic or environmental factors modifying the age of disease onset [114]. Therefore, exposure to pesticides may significantly affect the risk and onset age of PD in *G2019S LRRK2* mutation carriers, via the DNA hypermethylation of the promoter of *GSTP1* mediated by the mutant *LRRK2*.

The cytochrome P450 (CYP) genes play a crucial role in xenobiotic metabolism [115]; CYP2D6 activity is majorly affected by common genetic variants in the population, resulting in poor metabolizer phenotypes [87]. Hence, it has been hypothesized that the association between exposure to environmental toxins such as pesticides and PD risk might be affected by CYP2D6 gene variants. In this regard, exposure to pesticides has been demonstrated to potentially modify the effect of CYP2D6 variants on PD risk [116]. It has also been shown that the poor metabolizer genotype of CYP2D6 is associated with a higher DNA methylation status [53]. Hence, impaired DNA methylation might be at least partially responsible for the CYP2D6-dependent effects of pesticide exposure on PD risk.

Furthermore, as abovementioned, the promoter of the *CYP2E1* gene is hypomethylated in the brain of PD patients [24], and its protein product cytochrome P450 2E1 is involved in the formation of toxic metabolites contributing to dopaminergic degeneration [16]. Since paraquat-induced oxidative stress and ROS production is regulated by cytochrome P450 2E1 [117], pesticides might be implicated in PD pathogenesis via altered DNA methylation.

It has been indicated that environmental occupational exposure to the pesticides paraquat and maneb interacted with specific *DAT* gene variants to increase PD risk [118]. Global DNA hypermethylation has been associated with increased concentration of persistent organic pollutants (POPs) in the serum of elder individuals [119]. Given the fact that the *DAT* gene is found hypermethylated in PD patients, this epigenetic mechanism might be at least partially responsible for gene–environment interaction in PD and should be further investigated.

An interaction has been also detected between pesticides and specific SNPs of the *NOS1* gene regarding PD risk in two studies [120,121]. A recent clinical study indicated that the DNA methylation status was different in blood samples from PD patients compared to controls in mitochondria-related genes, including *LARS2*, *MIR1977*, and *DDAH2* [38]. Importantly, DDAH2 modulates the levels of ADMA, which in turn inhibits the activity of NOS [122]. Mitochondrial dysfunction is considered to be a key hallmark of environmental insults [123]. Therefore, a potential interaction between the methylation status of *DDAH2* or other mitochondria-related genes and pesticide exposure should be explored.

H3 and H4 hyperacetylation represents a crucial epigenetic mechanism in dopaminergic neurons upon their exposure to several neurotoxins, including dieldrin, paraquat, rotenone, and MPTP/MPP+ [124]. In particular, dieldrin has been demonstrated to increase H3 and H4 acetylation, potentially leading to proteosomal dysfunction and the accumulation of the cAMP response element-binding protein—aHAT—in mesencephalic dopaminergic neurons [59]. In this study, treatment with the HAT inhibitor anacardic acid prevented against dieldrin-induced histone hyper-acetylation, DNA fragmentation, and dopaminergic degeneration [59]. Similarly, exposure to the herbicide paraquat, which has been also linked with PD, could induce H3 acetylation in dopaminergic cells in vitro, and was also associated with reduced HDAC levels while anacardic acid also protected against these effects [60]. Another study indicated that in paraquat-treated mice α-synuclein was accumulated in the nucleus near acetylated H3, and α-synuclein was able to directly bind to H1 and form a 2:1 complex in vitro [61]. Rotenone has been also shown to promote H3K9 acetylation by downregulating SIRT1 and upregulating p53, thus enhancing neurodegeneration in vitro [62]. Therefore, pesticide induced H3 and H4 hyperacetylation represents a possible mechanism underlying the environmental effects of pesticides in PD, and it could be speculated that HAT inhibitors may inhibit this process.

Variants of *manganese-dependent superoxide dismutase* (*SOD2*), a mitochondrial antioxidant enzyme, have also been associated with an increased risk of PD in some but not all studies [125,126]. Sirtuin 3 (SIRT3), a HDAC, has been shown to deacetylate SOD2, resulting in protection against MPTP-induced ROS accumulation and dopaminergic neurodegeneration in vivo [63]. Another study has indicated that sirtuin 5 (SIRT5), another HDAC, was associated with increased SOD2 levels and improved mitochondrial function in MPTP-treated mice, thereby preventing nigrostriatal degeneration [64]. Therefore, HDACs, including SIRT3 and SIRT5, may act neuroprotectively in neurotoxin-induced PD-related pathology, opening the way for future research on the protection against PD in cases of pesticide exposures.

Several miRNA-related molecular pathways have been implicated in neurotoxin-induced neurodegeneration. For instance, miR-380-3p expression has been shown to be affected by the MPTP–Nrf2 interaction, and the miR-380-3p/Sp3-mRNA pathway was involved in MPTP-induced neurodegeneration [65]. Moreover, rotenone was associated with increased miR-26a and miR-34a levels and reduced miR-7 and let7a levels in rat models of PD [66]. MiR-34a, miR-141, and miR-9 were also differentially expressed in MPP+-treated PC12 cells [67]. MiR-384-5p, which targets and downregulates SIRT1 expression, was increased in rotenone-induced mice and SH-SY5Y cell models of PD [68]. MiR-34a-5p has been shown to play a key role in PD pathophysiology, and differential expression levels have been detected in the plasma of PD patients [69]. Thus, miRNA expression levels seem to be affected by environmental neurotoxins, highlighting their potential role in pesticide-induced PD pathology.

Collectively, this evidence strongly suggests that epigenetic mechanisms including DNA methylation and histone acetylation may underlie the effects of pesticides on PD development (Figure 2). The conflicting results between GWAS may at least be partially explained by the effects of pesticides or other environmental factors on the epigenetic regulation of PD-related susceptibility genes.

4.3. *Coffee Consumption*

PD risk has been shown to be lower in individuals who drink coffee [127,128]. However, there is also evidence that does not confirm this association [129]. Caffeine is considered to act as an adenosine A2A receptor antagonist, inhibiting neuroinflammation and oxidative stress. Recent evidence has also proposed that alterations in gut microbiota are implicated in the relationship between coffee and PD [130]. Concerning the epigenetic modifications underlying the effects of coffee, an association was observed between the DNA methylation status of CpG sites and coffee consumption in some genes causing familial PD, such as *GBA*, *Parkin*, and *PINK1* in the blood of non-PD individuals [70]. Since coffee consumption has recently been shown to possibly protect against early—but not late—onset PD [131], DNA methylation should be further explored especially in these genetic forms of PD.

It has been demonstrated that caffeine may increase the expression of *DAT*, *P450 1A2*, and the *adenosine A2A receptor* in the striatum of MPTP-treated mice [71]. *DAT, P450 1A2*, and *adenosine A2A receptor* gene variants have been already associated with PD risk in some studies, as already mentioned [132]. Although the exact molecular mechanisms remain unclear, epigenetic modifications could possibly underlie these effects, possibly by altering DNA methylation patterns.

Theacrine, a purine alkaloid derived from the Chinese tea "Kucha", is a chemical analogue to caffeine. A recent study demonstrated that theacrine protected against dopaminergic degeneration in in vitro and in vivo models of PD by directly activating SIRT3, resulting in SOD2 deacetylation, the prevention of apoptosis, the reduction in ROS accumulation, and the restoration of mitochondrial dysfunction [72]. Given the hypothesized similar molecular mechanism of action with caffeine [72], it could be proposed that caffeine might also exert its beneficial effects in PD at least partially through this epigenetic mechanism (Figure 2).

Mir-144 and miR-15b-5p have been shown to be upregulated following treatment with coffee compounds in fibrosis-associated hepatocarcinogenesis mouse models [73]. Furthermore, coffee has been demonstrated to upregulate miR-30 in Caco-2 human colon carcinoma cells in another study [74]. MiR-144, miR-15b-5p, and miR-30 have also been associated with PD pathophysiology [32]. Although innate differences in the gene expression patterns between different tissues should be taken into consideration, the effects of coffee consumption on the regulation of these three miRs should be further investigated as a potential mechanism underlying its effects on PD development.

4.4. Exposure to Heavy Metals

Pre-clinical evidence has shown that heavy metals can result in dopaminergic neurodegeneration via several mechanisms including mitochondrial damage, oxidative stress, excessive neuroinflammation, and epigenetic modifications [133]. It has been proposed that increased exposure to heavy metals, such as manganese, copper, mercury, zinc, lead, aluminum, arsenic, and iron increases the risk of PD [134], but the clinical evidence is inconclusive.

A large multicenter study did not indicate any relationship between manganese, iron, and copper and PD risk [135]. On the other hand, manganese in air pollution was demonstrated to be potentially associated with PD development in another Canadian study [136]. Manganese acts as a cofactor for several cellular enzymatic reactions [133], and chronic excessive manganese exposure results in parkinsonian symptoms. Manganese may lead to neurodegeneration by inducing mitochondrial dysfunction, an impairment of energy metabolism, neuroinflammation, and the disruption of synaptic transmission, as well as altering gene expression [137]. A recent in vitro study demonstrated that manganese chloride could inhibit H3 and H4 acetylation, increase HDAC3 and HDAC4 expression, and reduce HAT expression [75]. Furthermore, manganese-treated mice display increased DRD2 expression in their striatum, with unclear molecular mechanisms. Since the DRD2 variant is a susceptibility gene for PD, with PD patients displaying reduced methylation levels of DRD2 [102], it could be hypothesized that manganese could at least partially contribute to PD pathology by modulating *DRD2* gene methylation. Manganese has also been shown to be associated with lower levels of histone acetylation and expression levels of glutamate transporter 1 (GLT-1) and astrocytic glutamate aspartate transporter (GLAST), thereby promoting neurotoxicity [76]. In addition, the gene activities of *Parkin* and *PINK1* have been shown to be affected by increased DNA methylation in vitro upon exposure to manganese [77], suggesting that manganese may also trigger epigenetic modifications in causative PD genes. Thus, manganese may promote dopaminergic degeneration by altering DNA methylation and histone acetylation, although more evidence is needed to clarify this mechanism.

Lead exposure has been also associated with an increased risk for PD [138]. Reduced methylation levels of the promoter of *LINE-1* have been related to exposure to lead [78]. Given the fact that the amount of highly active retrotransposition competent (RC)-LINE-1 has been associated with PD risk and disease progression [139], this epigenetic mechanism could underlie the effects of lead on PD development.

An increased risk of PD has been also associated with exposure to high levels of farm soil arsenic [67]. Although the underlying molecular mechanisms of the effects of arsenic in PD have not been extensively studied, several arsenic-mediated epigenetic alterations have been identified in other conditions, including an impaired *LINE-1* methylation status [79].

5. Potential Implications for Diagnosis and Therapy

DNA methylation patterns already represent a promising biomarker for several types of human cancer [140]. Given the implication of impaired DNA methylation in neurodegenerative diseases, it has been proposed that it could also act as a diagnostic or prognostic biomarker for PD [141]. Importantly, DNA methylation alterations in peripheral leucocytes have been well-correlated with those in the brains of PD patients [42], suggesting that blood cells might successfully reflect DNA methylation patterns in the brains of patients with PD and they could be used for this purpose.

DNA methylation rates of *DRD2* in leukocytes were found to be reduced in PD patients but increased in Lewy body dementia (DLB) patients [102]. Given the association between *DRD2* variants and nicotine dependence mentioned before, smoking behavior should be considered as a potential factor contributing to the observed differences in this study.

Importantly, clinical studies have demonstrated that the levels of specific miRs in the blood may significantly differ between PD patients and controls, suggesting that they could be used as potential diagnostic biomarkers. For instance, miR-1, miR-22, and miR-29 were shown to be decreased in non-treated patients with PD compared to controls [142]. Differential expression levels of MiR-34a-5p have been detected in the plasma of PD patients and controls [69], and it has been suggested that it could serve as a potential diagnostic biomarker. However, given the diverse miR-34a levels induced by environmental toxins as mentioned above, exposure to pesticides or other environmental factors should be also considered for the interpretation of these results.

Some of the PD-related susceptibility genes have been also associated with levodopa-induced motor complications in PD patients; for instance, polymorphisms in *MAO-B* and *COMT* genes have been shown to increase the risk of developing dyskinesias and wearing-off [143]. Furthermore, specific *BDNF*, *DAT*, and *COMT* variants have been demonstrated to exert a synergistic effect on levodopa-induced motor complications [144]. Coffee consumption has been associated with a reduced risk of levodopa-induced dyskinesias in PD patients [145]. Given the potential epigenetic modifications mediated by environmental factors in these genes in PD, a possible effect of these toxins on the development of levodopa-induced motor complications via this mechanism should be further explored.

Pharmaceutical agents acting as epigenetic modulators have been successfully used in cancer. HDAC inhibitors, including vorinostat and agents causing DNA hypomethylation, including azacitidine, have been approved and used against cancer [146]. Novel pharmaceutical approaches targeting epigenetic modifications have also received increasing attention for the treatment of neurodegenerative disorders, including PD. In this regard, DNMT inhibitors exhibited conflicting results in pre-clinical models of PD; it has been indicated that 5-Aza-2′-Deoxycytidine (5-aza-dC), a DNMT inhibitor, can enhance the expression of *TH*—a gene implicated in the production of levodopa—but also the expression of *SNCA* and *UCHL1*, two genes contributing to PD pathogenesis [147].

In regard to potential treatment approaches related to DNA methylation in PD, it has recently been demonstrated that the use of β-naphthoflavone and ethanol, two CYP inducers, in MPP+-treated cells was associated with increased cell viability, lower levels of ROS, the rescue of mitochondrial membrane potential, and the protection of the activity of mitochondrial complex I against MPP+-induced effects [148]. Given the CYP2E1 hypomethylation in PD and its association with the metabolism of pesticides, it could be suggested that these therapeutic agents may prove beneficial in PD-related pesticides exposure and should be investigated. Furthermore, melatonin or silymarin treatment has been associated with the inhibition of paraquat- and maneb-induced dopaminergic degeneration and oxidative damage in PD mouse models, accompanied by decreased CYP2E1 expression [149]. Therefore, the pharmaceutical targeting of CYP2E1 expression via epigenetic modifications represents a promising approach against PD, especially in cases associated with pesticides exposure.

Atremorine, a novel bioproduct from *Vicia faba*, has been shown to act neuroprotectively and enhance dopamine production in PD patients [150]. Interestingly, a recent study revealed that the effects of atremorine as a dopamine enhancer largely depend on variants in several PD-related genes, including *SNCA*, *LRRK2*, *DRD2*, *CYP2D6*, *NAT2*, *DAT*, and *APOE* among others [150]. The underlying mechanism of atremorine activity is supposed to involve DNA hypermethylation, thereby regulating this extensive pharmaco-epigenetic network [150].

Furthermore, HDAC inhibitors, having been extensively studied in cancer, have also been investigated in neurodegenerative diseases, including PD [17]. Trichostatin A, an agent acting as a HDAC inhibitor leading to increased H3 acetylation, could prevent mitochondrial dysfunction and inhibit neuronal loss in in vitro models of PD, by upregulating (*mitofusin 2*) *MFN2* gene expression [151]. Suberoylanilide hydroxamic acid (SAHA), the first HDAC inhibitor approved for cancer therapy, has been demonstrated to protect against dopaminergic degeneration by enhancing the release of neurotrophic factors from astrocytes [152]. Valproic acid, another widely used anti-epileptic and mood-stabilizing pharmaceutical agent, can also act as a HDAC inhibitor, and promote H3 acetylation; pre-clinical evidence has shown that valproic acid could inhibit neuroinflammation and promote glial cell-derived factor (GDNF) and brain-derived neurotrophic factor (BDNF) expression, resulting in the protection of MPTP- and rotenone-induced dopaminergic neurotoxicity [40]. HDAC inhibitors could be used in case of PD related to pesticide exposure, such as dieldrin and paraquat, since these neurotoxins have been shown to be associated with histone acetylation in PD, as described above.

Moreover, trichostatin A has been demonstrated to protect against manganese-induced neuronal cell death in vitro [75]. Valproic acid could also prevent manganese-mediated decreased histone acetylation and inhibited manganese-induced dopaminergic neurodegeneration [76], highlighting the neuroprotective potential of HDAC inhibitors in manganese-related PD.

However, there is also evidence showing that HDAC inhibitors could be associated with adverse effects in pre-clinical PD models. Sodium butyrate-induced hyperacetylation of histone H4 in mice has been demonstrated to upregulate the protein kinase C δ (PKCδ) in the SN and striatum, enhancing the cellular sensitivity to oxidative stress, potentially resulting in dopaminergic degeneration [153]. Another study indicated that trichostatin A treatment was associated with the reduced survival of dopaminergic neurons [154]. Resveratrol, an agent activating SIRT1—a deacetylase enzyme—has been shown to inhibit rotenone-induced neuronal injury in vitro [62]. Given also the possible diverse responses of HDAC inhibitors in other cell types, further work is needed regarding the clinical effectiveness and safety of these approaches in human PD patients.

Curcumin, a phytochemical with pleiotropic functions, has been shown to protect against PD in several preclinical models [155]. Curcumin acts as a modulator of HATs, HDACs, DNMTs, and specific miRNAs, thereby being involved in several epigenetic modifications [155]. For instance, curcumin displays anti-cancer activities in tobacco smoke-induced lung cancer, by suppressing miR-19 transcription [156]. Hence, it could be speculated that curcumin and possibly other natural products might prevent against PD via epigenetic mechanisms particularly in cases of non-smoking.

Levodopa-induced dyskinesia (LID) represents a late complication of levodopa treatment in PD patients, characterized by involuntary movements often occurring at the peak of dose of levodopa therapy. LID has been associated with the reduced methylation of H3, as well as the deacetylation of histone H4 in the striatum of animal models of LID [157]. Given the protective role of coffee against LID development and PD progression [158], and the neuroprotective effects of caffeine on 6-OHDA-lesioned rat models of PD via histone deacetylase inhibition [159], histone deacetylation could be hypothesized to underlie the protective role of coffee in this case.

6. Future Perspectives

Although there are several pre-clinical studies investigating the impact of some environmental factors on epigenetic mechanisms in PD, the clinical evidence is limited. For a better clarification of this process in humans, it would be useful to explore specific epigenetic alterations between PD patients and controls in relation to their exposure to smoking, pesticides, heavy metals, and coffee. In addition, given the impact of environmental factors on the age of onset of the disease in some cases, and the well-known impact of ageing on the epigenome [160], this association requires clarification. Furthermore, given the diverse penetrance of some genetic causes of PD and the interaction between some environmental factors and PD-related genes, the specific impact of environmental exposure on familial cases of PD would be of interest. In this review, some possible mechanistic pathways have been proposed that could be further investigated for this purpose based on evidence from the current literature.

Epigenetic modulators represent an attractive novel approach for PD treatment. In contrast to the irreversible genetic mutations, small molecules targeting HATs, HDACs, DNMTs, miRNAs, and other ncRNAs are receiving increasing research interest. By elucidating the environmental impact on epigenetic alterations in PD, more personalized treatment and preventive approaches could be developed in the future. The deeper understanding of these mechanisms will also allow for earlier intervention at the preclinical stages of PD.

7. Conclusions

Collectively, although research on the epigenetic mechanisms underlying the effects of environmental factors on PD pathogenesis is still in its infancy, the combination and analysis of the existing results from relative studies has revealed several possible molecular links between gene–environment interactions that deserve further exploration.

Each PD patient carries a unique combination of various genetic factors that could increase (or decrease) PD susceptibility and is also exposed to a mixture of environmental factors that also affect this risk. Hence, PD is a multifactorial and pathophysiological heterogeneous disorder, in which genetic and environmental factors may rather interact differently in each patient. Further elucidation of the epigenetic mechanisms underlying this interaction will enable an understanding of PD pathogenesis, potentially leading to personalized and more effective treatment approaches.

Author Contributions: Conceptualization, E.A. and C.P.; methodology, Y.N.P.; software, E.A., Y.N.P.; validation, E.A., Y.N.P. and C.P.; formal analysis, E.A.; investigation, E.A.; resources, E.A.; data curation, S.G.P.; writing—original draft preparation, E.A.; writing—review and editing, C.P., S.G.P.; visualization, E.A.; supervision, C.P.; project administration, S.G.P., C.P.; funding acquisition, C.P. All authors have read and agreed to the published version of the manuscript.

Funding: This research received no external funding.

Informed Consent Statement: Not applicable.

Acknowledgments: Not applicable.

Conflicts of Interest: The authors declare no conflict of interest.

Abbreviations

PD	Parkinson's disease
SNpc	Substantia nigra pars compacta
GWAS	Genome wide association studies
MPTP	1-methyl-4-phenyl-1,2,3,6-tetrahydropyridine
miRNAs	microRNAs
CNS	Central nervous system
DNMTs	DNA methyltransferases
MBD	Methyl CpG-binding domain
CpGs	CpG dinucleotides
DAT	Dopamine transporter
HATs	Histone acetyltransferases
HDACs	Histone deacetylases
Nurr1	Nuclear receptor related 1 protein
lncRNAs	Long non-coding RNAs
DMRs	Differentially methylated regions
LINE-1	Long interspersed nucleotide element 1
PTBP1	Polypyrimidine Tract Binding Protein 1
GLP-1	Glucagon-like peptide 1
CYP	Cytochrome P450

References

1. de Lau, L.M.; Breteler, M.M. Epidemiology of Parkinson's disease. *Lancet Neurol.* **2006**, *5*, 525–535. [CrossRef]
2. Marino, B.L.B.; de Souza, L.R.; Sousa, K.P.A.; Ferreira, J.V.; Padilha, E.C.; da Silva, C.; Taft, C.A.; Hage-Melim, L.I.S. Parkinson's Disease: A Review from Pathophysiology to Treatment. *Mini Rev. Med. Chem.* **2020**, *20*, 754–767. [CrossRef] [PubMed]
3. Badanjak, K.; Fixemer, S.; Smajic, S.; Skupin, A.; Grunewald, A. The Contribution of Microglia to Neuroinflammation in Parkinson's Disease. *Int. J. Mol. Sci.* **2021**, *22*, 4676. [CrossRef] [PubMed]
4. Marogianni, C.; Sokratous, M.; Dardiotis, E.; Hadjigeorgiou, G.M.; Bogdanos, D.; Xiromerisiou, G. Neurodegeneration and Inflammation-An Interesting Interplay in Parkinson's Disease. *Int. J. Mol. Sci.* **2020**, *21*, 8421. [CrossRef]
5. Gruchot, J.; Kremer, D.; Kury, P. Neural Cell Responses Upon Exposure to Human Endogenous Retroviruses. *Front. Genet.* **2019**, *10*, 655. [CrossRef]
6. Kury, P.; Nath, A.; Creange, A.; Dolei, A.; Marche, P.; Gold, J.; Giovannoni, G.; Hartung, H.P.; Perron, H. Human Endogenous Retroviruses in Neurological Diseases. *Trends Mol. Med.* **2018**, *24*, 379–394. [CrossRef]
7. Sun, F.; Salinas, A.G.; Filser, S.; Blumenstock, S.; Medina-Luque, J.; Herms, J.; Sgobio, C. Impact of alpha-synuclein spreading on the nigrostriatal dopaminergic pathway depends on the onset of the pathology. *Brain Pathol.* **2021**, e13036. [CrossRef]
8. Postuma, R.B.; Berg, D.; Stern, M.; Poewe, W.; Olanow, C.W.; Oertel, W.; Obeso, J.; Marek, K.; Litvan, I.; Lang, A.E.; et al. MDS clinical diagnostic criteria for Parkinson's disease. *Mov. Disord. Off. J. Mov. Disord. Soc.* **2015**, *30*, 1591–1601. [CrossRef] [PubMed]
9. Pang, S.Y.; Ho, P.W.; Liu, H.F.; Leung, C.T.; Li, L.; Chang, E.E.S.; Ramsden, D.B.; Ho, S.L. The interplay of aging, genetics and environmental factors in the pathogenesis of Parkinson's disease. *Transl. Neurodegener.* **2019**, *8*, 23. [CrossRef]
10. Marques, S.C.; Oliveira, C.R.; Pereira, C.M.; Outeiro, T.F. Epigenetics in neurodegeneration: A new layer of complexity. *Prog. Neuro-Psychopharmacol. Biol. Psychiatry* **2011**, *35*, 348–355. [CrossRef]
11. Migliore, L.; Coppede, F. Genetics, environmental factors and the emerging role of epigenetics in neurodegenerative diseases. *Mutat. Res.* **2009**, *667*, 82–97. [CrossRef]
12. Singh, M.; Khan, A.J.; Shah, P.P.; Shukla, R.; Khanna, V.K.; Parmar, D. Polymorphism in environment responsive genes and association with Parkinson disease. *Mol. Cell. Biochem.* **2008**, *312*, 131–138. [CrossRef]
13. Dunn, A.R.; O'Connell, K.M.S.; Kaczorowski, C.C. Gene-by-environment interactions in Alzheimer's disease and Parkinson's disease. *Neurosci. Biobehav. Rev.* **2019**, *103*, 73–80. [CrossRef] [PubMed]
14. Bellou, V.; Belbasis, L.; Tzoulaki, I.; Evangelou, E.; Ioannidis, J.P. Environmental risk factors and Parkinson's disease: An umbrella review of meta-analyses. *Parkinsonism Relat. Disord.* **2016**, *23*, 1–9. [CrossRef]
15. Coppede, F. Genetics and epigenetics of Parkinson's disease. *Sci. World J.* **2012**, *2012*, 489830. [CrossRef] [PubMed]
16. Feng, Y.; Jankovic, J.; Wu, Y.C. Epigenetic mechanisms in Parkinson's disease. *J. Neurol. Sci.* **2015**, *349*, 3–9. [CrossRef]
17. Pavlou, M.A.S.; Outeiro, T.F. Epigenetics in Parkinson's Disease. *Adv. Exp. Med. Biol.* **2017**, *978*, 363–390. [CrossRef] [PubMed]
18. van Heesbeen, H.J.; Smidt, M.P. Entanglement of Genetics and Epigenetics in Parkinson's Disease. *Front. Neurosci.* **2019**, *13*, 277. [CrossRef] [PubMed]
19. Karakaidos, P.; Karagiannis, D.; Rampias, T. Resolving DNA Damage: Epigenetic Regulation of DNA Repair. *Molecules* **2020**, *25*, 2496. [CrossRef]
20. Creighton, S.D.; Stefanelli, G.; Reda, A.; Zovkic, I.B. Epigenetic Mechanisms of Learning and Memory: Implications for Aging. *Int. J. Mol. Sci.* **2020**, *21*, 6918. [CrossRef]
21. Moore, L.D.; Le, T.; Fan, G. DNA methylation and its basic function. *Neuropsychopharmacol. Off. Publ. Am. Coll. Neuropsychopharmacol.* **2013**, *38*, 23–38. [CrossRef]
22. Gowher, H.; Jeltsch, A. Mammalian DNA methyltransferases: New discoveries and open questions. *Biochem. Soc. Trans.* **2018**, *46*, 1191–1202. [CrossRef] [PubMed]
23. Masliah, E.; Dumaop, W.; Galasko, D.; Desplats, P. Distinctive patterns of DNA methylation associated with Parkinson disease: Identification of concordant epigenetic changes in brain and peripheral blood leukocytes. *Epigenetics* **2013**, *8*, 1030–1038. [CrossRef]
24. Desplats, P.; Spencer, B.; Coffee, E.; Patel, P.; Michael, S.; Patrick, C.; Adame, A.; Rockenstein, E.; Masliah, E. Alpha-synuclein sequesters Dnmt1 from the nucleus: A novel mechanism for epigenetic alterations in Lewy body diseases. *J. Biol. Chem.* **2011**, *286*, 9031–9037. [CrossRef] [PubMed]
25. International Parkinson's Disease Genomics, C.; Wellcome Trust Case Control, C. A two-stage meta-analysis identifies several new loci for Parkinson's disease. *PLoS Genet.* **2011**, *7*, e1002142. [CrossRef]
26. Rubino, A.; D'Addario, C.; Di Bartolomeo, M.; Michele Salamone, E.; Locuratolo, N.; Fattapposta, F.; Vanacore, N.; Pascale, E. DNA methylation of the 5′-UTR DAT 1 gene in Parkinson's disease patients. *Acta Neurol. Scand.* **2020**, *142*, 275–280. [CrossRef]
27. Pieper, H.C.; Evert, B.O.; Kaut, O.; Riederer, P.F.; Waha, A.; Wullner, U. Different methylation of the TNF-alpha promoter in cortex and substantia nigra: Implications for selective neuronal vulnerability. *Neurobiol. Dis.* **2008**, *32*, 521–527. [CrossRef]
28. Kaut, O.; Schmitt, I.; Wullner, U. Genome-scale methylation analysis of Parkinson's disease patients' brains reveals DNA hypomethylation and increased mRNA expression of cytochrome P450 2E1. *Neurogenetics* **2012**, *13*, 87–91. [CrossRef]
29. Shahabi, H.N.; Westberg, L.; Melke, J.; Hakansson, A.; Belin, A.C.; Sydow, O.; Olson, L.; Holmberg, B.; Nissbrandt, H. Cytochrome P450 2E1 gene polymorphisms/haplotypes and Parkinson's disease in a Swedish population. *J. Neural Transm.* **2009**, *116*, 567–573. [CrossRef]
30. Kakade, A.; Kumari, B.; Dholaniya, P.S. Feature selection using logistic regression in case-control DNA methylation data of Parkinson's disease: A comparative study. *J. Theor. Biol.* **2018**, *457*, 14–18. [CrossRef]

31. Jazvinscak Jembrek, M.; Orsolic, N.; Mandic, L.; Sadzak, A.; Segota, S. Anti-Oxidative, Anti-Inflammatory and Anti-Apoptotic Effects of Flavonols: Targeting Nrf2, NF-kappaB and p53 Pathways in Neurodegeneration. *Antioxidants* **2021**, *10*, 1628. [CrossRef] [PubMed]
32. Yang, Z.; Shao, Y.; Zhao, Y.; Li, Q.; Li, R.; Xiao, H.; Zhang, F.; Zhang, Y.; Chang, X.; Zhang, Y.; et al. Endoplasmic reticulum stress-related neuroinflammation and neural stem cells decrease in mice exposure to paraquat. *Sci. Rep.* **2020**, *10*, 17757. [CrossRef] [PubMed]
33. Coufal, N.G.; Garcia-Perez, J.L.; Peng, G.E.; Yeo, G.W.; Mu, Y.; Lovci, M.T.; Morell, M.; O'Shea, K.S.; Moran, J.V.; Gage, F.H. L1 retrotransposition in human neural progenitor cells. *Nature* **2009**, *460*, 1127–1131. [CrossRef] [PubMed]
34. Bartova, E.; Krejci, J.; Harnicarova, A.; Galiova, G.; Kozubek, S. Histone modifications and nuclear architecture: A review. *J. Histochem. Cytochem. Off. J. Histochem. Soc.* **2008**, *56*, 711–721. [CrossRef] [PubMed]
35. Chrun, E.S.; Modolo, F.; Daniel, F.I. Histone modifications: A review about the presence of this epigenetic phenomenon in carcinogenesis. *Pathol. Res. Pract.* **2017**, *213*, 1329–1339. [CrossRef] [PubMed]
36. van Heesbeen, H.J.; Mesman, S.; Veenvliet, J.V.; Smidt, M.P. Epigenetic mechanisms in the development and maintenance of dopaminergic neurons. *Development* **2013**, *140*, 1159–1169. [CrossRef] [PubMed]
37. Park, G.; Tan, J.; Garcia, G.; Kang, Y.; Salvesen, G.; Zhang, Z. Regulation of Histone Acetylation by Autophagy in Parkinson Disease. *J. Biol. Chem.* **2016**, *291*, 3531–3540. [CrossRef] [PubMed]
38. Renani, P.G.; Taheri, F.; Rostami, D.; Farahani, N.; Abdolkarimi, H.; Abdollahi, E.; Taghizadeh, E.; Gheibi Hayat, S.M. Involvement of aberrant regulation of epigenetic mechanisms in the pathogenesis of Parkinson's disease and epigenetic-based therapies. *J. Cell. Physiol.* **2019**, *234*, 19307–19319. [CrossRef]
39. Liu, Y.; Zhang, Y.; Zhu, K.; Chi, S.; Wang, C.; Xie, A. Emerging Role of Sirtuin 2 in Parkinson's Disease. *Front. Aging Neurosci.* **2019**, *11*, 372. [CrossRef]
40. Siddiqui, A.; Chinta, S.J.; Mallajosyula, J.K.; Rajagopolan, S.; Hanson, I.; Rane, A.; Melov, S.; Andersen, J.K. Selective binding of nuclear alpha-synuclein to the PGC1alpha promoter under conditions of oxidative stress may contribute to losses in mitochondrial function: Implications for Parkinson's disease. *Free. Radic. Biol. Med.* **2012**, *53*, 993–1003. [CrossRef]
41. Hoss, A.G.; Labadorf, A.; Beach, T.G.; Latourelle, J.C.; Myers, R.H. microRNA Profiles in Parkinson's Disease Prefrontal Cortex. *Front. Aging Neurosci.* **2016**, *8*, 36. [CrossRef]
42. Nies, Y.H.; Mohamad Najib, N.H.; Lim, W.L.; Kamaruzzaman, M.A.; Yahaya, M.F.; Teoh, S.L. MicroRNA Dysregulation in Parkinson's Disease: A Narrative Review. *Front. Neurosci.* **2021**, *15*, 660379. [CrossRef]
43. Jankovic, J.; Chen, S.; Le, W.D. The role of Nurr1 in the development of dopaminergic neurons and Parkinson's disease. *Prog. Neurobiol.* **2005**, *77*, 128–138. [CrossRef]
44. Kim, J.; Inoue, K.; Ishii, J.; Vanti, W.B.; Voronov, S.V.; Murchison, E.; Hannon, G.; Abeliovich, A. A MicroRNA feedback circuit in midbrain dopamine neurons. *Science* **2007**, *317*, 1220–1224. [CrossRef] [PubMed]
45. Cho, H.J.; Liu, G.; Jin, S.M.; Parisiadou, L.; Xie, C.; Yu, J.; Sun, L.; Ma, B.; Ding, J.; Vancraenenbroeck, R.; et al. MicroRNA-205 regulates the expression of Parkinson's disease-related leucine-rich repeat kinase 2 protein. *Hum. Mol. Genet.* **2013**, *22*, 608–620. [CrossRef]
46. Wang, Z.H.; Zhang, J.L.; Duan, Y.L.; Zhang, Q.S.; Li, G.F.; Zheng, D.L. MicroRNA-214 participates in the neuroprotective effect of Resveratrol via inhibiting alpha-synuclein expression in MPTP-induced Parkinson's disease mouse. *Biomed. Pharmacother. Biomed. Pharmacother.* **2015**, *74*, 252–256. [CrossRef] [PubMed]
47. Goh, S.Y.; Chao, Y.X.; Dheen, S.T.; Tan, E.K.; Tay, S.S. Role of MicroRNAs in Parkinson's Disease. *Int. J. Mol. Sci.* **2019**, *20*, 5649. [CrossRef]
48. Banerjee, A.; Waters, D.; Camacho, O.M.; Minet, E. Quantification of plasma microRNAs in a group of healthy smokers, ex-smokers and non-smokers and correlation to biomarkers of tobacco exposure. *Biomark. Biochem. Indic. Expo. Response Susceptibility Chem.* **2015**, *20*, 123–131. [CrossRef] [PubMed]
49. Aloizou, A.M.; Siokas, V.; Sapouni, E.M.; Sita, N.; Liampas, I.; Brotis, A.G.; Rakitskii, V.N.; Burykina, T.I.; Aschner, M.; Bogdanos, D.P.; et al. Parkinson's disease and pesticides: Are microRNAs the missing link? *Sci. Total Environ.* **2020**, *744*, 140591. [CrossRef] [PubMed]
50. Rasheed, M.; Liang, J.; Wang, C.; Deng, Y.; Chen, Z. Epigenetic Regulation of Neuroinflammation in Parkinson's Disease. *Int. J. Mol. Sci.* **2021**, *22*, 4956. [CrossRef]
51. Chuang, Y.-H.; Paul, K.C.; Bronstein, J.M.; Bordelon, Y.; Horvath, S.; Ritz, B. Parkinson's disease is associated with DNA methylation levels in human blood and saliva. *Genome Med.* **2017**, *9*, 1–12. [CrossRef] [PubMed]
52. Quik, M.; Bordia, T.; Zhang, D.; Perez, X.A. Nicotine and Nicotinic Receptor Drugs: Potential for Parkinson's Disease and Drug-Induced Movement Disorders. *Int. Rev. Neurobiol.* **2015**, *124*, 247–271. [CrossRef]
53. Lu, J.Y.D.; Su, P.; Barber, J.E.M.; Nash, J.E.; Le, A.D.; Liu, F.; Wong, A.H.C. The neuroprotective effect of nicotine in Parkinson's disease models is associated with inhibiting PARP-1 and caspase-3 cleavage. *PeerJ* **2017**, *5*, e3933. [CrossRef]
54. Mao, Y.; Huang, P.; Wang, Y.; Wang, M.; Li, M.D.; Yang, Z. Genome-wide methylation and expression analyses reveal the epigenetic landscape of immune-related diseases for tobacco smoking. *Clin. Epigenet.* **2021**, *13*, 215. [CrossRef]
55. Ding, J.; Li, F.; Cong, Y.; Miao, J.; Wu, D.; Liu, B.; Wang, L. Trichostatin A inhibits skeletal muscle atrophy induced by cigarette smoke exposure in mice. *Life Sci.* **2019**, *235*, 116800. [CrossRef]

56. Leus, N.G.; van den Bosch, T.; van der Wouden, P.E.; Krist, K.; Ourailidou, M.E.; Eleftheriadis, N.; Kistemaker, L.E.; Bos, S.; Gjaltema, R.A.; Mekonnen, S.A.; et al. HDAC1-3 inhibitor MS-275 enhances IL10 expression in RAW264.7 macrophages and reduces cigarette smoke-induced airway inflammation in mice. *Sci. Rep.* **2017**, *7*, 45047. [CrossRef]
57. Pace, E.; Di Vincenzo, S.; Ferraro, M.; Siena, L.; Chiappara, G.; Dino, P.; Vitulo, P.; Bertani, A.; Saibene, F.; Lanata, L.; et al. Effects of Carbocysteine and Beclomethasone on Histone Acetylation/Deacetylation Processes in Cigarette Smoke Exposed Bronchial Epithelial Cells. *J. Cell. Physiol.* **2017**, *232*, 2851–2859. [CrossRef] [PubMed]
58. Ganjali, M.; Kheirkhah, B.; Amini, K. Expression of miRNA-601 and PD-L1 among Iranian Patients with Lung Cancer and Their Relationship with Smoking and Mycoplasma Infection. *Cell J.* **2021**, *23*, 723–729. [CrossRef] [PubMed]
59. Kiyohara, C.; Miyake, Y.; Koyanagi, M.; Fujimoto, T.; Shirasawa, S.; Tanaka, K.; Fukushima, W.; Sasaki, S.; Tsuboi, Y.; Yamada, T.; et al. Genetic polymorphisms involved in dopaminergic neurotransmission and risk for Parkinson's disease in a Japanese population. *BMC Neurol.* **2011**, *11*, 89. [CrossRef] [PubMed]
60. Mellick, G.D. CYP450, genetics and Parkinson's disease: Gene x environment interactions hold the key. In *Parkinson's Disease and Related Disorders*; Springer: Vienna, Austria, 2006; pp. 159–165. [CrossRef]
61. Tiili, E.M.; Antikainen, M.S.; Mitiushkina, N.V.; Sukhovskaya, O.A.; Imyanitov, E.N.; Hirvonen, A.P. Effect of genotype and methylation of CYP2D6 on smoking behaviour. *Pharm. Genom.* **2015**, *25*, 531–540. [CrossRef]
62. Kiyohara, C.; Miyake, Y.; Koyanagi, M.; Fujimoto, T.; Shirasawa, S.; Tanaka, K.; Fukushima, W.; Sasaki, S.; Tsuboi, Y.; Yamada, T.; et al. GST polymorphisms, interaction with smoking and pesticide use, and risk for Parkinson's disease in a Japanese population. *Parkinsonism Relat. Disord.* **2010**, *16*, 447–452. [CrossRef] [PubMed]
63. Deng, Y.; Newman, B.; Dunne, M.P.; Silburn, P.A.; Mellick, G.D. Case-only study of interactions between genetic polymorphisms of GSTM1, P1, T1 and Z1 and smoking in Parkinson's disease. *Neurosci. Lett.* **2004**, *366*, 326–331. [CrossRef] [PubMed]
64. De Palma, G.; Dick, F.D.; Calzetti, S.; Scott, N.W.; Prescott, G.J.; Osborne, A.; Haites, N.; Mozzoni, P.; Negrotti, A.; Scaglioni, A.; et al. A case-control study of Parkinson's disease and tobacco use: Gene-tobacco interactions. *Mov. Disord. Off. J. Mov. Disord. Soc.* **2010**, *25*, 912–919. [CrossRef]
65. Korff, A.; Pfeiffer, B.; Smeyne, M.; Kocak, M.; Pfeiffer, R.F.; Smeyne, R.J. Alterations in glutathione S-transferase pi expression following exposure to MPP+ -induced oxidative stress in the blood of Parkinson's disease patients. *Parkinsonism Relat. Disord.* **2011**, *17*, 765–768. [CrossRef]
66. Hernandez, H.G.; Sandoval-Hernandez, A.G.; Garrido-Gil, P.; Labandeira-Garcia, J.L.; Zelaya, M.V.; Bayon, G.F.; Fernandez, A.F.; Fraga, M.F.; Arboleda, G.; Arboleda, H. Alzheimer's disease DNA methylome of pyramidal layers in frontal cortex: Laser-assisted microdissection study. *Epigenomics* **2018**, *10*, 1365–1382. [CrossRef]
67. Searles Nielsen, S.; Checkoway, H.; Butler, R.A.; Nelson, H.H.; Farin, F.M.; Longstreth, W.T., Jr.; Franklin, G.M.; Swanson, P.D.; Kelsey, K.T. LINE-1 DNA methylation, smoking and risk of Parkinson's disease. *J. Parkinson's Dis.* **2012**, *2*, 303–308. [CrossRef]
68. Patchsung, M.; Boonla, C.; Amnattrakul, P.; Dissayabutra, T.; Mutirangura, A.; Tosukhowong, P. Long interspersed nuclear element-1 hypomethylation and oxidative stress: Correlation and bladder cancer diagnostic potential. *PLoS ONE* **2012**, *7*, e37009. [CrossRef]
69. Rusiecki, J.A.; Baccarelli, A.; Bollati, V.; Tarantini, L.; Moore, L.E.; Bonefeld-Jorgensen, E.C. Global DNA hypomethylation is associated with high serum-persistent organic pollutants in Greenlandic Inuit. *Environ. Health Perspect.* **2008**, *116*, 1547–1552. [CrossRef] [PubMed]
70. Lambrou, A.; Baccarelli, A.; Wright, R.O.; Weisskopf, M.; Bollati, V.; Amarasiriwardena, C.; Vokonas, P.; Schwartz, J. Arsenic exposure and DNA methylation among elderly men. *Epidemiology* **2012**, *23*, 668–676. [CrossRef] [PubMed]
71. Santiago, J.A.; Potashkin, J.A. Network-based metaanalysis identifies HNF4A and PTBP1 as longitudinally dynamic biomarkers for Parkinson's disease. *Proc. Natl. Acad. Sci. USA* **2015**, *112*, 2257–2262. [CrossRef] [PubMed]
72. Knoch, K.P.; Meisterfeld, R.; Kersting, S.; Bergert, H.; Altkruger, A.; Wegbrod, C.; Jager, M.; Saeger, H.D.; Solimena, M. cAMP-dependent phosphorylation of PTB1 promotes the expression of insulin secretory granule proteins in beta cells. *Cell Metab.* **2006**, *3*, 123–134. [CrossRef] [PubMed]
73. Foltynie, T.; Aviles-Olmos, I. Exenatide as a potential treatment for patients with Parkinson's disease: First steps into the clinic. *Alzheimer's Dement. J. Alzheimer's Assoc.* **2014**, *10*, S38–S46. [CrossRef] [PubMed]
74. Wang, Z.; Zhou, S.; Zhao, J.; Nie, S.; Sun, J.; Gao, X.; Lenahan, C.; Lin, Z.; Huang, Y.; Chen, G. Tobacco Smoking Increases Methylation of Polypyrimidine Tract Binding Protein 1 Promoter in Intracranial Aneurysms. *Front. Aging Neurosci.* **2021**, *13*, 688179. [CrossRef] [PubMed]
75. Zhai, D.; Li, S.; Zhao, Y.; Lin, Z. SLC6A3 is a risk factor for Parkinson's disease: A meta-analysis of sixteen years' studies. *Neurosci. Lett.* **2014**, *564*, 99–104. [CrossRef]
76. Tiili, E.M.; Mitiushkina, N.V.; Sukhovskaya, O.A.; Imyanitov, E.N.; Hirvonen, A.P. The effect of SLC6A3 variable number of tandem repeats and methylation levels on individual susceptibility to start tobacco smoking and on the ability of smokers to quit smoking. *Pharm. Genom.* **2020**, *30*, 117–123. [CrossRef]
77. Ozaki, Y.; Yoshino, Y.; Yamazaki, K.; Ochi, S.; Iga, J.I.; Nagai, M.; Nomoto, M.; Ueno, S.I. DRD2 methylation to differentiate dementia with Lewy bodies from Parkinson's disease. *Acta Neurol. Scand.* **2020**, *141*, 177–182. [CrossRef]
78. Liu, Q.; Xu, Y.; Mao, Y.; Ma, Y.; Wang, M.; Han, H.; Cui, W.; Yuan, W.; Payne, T.J.; Xu, Y.; et al. Genetic and Epigenetic Analysis Revealing Variants in the NCAM1-TTC12-ANKK1-DRD2 Cluster Associated Significantly With Nicotine Dependence in Chinese Han Smokers. *Nicotine Tob. Res. Off. J. Soc. Res. Nicotine Tob.* **2020**, *22*, 1301–1309. [CrossRef]

79. Qin, Z.; Wan, J.J.; Sun, Y.; Wu, T.; Wang, P.Y.; Du, P.; Su, D.F.; Yang, Y.; Liu, X. Nicotine protects against DSS colitis through regulating microRNA-124 and STAT3. *J. Mol. Med.* **2017**, *95*, 221–233. [CrossRef]
80. Angelopoulou, E.; Paudel, Y.N.; Piperi, C. miR-124 and Parkinson's disease: A biomarker with therapeutic potential. *Pharmacol. Res.* **2019**, *150*, 104515. [CrossRef]
81. Izzotti, A.; Calin, G.A.; Arrigo, P.; Steele, V.E.; Croce, C.M.; De Flora, S. Downregulation of microRNA expression in the lungs of rats exposed to cigarette smoke. *FASEB J. Off. Publ. Fed. Am. Soc. Exp. Biol.* **2009**, *23*, 806–812. [CrossRef]
82. Chambers-Richards, T.; Su, Y.; Chireh, B.; D'Arcy, C. Exposure to toxic occupations and their association with Parkinson's disease: A systematic review with meta-analysis. *Rev. Environ. Health* **2021**. [CrossRef]
83. Cao, F.; Souders Ii, C.L.; Perez-Rodriguez, V.; Martyniuk, C.J. Elucidating Conserved Transcriptional Networks Underlying Pesticide Exposure and Parkinson's Disease: A Focus on Chemicals of Epidemiological Relevance. *Front. Genet.* **2018**, *9*, 701. [CrossRef]
84. Kanthasamy, A.G.; Kitazawa, M.; Yang, Y.; Anantharam, V.; Kanthasamy, A. Environmental neurotoxin dieldrin induces apoptosis via caspase-3-dependent proteolytic activation of protein kinase C delta (PKCdelta): Implications for neurodegeneration in Parkinson's disease. *Mol. Brain* **2008**, *1*, 12. [CrossRef] [PubMed]
85. Tanner, C.M.; Kamel, F.; Ross, G.W.; Hoppin, J.A.; Goldman, S.M.; Korell, M.; Marras, C.; Bhudhikanok, G.S.; Kasten, M.; Chade, A.R.; et al. Rotenone, paraquat, and Parkinson's disease. *Environ. Health Perspect.* **2011**, *119*, 866–872. [CrossRef]
86. Giambo, F.; Leone, G.M.; Gattuso, G.; Rizzo, R.; Cosentino, A.; Cina, D.; Teodoro, M.; Costa, C.; Tsatsakis, A.; Fenga, C.; et al. Genetic and Epigenetic Alterations Induced by Pesticide Exposure: Integrated Analysis of Gene Expression, microRNA Expression, and DNA Methylation Datasets. *Int. J. Environ. Res. Public Health* **2021**, *18*, 8697. [CrossRef]
87. Sabarwal, A.; Kumar, K.; Singh, R.P. Hazardous effects of chemical pesticides on human health-Cancer and other associated disorders. *Environ. Toxicol. Pharmacol.* **2018**, *63*, 103–114. [CrossRef]
88. Menegon, A.; Board, P.G.; Blackburn, A.C.; Mellick, G.D.; Le Couteur, D.G. Parkinson's disease, pesticides, and glutathione transferase polymorphisms. *Lancet* **1998**, *352*, 1344–1346. [CrossRef]
89. Wilk, J.B.; Tobin, J.E.; Suchowersky, O.; Shill, H.A.; Klein, C.; Wooten, G.F.; Lew, M.F.; Mark, M.H.; Guttman, M.; Watts, R.L.; et al. Herbicide exposure modifies GSTP1 haplotype association to Parkinson onset age: The GenePD Study. *Neurology* **2006**, *67*, 2206–2210. [CrossRef]
90. Chen, J.; Liou, A.; Zhang, L.; Weng, Z.; Gao, Y.; Cao, G.; Zigmond, M.J.; Chen, J. GST P1, a novel downstream regulator of LRRK2, G2019S-induced neuronal cell death. *Front. Biosci.* **2012**, *4*, 2365–2377. [CrossRef]
91. Rudyk, C.; Dwyer, Z.; Hayley, S.; membership, C. Leucine-rich repeat kinase-2 (LRRK2) modulates paraquat-induced inflammatory sickness and stress phenotype. *J. Neuroinflamm.* **2019**, *16*, 120. [CrossRef] [PubMed]
92. Marder, K.; Wang, Y.; Alcalay, R.N.; Mejia-Santana, H.; Tang, M.X.; Lee, A.; Raymond, D.; Mirelman, A.; Saunders-Pullman, R.; Clark, L.; et al. Age-specific penetrance of LRRK2 G2019S in the Michael, J. Fox Ashkenazi Jewish LRRK2 Consortium. *Neurology* **2015**, *85*, 89–95. [CrossRef] [PubMed]
93. Zhao, M.; Ma, J.; Li, M.; Zhang, Y.; Jiang, B.; Zhao, X.; Huai, C.; Shen, L.; Zhang, N.; He, L.; et al. Cytochrome P450 Enzymes and Drug Metabolism in Humans. *Int. J. Mol. Sci.* **2021**, *22*, 12808. [CrossRef] [PubMed]
94. Elbaz, A.; Levecque, C.; Clavel, J.; Vidal, J.S.; Richard, F.; Amouyel, P.; Alperovitch, A.; Chartier-Harlin, M.C.; Tzourio, C. CYP2D6 polymorphism, pesticide exposure, and Parkinson's disease. *Ann. Neurol.* **2004**, *55*, 430–434. [CrossRef] [PubMed]
95. Singh, D.; Yadav, A.; Singh, C. Autonomous regulation of inducible nitric oxide synthase and cytochrome P450 2E1-mediated oxidative stress in maneb- and paraquat-treated rat polymorphs. *Pestic. Biochem. Physiol.* **2021**, *178*, 104944. [CrossRef] [PubMed]
96. Ritz, B.R.; Manthripragada, A.D.; Costello, S.; Lincoln, S.J.; Farrer, M.J.; Cockburn, M.; Bronstein, J. Dopamine transporter genetic variants and pesticides in Parkinson's disease. *Environ. Health Perspect.* **2009**, *117*, 964–969. [CrossRef] [PubMed]
97. Lind, L.; Penell, J.; Luttropp, K.; Nordfors, L.; Syvanen, A.C.; Axelsson, T.; Salihovic, S.; van Bavel, B.; Fall, T.; Ingelsson, E.; et al. Global DNA hypermethylation is associated with high serum levels of persistent organic pollutants in an elderly population. *Environ. Int.* **2013**, *59*, 456–461. [CrossRef] [PubMed]
98. Hancock, D.B.; Martin, E.R.; Vance, J.M.; Scott, W.K. Nitric oxide synthase genes and their interactions with environmental factors in Parkinson's disease. *Neurogenetics* **2008**, *9*, 249–262. [CrossRef]
99. Paul, K.C.; Sinsheimer, J.S.; Rhodes, S.L.; Cockburn, M.; Bronstein, J.; Ritz, B. Organophosphate Pesticide Exposures, Nitric Oxide Synthase Gene Variants, and Gene-Pesticide Interactions in a Case-Control Study of Parkinson's Disease, California (USA). *Environ. Health Perspect.* **2016**, *124*, 570–577. [CrossRef]
100. Fiedler, L. The DDAH/ADMA pathway is a critical regulator of NO signalling in vascular homeostasis. *Cell Adhes. Migr.* **2008**, *2*, 149–150. [CrossRef]
101. Duarte-Hospital, C.; Tete, A.; Brial, F.; Benoit, L.; Koual, M.; Tomkiewicz, C.; Kim, M.J.; Blanc, E.B.; Coumoul, X.; Bortoli, S. Mitochondrial Dysfunction as a Hallmark of Environmental Injury. *Cells* **2021**, *11*, 110. [CrossRef]
102. Wang, R.; Sun, H.; Wang, G.; Ren, H. Imbalance of Lysine Acetylation Contributes to the Pathogenesis of Parkinson's Disease. *Int. J. Mol. Sci.* **2020**, *21*, 7182. [CrossRef]
103. Song, C.; Kanthasamy, A.; Anantharam, V.; Sun, F.; Kanthasamy, A.G. Environmental neurotoxic pesticide increases histone acetylation to promote apoptosis in dopaminergic neuronal cells: Relevance to epigenetic mechanisms of neurodegeneration. *Mol. Pharmacol.* **2010**, *77*, 621–632. [CrossRef]

104. Song, C.; Kanthasamy, A.; Jin, H.; Anantharam, V.; Kanthasamy, A.G. Paraquat induces epigenetic changes by promoting histone acetylation in cell culture models of dopaminergic degeneration. *Neurotoxicology* **2011**, *32*, 586–595. [CrossRef]
105. Goers, J.; Manning-Bog, A.B.; McCormack, A.L.; Millett, I.S.; Doniach, S.; Di Monte, D.A.; Uversky, V.N.; Fink, A.L. Nuclear localization of alpha-synuclein and its interaction with histones. *Biochemistry* **2003**, *42*, 8465–8471. [CrossRef]
106. Feng, Y.; Liu, T.; Dong, S.Y.; Guo, Y.J.; Jankovic, J.; Xu, H.; Wu, Y.C. Rotenone affects p53 transcriptional activity and apoptosis via targeting SIRT1 and H3K9 acetylation in SH-SY5Y cells. *J. Neurochem.* **2015**, *134*, 668–676. [CrossRef] [PubMed]
107. Liu, C.; Fang, J.; Liu, W. Superoxide dismutase coding of gene polymorphisms associated with susceptibility to Parkinson's disease. *J. Integr. Neurosci.* **2019**, *18*, 299–303. [CrossRef]
108. Farin, F.M.; Hitosis, Y.; Hallagan, S.E.; Kushleika, J.; Woods, J.S.; Janssen, P.S.; Smith-Weller, T.; Franklin, G.M.; Swanson, P.D.; Checkoway, H. Genetic polymorphisms of superoxide dismutase in Parkinson's disease. *Mov. Disord. Off. J. Mov. Disord. Soc.* **2001**, *16*, 705–707. [CrossRef]
109. Zhang, X.; Ren, X.; Zhang, Q.; Li, Z.; Ma, S.; Bao, J.; Li, Z.; Bai, X.; Zheng, L.; Zhang, Z.; et al. PGC-1alpha/ERRalpha-Sirt3 Pathway Regulates DAergic Neuronal Death by Directly Deacetylating SOD2 and ATP Synthase beta. *Antioxid. Redox Signal.* **2016**, *24*, 312–328. [CrossRef] [PubMed]
110. Liu, L.; Peritore, C.; Ginsberg, J.; Shih, J.; Arun, S.; Donmez, G. Protective role of SIRT5 against motor deficit and dopaminergic degeneration in MPTP-induced mice model of Parkinson's disease. *Behav. Brain Res.* **2015**, *281*, 215–221. [CrossRef] [PubMed]
111. Wang, Q.; Ren, N.; Cai, Z.; Lin, Q.; Wang, Z.; Zhang, Q.; Wu, S.; Li, H. Paraquat and MPTP induce neurodegeneration and alteration in the expression profile of microRNAs: The role of transcription factor Nrf2. *NPJ Parkinsons Dis.* **2017**, *3*, 31. [CrossRef]
112. Horst, C.H.; Schlemmer, F.; de Aguiar Montenegro, N.; Domingues, A.C.M.; Ferreira, G.G.; da Silva Ribeiro, C.Y.; de Andrade, R.R.; Del Bel Guimaraes, E.; Titze-de-Almeida, S.S.; Titze-de-Almeida, R. Signature of Aberrantly Expressed microRNAs in the Striatum of Rotenone-Induced Parkinsonian Rats. *Neurochem. Res.* **2018**, *43*, 2132–2140. [CrossRef]
113. Rostamian Delavar, M.; Baghi, M.; Safaeinejad, Z.; Kiani-Esfahani, A.; Ghaedi, K.; Nasr-Esfahani, M.H. Differential expression of miR-34a, miR-141, and miR-9 in MPP+-treated differentiated PC12 cells as a model of Parkinson's disease. *Gene* **2018**, *662*, 54–65. [CrossRef]
114. Tao, H.; Liu, Y.; Hou, Y. miRNA3845p regulates the progression of Parkinson's disease by targeting SIRT1 in mice and SHSY5Y cell. *Int. J. Mol. Med.* **2020**, *45*, 441–450. [CrossRef]
115. Grossi, I.; Radeghieri, A.; Paolini, L.; Porrini, V.; Pilotto, A.; Padovani, A.; Marengoni, A.; Barbon, A.; Bellucci, A.; Pizzi, M.; et al. MicroRNA34a5p expression in the plasma and in its extracellular vesicle fractions in subjects with Parkinson's disease: An exploratory study. *Int. J. Mol. Med.* **2021**, *47*, 533–546. [CrossRef]
116. Chen, Y.; Sun, X.; Lin, Y.; Zhang, Z.; Gao, Y.; Wu, I.X.Y. Non-Genetic Risk Factors for Parkinson's Disease: An Overview of 46 Systematic Reviews. *J. Parkinsons Dis.* **2021**, *11*, 919–935. [CrossRef]
117. Socala, K.; Szopa, A.; Serefko, A.; Poleszak, E.; Wlaz, P. Neuroprotective Effects of Coffee Bioactive Compounds: A Review. *Int. J. Mol. Sci.* **2020**, *22*, 107. [CrossRef] [PubMed]
118. Domenighetti, C.; Sugier, P.E.; Sreelatha, A.A.K.; Schulte, C.; Grover, S.; Mohamed, O.; Portugal, B.; May, P.; Bobbili, D.R.; Radivojkov-Blagojevic, M.; et al. Mendelian Randomisation Study of Smoking, Alcohol, and Coffee Drinking in Relation to Parkinson's Disease. *J. Parkinsons Dis.* **2021**, 1–16. [CrossRef] [PubMed]
119. Liu, J.; Zhang, Y.; Ye, T.; Yu, Q.; Yu, J.; Yuan, S.; Gao, X.; Wan, X.; Zhang, R.; Han, W.; et al. Effect of Coffee against MPTP-Induced Motor Deficits and Neurodegeneration in Mice Via Regulating Gut Microbiota. *J. Agric. Food Chem.* **2022**, *70*, 184–195. [CrossRef] [PubMed]
120. Chuang, Y.H.; Quach, A.; Absher, D.; Assimes, T.; Horvath, S.; Ritz, B. Coffee consumption is associated with DNA methylation levels of human blood. *Eur. J. Hum. Genet. EJHG* **2017**, *25*, 608–616. [CrossRef]
121. Angelopoulou, E.; Bozi, M.; Simitsi, A.M.; Koros, C.; Antonelou, R.; Papagiannakis, N.; Maniati, M.; Poula, D.; Stamelou, M.; Vassilatis, D.K.; et al. The relationship between environmental factors and different Parkinson's disease subtypes in Greece: Data analysis of the Hellenic Biobank of Parkinson's disease. *Parkinsonism Relat. Disord.* **2019**, *67*, 105–112. [CrossRef]
122. Singh, S.; Singh, K.; Gupta, S.P.; Patel, D.K.; Singh, V.K.; Singh, R.K.; Singh, M.P. Effect of caffeine on the expression of cytochrome P450 1A2, adenosine A2A receptor and dopamine transporter in control and 1-methyl 4-phenyl 1, 2, 3, 6-tetrahydropyridine treated mouse striatum. *Brain Res.* **2009**, *1283*, 115–126. [CrossRef] [PubMed]
123. Siokas, V.; Aloizou, A.M.; Tsouris, Z.; Liampas, I.; Liakos, P.; Calina, D.; Docea, A.O.; Tsatsakis, A.; Bogdanos, D.P.; Hadjigeorgiou, G.M.; et al. ADORA2A rs5760423 and CYP1A2 rs762551 Polymorphisms as Risk Factors for Parkinson's Disease. *J. Clin. Med.* **2021**, *10*, 381. [CrossRef]
124. Duan, W.J.; Liang, L.; Pan, M.H.; Lu, D.H.; Wang, T.M.; Li, S.B.; Zhong, H.B.; Yang, X.J.; Cheng, Y.; Liu, B.; et al. Theacrine, a purine alkaloid from kucha, protects against Parkinson's disease through SIRT3 activation. *Phytomed. Int. J. Phytother. Phytopharm.* **2020**, *77*, 153281. [CrossRef]
125. Romualdo, G.R.; Prata, G.B.; da Silva, T.C.; Evangelista, A.F.; Reis, R.M.; Vinken, M.; Moreno, F.S.; Cogliati, B.; Barbisan, L.F. The combination of coffee compounds attenuates early fibrosis-associated hepatocarcinogenesis in mice: Involvement of miRNA profile modulation. *J. Nutr. Biochem.* **2020**, *85*, 108479. [CrossRef]
126. Nakayama, T.; Funakoshi-Tago, M.; Tamura, H. Coffee reduces KRAS expression in Caco-2 human colon carcinoma cells via regulation of miRNAs. *Oncol. Lett.* **2017**, *14*, 1109–1114. [CrossRef] [PubMed]
127. Kulshreshtha, D.; Ganguly, J.; Jog, M. Manganese and Movement Disorders: A Review. *J. Mov. Disord.* **2021**, *14*, 93–102. [CrossRef]

128. Raj, K.; Kaur, P.; Gupta, G.D.; Singh, S. Metals associated neurodegeneration in Parkinson's disease: Insight to physiological, pathological mechanisms and management. *Neurosci. Lett.* **2021**, *753*, 135873. [CrossRef]
129. Dick, F.D.; De Palma, G.; Ahmadi, A.; Scott, N.W.; Prescott, G.J.; Bennett, J.; Semple, S.; Dick, S.; Counsell, C.; Mozzoni, P.; et al. Environmental risk factors for Parkinson's disease and parkinsonism: The Geoparkinson study. *Occup. Environ. Med.* **2007**, *64*, 666–672. [CrossRef] [PubMed]
130. Finkelstein, M.M.; Jerrett, M. A study of the relationships between Parkinson's disease and markers of traffic-derived and environmental manganese air pollution in two Canadian cities. *Environ. Res.* **2007**, *104*, 420–432. [CrossRef]
131. Aschner, M.; Erikson, K.M.; Herrero Hernandez, E.; Tjalkens, R. Manganese and its role in Parkinson's disease: From transport to neuropathology. *Neuromol. Med.* **2009**, *11*, 252–266. [CrossRef]
132. Guo, Z.; Zhang, Z.; Wang, Q.; Zhang, J.; Wang, L.; Zhang, Q.; Li, H.; Wu, S. Manganese chloride induces histone acetylation changes in neuronal cells: Its role in manganese-induced damage. *Neurotoxicology* **2018**, *65*, 255–263. [CrossRef] [PubMed]
133. Johnson, J., Jr.; Pajarillo, E.; Karki, P.; Kim, J.; Son, D.S.; Aschner, M.; Lee, E. Valproic acid attenuates manganese-induced reduction in expression of GLT-1 and GLAST with concomitant changes in murine dopaminergic neurotoxicity. *Neurotoxicology* **2018**, *67*, 112–120. [CrossRef]
134. Tarale, P.; Sivanesan, S.; Daiwile, A.P.; Stoger, R.; Bafana, A.; Naoghare, P.K.; Parmar, D.; Chakrabarti, T.; Kannan, K. Global DNA methylation profiling of manganese-exposed human neuroblastoma SH-SY5Y cells reveals epigenetic alterations in Parkinson's disease-associated genes. *Arch. Toxicol.* **2017**, *91*, 2629–2641. [CrossRef] [PubMed]
135. Coon, S.; Stark, A.; Peterson, E.; Gloi, A.; Kortsha, G.; Pounds, J.; Chettle, D.; Gorell, J. Whole-body lifetime occupational lead exposure and risk of Parkinson's disease. *Environ. Health Perspect.* **2006**, *114*, 1872–1876. [CrossRef] [PubMed]
136. Li, C.; Yang, X.; Xu, M.; Zhang, J.; Sun, N. Epigenetic marker (LINE-1 promoter) methylation level was associated with occupational lead exposure. *Clin. Toxicol.* **2013**, *51*, 225–229. [CrossRef]
137. Pfaff, A.L.; Bubb, V.J.; Quinn, J.P.; Koks, S. An Increased Burden of Highly Active Retrotransposition Competent L1s Is Associated with Parkinson's Disease Risk and Progression in the PPMI Cohort. *Int. J. Mol. Sci.* **2020**, *21*, 6562. [CrossRef]
138. Lee, C.P.; Zhu, C.H.; Su, C.C. Increased prevalence of Parkinson's disease in soils with high arsenic levels. *Parkinsonism Relat. Disord.* **2021**, *88*, 19–23. [CrossRef]
139. Bailey, K.A.; Fry, R.C. Arsenic-Associated Changes to the Epigenome: What Are the Functional Consequences? *Curr. Environ. Health Rep.* **2014**, *1*, 22–34. [CrossRef] [PubMed]
140. Huo, M.; Zhang, J.; Huang, W.; Wang, Y. Interplay Among Metabolism, Epigenetic Modifications, and Gene Expression in Cancer. *Front. Cell Dev. Biol.* **2021**, *9*, 793428. [CrossRef]
141. Martinez-Iglesias, O.; Naidoo, V.; Cacabelos, N.; Cacabelos, R. Epigenetic Biomarkers as Diagnostic Tools for Neurodegenerative Disorders. *Int. J. Mol. Sci.* **2021**, *23*, 13. [CrossRef]
142. Margis, R.; Margis, R.; Rieder, C.R. Identification of blood microRNAs associated to Parkinsonis disease. *J. Biotechnol.* **2011**, *152*, 96–101. [CrossRef]
143. Hao, H.; Shao, M.; An, J.; Chen, C.; Feng, X.; Xie, S.; Gu, Z.; Chan, P.; Chinese Parkinson Study, G. Association of Catechol-O-Methyltransferase and monoamine oxidase B gene polymorphisms with motor complications in parkinson's disease in a Chinese population. *Parkinsonism Relat. Disord.* **2014**, *20*, 1041–1045. [CrossRef]
144. Michalowska, M.; Chalimoniuk, M.; Jowko, E.; Przybylska, I.; Langfort, J.; Toczylowska, B.; Krygowska-Wajs, A.; Fiszer, U. Gene polymorphisms and motor levodopa-induced complications in Parkinson's disease. *Brain Behav.* **2020**, *10*, e01537. [CrossRef] [PubMed]
145. Wills, A.M.; Eberly, S.; Tennis, M.; Lang, A.E.; Messing, S.; Togasaki, D.; Tanner, C.M.; Kamp, C.; Chen, J.F.; Oakes, D.; et al. Caffeine consumption and risk of dyskinesia in CALM-PD. *Mov. Disord. Off. J. Mov. Disord. Soc.* **2013**, *28*, 380–383. [CrossRef] [PubMed]
146. Fardi, M.; Solali, S.; Farshdousti Hagh, M. Epigenetic mechanisms as a new approach in cancer treatment: An updated review. *Genes Dis.* **2018**, *5*, 304–311. [CrossRef] [PubMed]
147. Wang, Y.; Wang, X.; Li, R.; Yang, Z.F.; Wang, Y.Z.; Gong, X.L.; Wang, X.M. A DNA methyltransferase inhibitor, 5-aza-2′-deoxycytidine, exacerbates neurotoxicity and upregulates Parkinson's disease-related genes in dopaminergic neurons. *CNS Neurosci. Ther.* **2013**, *19*, 183–190. [CrossRef] [PubMed]
148. Fernandez-Abascal, J.; Chiaino, E.; Frosini, M.; Davey, G.P.; Valoti, M. beta-Naphthoflavone and Ethanol Reverse Mitochondrial Dysfunction in A Parkinsonian Model of Neurodegeneration. *Int. J. Mol. Sci.* **2020**, *21*, 3955. [CrossRef]
149. Corrigendum. *J. Pineal Res.* **2019**, *66*, e12529. [CrossRef]
150. Cacabelos, R.; Carrera, I.; Martinez, O.; Alejo, R.; Fernandez-Novoa, L.; Cacabelos, P.; Corzo, L.; Rodriguez, S.; Alcaraz, M.; Nebril, L.; et al. Atremorine in Parkinson's disease: From dopaminergic neuroprotection to pharmacogenomics. *Med. Res. Rev.* **2021**, *41*, 2841–2886. [CrossRef] [PubMed]
151. Zhu, M.; Li, W.W.; Lu, C.Z. Histone decacetylase inhibitors prevent mitochondrial fragmentation and elicit early neuroprotection against MPP+. *CNS Neurosci. Ther.* **2014**, *20*, 308–316. [CrossRef] [PubMed]
152. Chen, S.H.; Wu, H.M.; Ossola, B.; Schendzielorz, N.; Wilson, B.C.; Chu, C.H.; Chen, S.L.; Wang, Q.; Zhang, D.; Qian, L.; et al. Suberoylanilide hydroxamic acid, a histone deacetylase inhibitor, protects dopaminergic neurons from neurotoxin-induced damage. *Br. J. Pharmacol.* **2012**, *165*, 494–505. [CrossRef]

153. Jin, H.; Kanthasamy, A.; Harischandra, D.S.; Kondru, N.; Ghosh, A.; Panicker, N.; Anantharam, V.; Rana, A.; Kanthasamy, A.G. Histone hyperacetylation up-regulates protein kinase Cdelta in dopaminergic neurons to induce cell death: Relevance to epigenetic mechanisms of neurodegeneration in Parkinson disease. *J. Biol. Chem.* **2014**, *289*, 34743–34767. [CrossRef] [PubMed]
154. Wang, Y.; Wang, X.; Liu, L.; Wang, X. HDAC inhibitor trichostatin A-inhibited survival of dopaminergic neuronal cells. *Neurosci. Lett.* **2009**, *467*, 212–216. [CrossRef]
155. Cortes, H.; Reyes-Hernandez, O.D.; Gonzalez-Torres, M.; Vizcaino-Dorado, P.A.; Del Prado-Audelo, M.L.; Alcala-Alcala, S.; Sharifi-Rad, J.; Figueroa-Gonzalez, G.; Gonzalez-Del Carmen, M.; Floran, B.N.; et al. Curcumin for parkinson s disease: Potential therapeutic effects, molecular mechanisms, and nanoformulations to enhance its efficacy. *Cell. Mol. Biol.* **2021**, *67*, 101–105. [CrossRef]
156. Xie, C.; Zhu, J.; Yang, X.; Huang, C.; Zhou, L.; Meng, Z.; Li, X.; Zhong, C. TAp63alpha Is Involved in Tobacco Smoke-Induced Lung Cancer EMT and the Anti-cancer Activity of Curcumin via miR-19 Transcriptional Suppression. *Front. Cell Dev. Biol.* **2021**, *9*, 645402. [CrossRef]
157. Nicholas, A.P.; Lubin, F.D.; Hallett, P.J.; Vattem, P.; Ravenscroft, P.; Bezard, E.; Zhou, S.; Fox, S.H.; Brotchie, J.M.; Sweatt, J.D.; et al. Striatal histone modifications in models of levodopa-induced dyskinesia. *J. Neurochem.* **2008**, *106*, 486–494. [CrossRef]
158. Hong, C.T.; Chan, L.; Bai, C.H. The Effect of Caffeine on the Risk and Progression of Parkinson's Disease: A Meta-Analysis. *Nutrients* **2020**, *12*, 1860. [CrossRef]
159. Machado-Filho, J.A.; Correia, A.O.; Montenegro, A.B.; Nobre, M.E.; Cerqueira, G.S.; Neves, K.R.; Naffah-Mazzacoratti Mda, G.; Cavalheiro, E.A.; de Castro Brito, G.A.; de Barros Viana, G.S. Caffeine neuroprotective effects on 6-OHDA-lesioned rats are mediated by several factors, including pro-inflammatory cytokines and histone deacetylase inhibitions. *Behav. Brain Res.* **2014**, *264*, 116–125. [CrossRef] [PubMed]
160. Sen, P.; Shah, P.P.; Nativio, R.; Berger, S.L. Epigenetic Mechanisms of Longevity and Aging. *Cell* **2016**, *166*, 822–839. [CrossRef] [PubMed]

MDPI

Review

Brain Connectivity and Graph Theory Analysis in Alzheimer's and Parkinson's Disease: The Contribution of Electrophysiological Techniques

Francesca Miraglia [1,2,*], Fabrizio Vecchio [1,2], Chiara Pappalettera [1,2], Lorenzo Nucci [1], Maria Cotelli [3], Elda Judica [4], Florinda Ferreri [5,6] and Paolo Maria Rossini [1]

1 Brain Connectivity Laboratory, Department of Neuroscience and Neurorehabilitation, IRCCS San Raffaele Roma, 00163 Rome, Italy; fabrizio.vecchio@uniecampus.it (F.V.); pappaletterachiara@gmail.com (C.P.); lorenzo.nucci.ln@gmail.com (L.N.); paolomaria.rossini@sanraffaele.it (P.M.R.)
2 Department of Theoretical and Applied Sciences, eCampus University, 22060 Como, Italy
3 Neuropsychology Unit, IRCCS Istituto Centro San Giovanni di Dio Fatebenefratelli, 25125 Brescia, Italy; mcotelli@fatebenefratelli.eu
4 Department of Neurorehabilitation Sciences, Casa Cura Policlinico, 20144 Milan, Italy; e.judica@ccppdezza.it
5 Unit of Neurology, Unit of Clinical Neurophysiology and Study Center of Neurodegeneration (CESNE), Department of Neuroscience, University of Padua, 35122 Padua, Italy; florinda.ferreri@unipd.it
6 Department of Clinical Neurophysiology, Kuopio University Hospital, University of Eastern Finland, 70210 Kuopio, Finland
* Correspondence: fra.miraglia@gmail.com

Abstract: In recent years, applications of the network science to electrophysiological data have increased as electrophysiological techniques are not only relatively low cost, largely available on the territory and non-invasive, but also potential tools for large population screening. One of the emergent methods for the study of functional connectivity in electrophysiological recordings is graph theory: it allows to describe the brain through a mathematic model, the graph, and provides a simple representation of a complex system. As Alzheimer's and Parkinson's disease are associated with synaptic disruptions and changes in the strength of functional connectivity, they can be well described by functional connectivity analysis computed via graph theory. The aim of the present review is to provide an overview of the most recent applications of the graph theory to electrophysiological data in the two by far most frequent neurodegenerative disorders, Alzheimer's and Parkinson's diseases.

Keywords: EEG; MEG; graph theory; Alzheimer; Parkinson

Citation: Miraglia, F.; Vecchio, F.; Pappalettera, C.; Nucci, L.; Cotelli, M.; Judica, E.; Ferreri, F.; Rossini, P.M. Brain Connectivity and Graph Theory Analysis in Alzheimer's and Parkinson's Disease: The Contribution of Electrophysiological Techniques. *Brain Sci.* **2022**, *12*, 402. https://doi.org/10.3390/brainsci12030402

Academic Editors: Christina Piperi, Chiara Villa and Yam Nath Paudel

Received: 31 January 2022
Accepted: 16 March 2022
Published: 18 March 2022

Publisher's Note: MDPI stays neutral with regard to jurisdictional claims in published maps and institutional affiliations.

Copyright: © 2022 by the authors. Licensee MDPI, Basel, Switzerland. This article is an open access article distributed under the terms and conditions of the Creative Commons Attribution (CC BY) license (https://creativecommons.org/licenses/by/4.0/).

1. Introduction

The brain is one of the most complex and less explored systems of the human body. It consists of 100 billions of neurons that reciprocally communicate through networks of connections. In order to explain the mechanisms of brain networks, the "brain connectivity analysis" was created recently. Theoretically speaking, the analysis consists of three main types of connectivity: structural, functional and effective connectivity. Structural connectivity is based on anatomical constraints, that is, the set of physical (fibers) and structural (synaptic) connections linking neuronal units at a given time. Anatomical connectivity refers to a network of synaptic connections linking sets of neurons or neuronal elements, as well as their associated structural biophysical attributes condensed in parameters, such as synaptic strength or effectiveness [1,2]. The "functional connectivity space" is defined as the physical substrate in which all neural information processes happen, thus providing plausible biological boundaries for theories and inferences about neural interactions when analyzing functional neuroimaging data and developing computer simulations. In fact, because the structural/anatomical input/output connections of a given brain region are the main constraints for its functional properties, structural brain connectivity does not rigidly determine neural interactions, but acts instead as a dynamic support that reduces the dimensionality of the neural network state space. Meanwhile, functional interactions contribute to modify the underlying structural substrate by modifying the synaptic connections (i.e., enlarging/reducing the synaptic knob area, forming new synapses, and pruning preexisting synapses) [3].

In fact, functional connectivity is time dependent and captures the patterns of deviations between distributed and often spatially remote neuronal units, measuring the statistical correlation or their time-dependent activity. Effective connectivity describes the set of causal/modulating effects of one neural assembly activity over another and defines their inner hierarchy. Structural connectivity has been usually assessed by high spatial resolution technologies, such as magnetic resonance imaging (MRI-tractography); functional and effective connectivity are largely dependent on calculating the correspondence of neural signals over time, using electrophysiological techniques, such as EEG, TMS-EEG and MEG, that have an excellent temporal resolution and are optimal for calculating such connectivity [4,5]. Moreover, EEG recordings can be carried out in more ecological conditions since they do not need any specific environment, in opposition to the need of a shielded room for the fMRI and MEG recordings.

In the human brain, the connectome concept strongly relies on the evidence that neuronal populations interact with each other by means of their connections as well as of their temporally related dynamics. This is particularly evident when considering the innumerable brain dynamic states, which vary instantly and continuously because of changing sensory inputs from internal and external environments [2]. According to the principles of segregation and integration [6] in the human nervous systems, brain anatomical connections are both specific and variable. Specificity depends on the arrangement of individual synaptic connections between morphologically and physiologically different neuronal types, in the organization of axonal arborizations and long-range connectivity between separate cell nuclei or brain regions [1,2].

Recently, the study of brain connectivity was investigated in two of the main neurodegenerative diseases, in particular Alzheimer's (AD) and Parkinson's (PD) disease. In fact, AD is histopathologically defined by the presence of amyloid-beta plaques and tau-related neurofibrillary tangles, which have been associated with local synaptic disruptions, loss of fibers and neuronal death: this evidence suggests that AD is a dysconnectivity disease [7–12] whose early stages are due to synaptic failure. In addition, previous studies on PD have shown changes in the strength of functional connectivity between distributed brain regions associated with clinical symptoms, such as motor features [13–15], as well as a variety of non-motor disturbances, including cognitive impairment [16].

As already mentioned, one of the emergent tools for the study of functional connectivity is graph theory, which allows describing the brain through a mathematic model, the graph, which provides a simple representation of a complex system. The origins of graph theory and network science are to be found in the distant past, but their application in neuroscience is definitely more recent [17]. With the graph model, the brain is shaped as a network composed by nodes linked by directed or undirected, weighted or unweighted edges [18,19]. The characteristics of the graph are measurable through several parameters; the most explored ones are reported in Table 1.

Within this theoretical framework, the current review provides an overview of the most recent applications of graph theory analysis to electrophysiological data for the study of brain functional connectivity in two of the main neurodegenerative diseases, that is, AD and PD.

Table 1. Description of the main graph theory parameters.

Parameters	Description
Clustering Coefficient, C	The number of connections that exist between the nearest neighbors of a node as a proportion of maximum number of possible connections. It reflects the tendency of a network to form topologically organized circuits and it is often interpreted as a metric of information segregation in networks [20].
Path Length, PL	The minimum number of edges that must be traversed to go from one node to another. It is used as a measure of global integration of the network [20].
Small-world, SW	The ratio of the normalized clustering coefficient and normalized path length. It describes a balance between segregation and integration network properties integrating the information of global and local network characteristics [21].
Divergence	Measure of the broadness of the weighted degree distribution, where weighted degree is the summed weights of all edges connected to a node [22].
Modularity	Ratio of the intra- and intermodular connectivity strength where modules are subgraphs containing nodes that are more strongly connected to themselves than to other nodes. Modularity is a measure of the strength of the modules [22].
Efficiency	The ability of information exchange within the network [23].
Global efficiency	Measure of network integration and its overall performance for information transferring. This measure is inversely related to the average shortest path length [24].
Local efficiency	Local efficiency, which has a similar interpretation as clustering coefficient, measures the compactness of the subnetwork [25].
Centrality	The importance of a node and its direct impact on adjacent brain areas [23].
Betweenness	Used to investigate the contribution of each node to all other node pairs on the shortest path. It measures not only the importance of the nodes, but also the amount of information flowing through the node [25].
Strength	The sum of weights of connections (edges) of node. The strength can be averaged over the whole network to obtain a global measure of connection weights [26].
Degree	The degree of a node is the sum of its incoming (afferent) and outgoing (efferent) edges [27].
In-degree	Number of afferent connections to the node [27].
Out-degree	Number of efferent connections to the node [27].
Assortativity coefficient	The assortativity coefficient represents a measure of a network's resilience. It is a correlation coefficient between the degrees of all vertices on two opposite ends of an edge [27].

2. Alzheimer's Disease and Graph Theory

AD is the most common progressive and multifactorial, neurodegenerative disease in the elderly population and the main cause of cognitive impairment. The histopathological hallmarks of AD are the accumulation of the protein fragment beta-amyloid (plaques) outside neurons and of the protein tau (tangles) inside neurons. These changes are accompanied by the death of neurons and consequently by the damage of brain tissue [28].Over the years, as AD has been increasingly considered as a synaptic disconnection syndrome in its early stages—the pre-symptomatic stage of the disease can last many years and does not manifest due to the "neural reserve" that vicariate the lost functions—its complex brain dynamics have been studied by network approach. In fact, functional brain abnormalities can be reflected in changes of connectivity and networks architecture: this can be useful for the characterization of the brain condition in advance of symptom onset and disease progression.

Graph theory analysis provides tools to quantify networks properties and to understand the association between various pathological processes at the basis of AD. In recent years, researchers have advanced the idea of interpreting neurophysiological data (from EEG/MEG) via graph theory. For the first time, in 2007, Stam and colleagues applied graph theory methods to the EEG data of AD patients, comparing them to a group of control subjects and using synchronization likelihood (SL) as a measure to weight the graph. The authors demonstrated that the PL measure was higher in AD patients, whereas the C showed no significant alterations between the two groups [29]. The authors concluded that AD patients showed a loss of SW network characteristics, indicating less complexity and organization of the brain.

Since then, numerous scientists have explored the modulation of both local and global connectivity as computed main indexes, such as PL, C and SW, in the M/EEG frequency bands and over the years, and various reviews have been produced [30–32]. More recently, the distinctive features of physiological and pathological brain aging [33] were explored in order to describe the modulation of graph theory parameters in AD compared to healthy elderly people as well as to mild cognitive impairment (MCI) subjects. Indeed, MCI subjects do not yet meet the diagnosis of dementia, but carry a remarkably high risk, since about half of them become demented during a 3 to 5 years follow-up. Furthermore, studies of the graph derived from EEGs of AD patients were increasingly published, thanks to the low cost, large diffusion on the territory and non-invasiveness of the technique. Because of such characteristics, the EEG advanced analysis with graph theory might become a tool for large population screening in the near future [34,35]. In recent reviews, Rossini and collaborators [36] and Hallett and collaborators [37] summarized the results obtained from measures of brain connectivity (including graph theory) and their application in neurological diseases, such as AD, across MRI, EEG and MEG techniques.

However, some consistent results are available. In general, a decrease in PL was found in the lower alpha [38,39] and gamma bands [38], whereas an increase was found in the theta band [40] and in both the delta and theta bands [41].

Moreover, the C coefficient in AD patients have shown consistently a lower value in the alpha1 and beta bands [38], while a higher value of C was found in the theta and alpha1 bands [40,41], and in the alpha and beta bands [24,41].

Regarding the SW, the results seem less solid; however, some conclusions can be drawn. Several studies have reported a disruptive reorganization in the brain networks of AD patients in some of the frequency bands analyzed; in particular, the SW values seem to decrease in the delta, theta and in beta bands [33,38,42] and increase in the alpha one [43]. Further studies reported a significant reduction in the SW brain architecture in all the EEG frequency bands computed in mild AD patients compared to healthy controls [25,44]. Other researchers have adopted the SW index as a biomarker of the pathologic conversion of the MCI subject to AD patients, showing a high level of accuracy in combination with other biomarkers, such as Apo-E allele testing [45], as shown in Figure 1.

A pivotal aspect of graph measures is their potential as a prognostic tool in the conversion to AD status. In this regard, Miraglia and collaborators [46] deepened the analysis of SW in the Default Mode Network in a cohort of MCI subjects, discovering that SW index decreased in the gamma band in converted MCI subjects compared to stable MCI subjects. Moreover, in converted MCI subjects with impairment in linguistic domain, the SW index significantly decreased in the delta band, while in those converted MCI subjects with impairment in the executive domain, the SW index decreased in the delta and gamma bands and increased in the alpha 1 band (Table 1).

The most recent studies have intensified the research of the changes of the graph theory's measures, by analyzing new parameters and correlating them with neuropsychological tests [27,47] (Figure 2) and other biomarkers of AD, such as the hippocampal volume [43].

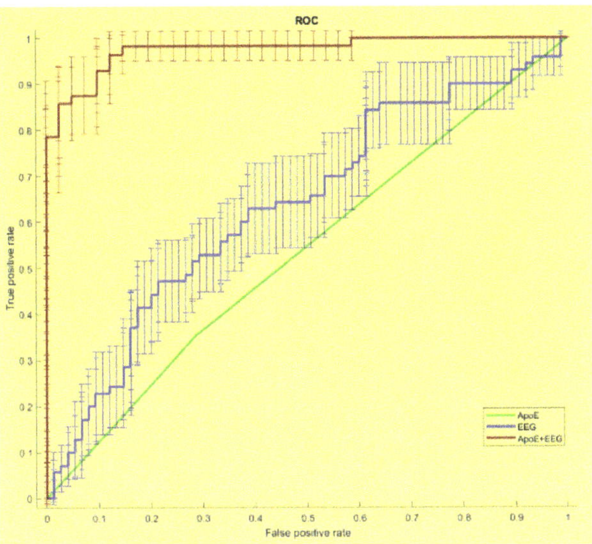

Figure 1. The receiver operating characteristic (ROC) curves illustrating the classification of the Stable and Converted MCI individuals based on the Apo-E (red line), SW (green line) and Apo-E and SW (blue line) values. The area under the (ROC) curve (AUC) was, respectively, 0.52, 0.64 and 0.97. Adapted from [45].

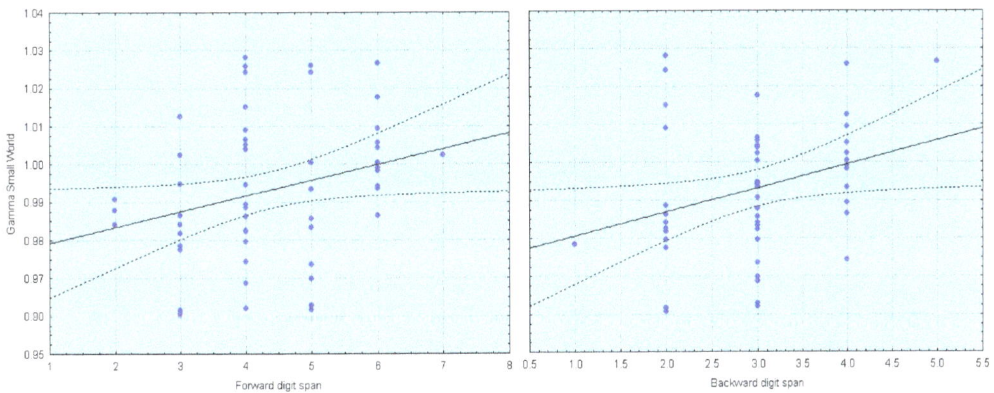

Figure 2. Scatterplots of SW index correlation with memory tasks. Less ordered brain network (as reflected by increasing value of SW) in the gamma band is associated with better memory performance. Adapted from [48].

Several research groups are working on this. In a recent study, the number of edges (degree), of inward edges (in-degree), and of outgoing edges (out-degree) were compared among healthy controls, MCI and AD patients with mild dementia by Franciotti and colleagues [27] to evaluate if degree anomaly could involve the measure of degree vertices, called hubs, in both prodromal and over AD stage. Degree, in-degree and out-degree values were smaller in MCI and mild AD than the control group for all hubs, confirming the hypothesis of an affected pattern of information flow in the brain networks. In the same study, the assortative coefficient, a correlation between the degrees of vertices on two opposite ends of an edge, was computed; however, not significant results emerged.

Networks with a positive assortative coefficient are resilient high-degree hubs. To the contrary, networks with a negative assortative coefficient are more vulnerable hubs.

Majd Abazid and colleagues [49] used a further innovative approach analyzing a dataset of EEGs of subjective cognitive impairment patients, MCI and AD patients. They quantified the graph links by weighing them by an entropy measure and comparing the accuracy of disease classification to other more used weight measures (i.e., phase lag index and coherence). They demonstrated the higher effectiveness of the entropy measure to analyze the brain network in patients with different stages of cognitive dysfunction.

Furthermore, Kocagoncu and colleagues [50] demonstrated the presence of a correlation between the SW index in the beta and gamma bands and the deposition of the protein tau, meaning that the higher tau burden in early AD's disease was associated with a shift away from the optimal SW organization and a more fragmented network, especially in the beta and gamma bands. Additionally, several studies have described a link of correlation of the graph parameters with the participant's generic cognitive status, evaluated by the mini-mental state examination (MMSE) test and memory assessment [27,42,51]. In particular, Franciotti and colleagues [27] found a positive correlation between the clustering coefficient and MMSE inpatients' groups, namely a higher clustering and higher MMSE, suggesting that high clustering is associated with the robustness of a network and resilience against damage.

Tait and collaborators [51] found a positive correlation between SW calculated in the temporal lobe and the language sub-score of MMSE, indicating that disruption to temporal lobe connectivity plays an important role in the language impairments of AD subjects. In a recent study, the SW measures were used as biomarkers to evaluate the effects of a repetitive transcranial magnetic stimulation and cognitive rehabilitation therapy for AD patients, recording the EEG before and after the treatment [52]. This study showed that the graph parameters can be awarded the role of diagnostic and evaluation biomarkers of AD stages and treatments (Figure 3).

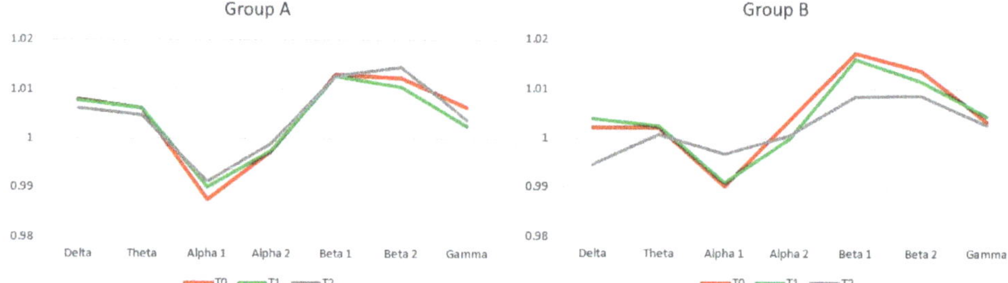

Figure 3. SW evaluation for two AD groups of two type of rehabilitation (repetitive Transcranial Magnetic Stimulation (rTMS) and Cognitive Training (Cog) for Group A and sham rTMS and Cog for Group B) for the evaluation of the effectiveness of the rTMS treatment before (T0), after the rehabilitation (T1) and at the 40 week follow up (T2). Adapted from [52].

The study of graph theory was applied to explore brain connectivity differences between AD and other dementia as well, such as vascular dementia patients compared to mild cognitive impairment (MCI) and normal elderly (Nold) subjects. It was confirmed that AD patients presented more ordered delta and theta SW organization (lower values), and more random alpha SW (higher values) than Nold and MCI subjects [53].

Finally, Li and collaborators [54] proposed a new combined approach based on an integrative graph analysis, by recording EEG and fNIRS signals in AD and controls subjects during a cognitive task. In particular, the graph indices were calculated from reconstructed EEG sources by using fNIRS localization to assess differences due to the disease. The results revealed lower values of all graph parameters (i.e., degree, C, and centrality) in the alpha and beta bands to the orbitofrontal and parietal regions and across all frequency bands in the frontal pole and medial orbitofrontal frequency, and higher values for the superior temporal sulcus. These findings suggest that the integration of EEG-fNIRS in the graph approaches could be useful to understand the spatiotemporal dynamics of brain activity better.

The main results just described are summarized in Table 2. Clearly, much work has been performed in the description of AD brain characteristics and network architecture, but there is still a lot to achieve in the definition of consensus biomarkers able to intercept the disease at its early stage, as this could provide treatment and rehabilitation strategies to improve the clinical condition of the patients.

Table 2. Summary of the main results of AD studies reported in the present review.

Authors	Recording Type	Graph Parameters	Main Results (All Results Refer to AD vs. Healthy)
Stam et al., 2007 [29]	EEG	PL C	• Beta PL ↑ • Beta C ↓ Pearson's correlation: • Beta PL ↑ MMSE ↓
Stam et al., 2009 [39]	MEG	PL C	• Alpha 1 PL↓ • Alpha 1 C ↓ Pearson's correlation: • Alpha 1 C ↓ MMSE ↓
de Haan et al., 2009 [38]	EEG	PL C	• Alpha-1 and beta C ↓ • Alpha 1 and gamma PL ↓
Poza et al., 2013 [41]	EEG	PL C	• Delta e theta PL ↑ • Alpha 2 and beta PL ↓ • Delta and theta C ↓ • Alpha 2 and beta C ↑
Wang et al., 2014 [25]	EEG	PL C Global Efficiency Local Efficiency SW	• PL↑ in all frequency bands (except delta) • C ↓ in all frequency bands (except delta) • Global Efficiency ↓ in all frequency bands • Local Efficiency ↓ in all frequency bands • SW ↓ in all frequency bands
Vecchio et al., 2014 [40]	EEG	PL C	• Theta PL ↑ • Delta, theta and alpha-1 C ↑
Frantzidis et al., 2014 [44]	EEG	SW	• SW ↓ Pearson's correlation: -SW ↓ MMSE ↓; SW ↓ MoCA ↓
Vecchio et al., 2016 [42]	EEG	SW	Pearson's correlation: • Gamma SW ↓ Digit Span Test ↓
Miraglia et al., 2017 [33]	EEG	SW	• EO: delta and theta SW Nold>MCI>AD • EC: delta SW Nold and MCI > AD
Vecchio et al., 2017 [34]	EEG	SW	Pearson's correlations: • Alpha SW ↓ hippocampal volume ↑ • Delta, beta, and gamma SW ↑ hippocampal volume ↑

Table 2. *Cont.*

Authors	Recording Type	Graph Parameters	Main Results (All Results Refer to AD vs. Healthy)
Saeedeh Afshari and Mahdi Jalili, 2017 [24]	EEG	Global efficiency Local efficiency	• Beta global efficiency ↓ • Alpha local efficiency↑
Vecchio et al., 2018 [45]	EEG	SW	• ROC curve accuracy 97%
Franciotti et al., 2019 [27]	EEG	Degree In-degree Out-degree Assortative Coefficient	• Degree, in-degree, out-degree ↓
Li et al., 2019 [54]	EEG	Degree C Centrality	• Alpha 2 and beta degree, C, centrality ↓ in orbitofrontal and parietal regions • All frequency degree, C, centrality ↓ in frontal pole and medial orbitofrontal regions • All frequency degree, C, centrality ↑ in the temporal sulcus
Vecchio et al., 2020 [55]	EEG	SW	• ROC curve accuracy 95%
Miraglia et al., 2020 [46]	EEG	SW	• Gamma SW ↓ in converted MCI vs. stable MCI • Delta SW ↓ in converted MCI in linguistic domain • Delta and gamma SW ↓ and alpha 1 SW ↑ in converted MCI in executive domain
Cecchetti et al., 2021 [47]	EEG	PL C	• Theta PL ↓ • Alpha 2 PL ↑ • Theta C ↑ • Alpha 2 C ↓
Majd Abazid et al., 2021 [49]	EEG	PL C Degree Efficiency Betweenness	• Higher accuracy of classification of AD for the graph parameters
Kocagoncu 2020 [50]	E/MEG	SW	Pearson's correlation: • Beta and gamma SW ↑protein Tau ↑
Tait et al., 2019 [51]	EEG	SW	Pearson's correlation: • Temporal lobe SW ↑ language sub-score ↑
Vecchio et al., 2021 [52]	EEG	SW	Pearson's correlations: • Delta SW ↓ ADAS-Cog ↑ • Alpha 1 ↑ ADAS-Cog ↑
Vecchio et al., 2021 [53]	EEG	SW	• Delta and theta SW ↓ • Alpha 2 SW↑

The arrows refer to an increase (↑) or a decrease (↓) of the indicated parameters in AD patients. All results in the table refer to AD patients compared to elderly healthy controls, except when differently indicated. Abbreviations: EEG: electroencephalography; MEG: magnetoencephalography; PL: path length; C: clustering coefficient; SW: small-world index; MMSE: mini-mental state examination; MoCA: Montreal Cognitive Assessment; EO: eyes open; EC: eyes closed; NOLD: NOrmal eLDerly; MCI: Mild Cognitive Impairment; ROC: received operating characteristics.

3. Parkinson's Disease and Graph Theory

PD is a neurodegenerative disease mainly characterized by movement impairment. Non-motor disturbances, including cognitive impairment, mood changes and dementia, are commonly present during the progression of PD [16]. The mechanisms underlying the development of motor and cognitive disorders are not completely understood. PD disorders were largely studied by functional imaging methods that have shown changes in brain functional connectivity. In spite of the promising findings of some studies [56], there are still few applications of graph theory to the M/EEG data of PD patients, probably because it is still assumed that PD is mainly due to subcortical relays degeneration (namely basal ganglia) that could be not captured by scalp recordings, which are dominated by the cortical EEG activity. However, it should be noted that such subcortical relays are heavily and mutually connected to cortical areas and that the clinical symptoms that characterize the disease mainly stem from the disruption of these connections. Therefore, M/EEG data recordings are suitable tools for PD connectivity studies. Figure 4 shows a representation of these relays, that is, the dopaminergic pathways linking the basal ganglia and the cortex. This picture illustrates the link between the cited subcortical structures and the activity of neurons in the cortex, justifying the use of scalp EEG recordings for the evaluation of brain network modulations due to subcortical network deregulations in PD.

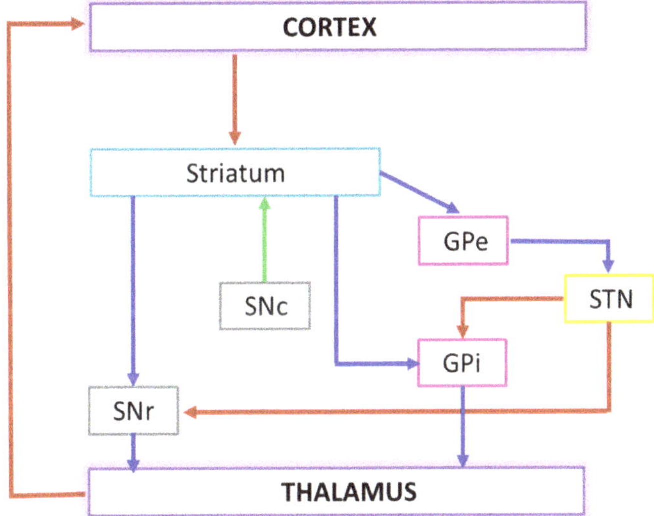

Figure 4. Dopaminergic pathways linking the basal ganglia and the cortex. Connectivity diagram showing excitatory glutamatergic pathways as red, inhibitory GABAergic pathways as blue, and modulatory dopaminergic as green. Their final functional output is the modulation of the cortical activity, mainly for motor-related circuits. Abbreviations: STN: subthalamic nucleus; SNr: substantia nigra pars reticulata; SNc: substantia nigra pars compacta; GPe: external segment of the globus pallidus; GPi: internal segment of the globus pallidus.

Notably, most of the relatively few studies in this disease have used the C and the PL as network measures, and of those selected, only one adds to them modularity and divergence [22] parameters study, whereas another study considers the SW index [21].

The results reported in this review include the most interesting studies concerning the investigation of graph theory application on the electrophysiological data of cognitively normal PD patients, PD subjects with Mild Cognitive Impairments and PD patients with dementia.

In a MEG study [57], cognitively normal PD subjects were compared to the control group and showed a decreased in the C in the delta band, whereas no differences in the PL compared to healthy subjects were reported, indicating the reduced local integration with preserved global efficiency of the brain network.

In a more recent EEG study, Suarez and colleagues [58] reported a decrease in the C in different bands from a previous work, theta and beta, and in addition, in the same bands, they also found a decrease in PL, highlighting a reduction in functional segregation and an increase in functional integration in both bands.

Moreover, in a more recent study on resting state EEG [21], it was found that SW index decreased in the theta and increased in the alpha band, describing a more ordered structure for the lower frequencies and a more random organization for the higher alpha frequencies for PD compared to healthy subjects, as described in Figure 5.

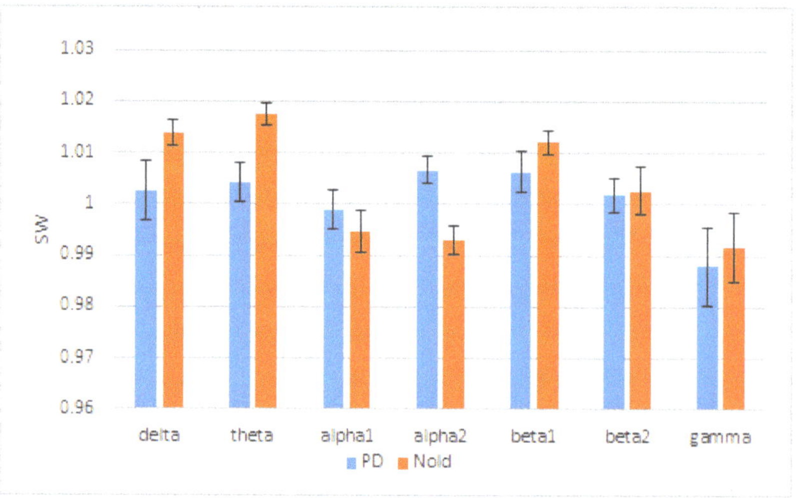

Figure 5. SW index trend in PD patients (blue line) and healthy elderly subjects (red line) in EEG frequency bands (delta, theta, alpha 1, alpha 2, beta 1, beta 2, and gamma). Adapted from [21].

Other studies have followed the temporal development of PD through a follow up or through the assessment of various cognitive impairment stages (MCI and dementia). Over a 4 year period analysis, Olde Dubbelink and colleagues [57] reported, within the same Parkinsonian group, a reduction in the PL in the alpha frequency range and in the C in the theta and alpha bands, showing a progressive impairment in the local integration and an additional loss of global efficiency as reflected in the alpha frequency band along the time.

Regarding Parkinsonian MCI subjects compared to healthy subjects, Suarez and colleagues [58] reported a reduction in the C in the alpha band and a decrement in the PL in the delta and theta bands. Furthermore, a similar reduction in the C was found in the alpha band in the work of Utianski and colleagues [22], who compared MCI PD subjects to PD subjects without cognitive impairment. In a recent study [26], through a dense-EEG source connectivity analysis, it was observed a decreasing tendency of global topological graph features (PL, C, modularity and strength) in the alpha frequency band, from PD patients with normal cognitive profile to PD subjects with dementia. Among functional connectivity studies in PD, the decrease in the patterns in the alpha band seems to be associated with cognitive impairment development. Moreover, the study of Mehraram and colleagues [59] of PD demented subjects showed a decreased C and node degree in the alpha band, an increase in the PL in the alpha band, and of modularity in the theta and alpha bands, compared to healthy subjects, demonstrating that the network measures in the alpha band were significantly affected in demented subjects.

Changes in brain networks in the alpha band can be found even in PD subjects with dementia compared to the ones with a normal cognitive profile. More specifically, a reduction in the C, the PL and of the divergence parameter in the alpha band, and an increase in the modularity index in the alpha and bands were reported [22], demonstrating an impairment of functional connectivity in the background frequency band (alpha).

Other studies have evaluated graph theory coefficients during the execution of a task. One of them has analyzed the performance of a local contextual processing paradigm [60] in order to demonstrate that functional disconnection is involved with contextual processing deficits in PD subjects. In particular, PD subjects showed a larger C and a longer PL compared to control subjects for predicted targets in the alpha and theta frequency bands, underlining a more structured functional connection in the detection of the predicted target and suggesting that the deficits observed in the process of context in PD may be due to ineffective interactions across cortical regions.

In another study [23], the pre-stimulus network abnormalities of PD patients experiencing pareidolias were investigated during a visual task. A higher global C and parietal efficiency and a decreased frontal degree centrality were found compared to normal PD and healthy subjects in the low-alpha band, suggesting an efficient information transfer within the parietal network and a reduced disengagement of the posterior cortex.

Finally, the latest evidence suggested that graph theory parameters are able to show different modulation of brain rhythms in deep brain stimulation (DBS). Li and colleagues [20] demonstrated that PD subjects with DBS-ON and DBS-OFF reported a lower C and local efficiency in the alpha and beta bands, compared to healthy subjects. Moreover, no evidence was found within patients with PD in DBS-ON and DBS-OFF in any frequency bands. Although significant changes in PL and global efficiency were not found between PD and healthy subjects, the study results indicated decreased brain network segregation in PD subjects and the moving forward to their more random organization, as highlighted in Vecchio and colleagues' study [21].

Bočkovà and collaborators [61] analyzed graph theory coefficients between normal cognitive PD groups treated by subthalamic brain stimulation (STN-DBS) in OFF and ON stimulation states during a visual task. They found that, in the 1–8 Hz band, subjects responding faster with DBS-OFF demonstrated a decrease in the C and node strength, while the PL was increased in the DBS-ON state compared to DBS-OFF after target stimuli. This network analysis highlights a dysfunction of the large-scale cerebral networks in subjects responding faster with DBS-OFF rather than DBS-ON, reporting global connectivity and slower communication within the frontal network in low frequencies bands (1–8 Hz). These findings suggest that the subjects with such reductions in low frequencies are more likely to develop cognitive deterioration.

The main evidence highlighted in the current review are reported in Table 3. Although only few studies are available on the application of graph theory to the electrophysiological data of PD patients, their results are encouraging for the characterization of the disease and for the response to treatments, suggesting network analysis as a promising tool and interesting field of research to explore.

Table 3. Summary of the main results of PD studies reported in the present review.

Authors	Recording Type	Graph Parameters	Main Results
Fogelson et al., 2013 [60]	EEG	C PL	• C ↑ in theta and alpha bands • PL ↑ in theta band for PD vs. Healthy subjects.
Olde Dubbelink et al., 2014 [57]	MEG	C PL	• C↓ in delta band for PD vs. Healthy subjects • C ↓ in theta and alpha 2 bands, PL ↓ in alpha 2 band for PD subjects

Table 3. *Cont.*

Authors	Recording Type	Graph Parameters	Main Results
Utianski et al., 2016 [22]	EEG	C PL Divergence Modularity	• C and PL ↑ in all frequency bands, Divergence ↑ in theta and beta bands and ↓ in delta and alpha bands, Modularity ↑ in all frequency bands, for normally cognitive PD vs. Healthy subjects. • C ↓ in alpha 1 band for PD-MCI vs. normally cognitive PD subjects. • C, PL and Divergence ↓ in alpha 1, Modularity ↑ in alpha 1 and 2 frequency bands, for demented PD vs. normally cognitive PD subjects.
Hassan et al., 2017 [26]	EEG	C PL Modularity Strength	• PL, C, Modularity and Strength ↓ in alpha frequency band for demented PD vs. normally cognitive PD subjects.
Mehraram et al., 2020 [59]	EEG	Node degree C PL SW Modularity	• C and Node degree↓ in alpha band, PL ↑ in alpha band and Modularity↑ in theta and alpha bands, for PD demented vs. Healthy subjects. • PL ↑ in alpha band, Modularity↑ in theta and alpha bands and SW ↑in theta band, for PD demented vs. AD subjects.
Bočková et al., 2021 [61]	EEG	Node strength C PL Modularity	• Node strength ↓, C ↓ and PL ↑ in 1–8 Hz frequencies band for DBS-ON compared to DBS-OFF for subjects responding faster with DBS-OFF rather than DBS-ON.
Suárez et al., 2021 [58]	EEG	C PL Local efficiency Global connectivity	• C and PL ↓ in theta and beta bands for normally cognitive PD vs. Healthy subjects. • C ↓ in alpha band, PL ↓ in delta and theta bands in PD-MCI vs. Healthy subjects.
Vecchio et al., 2021 [21]	EEG	SW	• SW ↓ in theta band and ↑ in alpha 2 band.
Li et al., 2021 [20]	EEG	C PL Global efficiency Local efficiency	• C and Local efficiency ↓ in alpha, beta 1 and beta 2 bands for PD subjects in DBS-OFF and DBS-ON vs. healthy subjects.
Revankar et al., 2021 [23]	EEG	C PL Efficiency Centrality	• C and parietal Efficiency ↑ in alpha 1 band, frontal Centrality ↓ for PD with pareidolias vs. normal PD and Healthy subjects.

The arrows refer to an increase (↑) or a decrease (↓) of the indicated parameters. Abbreviations: EEG: electroencephalography; MEG: magnetoencephalography; PL: path length; C: clustering coefficient; SW: small-world index; PD-MCI: Parkinson disease with mild cognitive impairments.

4. Conclusions

The aim of the present review was to provide an overview of the most recent applications of the graph theory to electrophysiological data over two of the main neurodegenerative disorders, that is, AD and PD.

In recent years, the applications of the networks science to electrophysiological data have increased as electrophysiological techniques are low cost, largely available on the territory and non-invasive with the potential to become a tool for large population screening [34,35]. Indeed, health systems are actually looking for a combination f biomarkers characterized by high accuracy, specificity, sensitivity as well as reasonable costs, non-invasiveness, and large availability. We can confidently conclude that the neurophysiological techniques described in this review embody all the required characteristics and are promising as optimal candidates for a first level screening. All of the reported evidence confirms the role of graph analysis as a promising tool in the characterization of the modulation of brain mechanisms of local and global integration in AD and PD as computed by main indexes (such as PL, C and SW in the M/EEG frequency bands) and over the years, and as demonstrated by new parameters (such as assortative coefficient, degree, modularity and divergence). Likewise, the optimal approach for quantifying functional connectivity is an open question [62] still in the absence of methodological convergence. In fact, a graph could be constructed with weighted or unweighted, directed or undirected edges. Moreover, in the case of a weighted graph, there are different weights of the edges [15]. Accordingly, examining the past and more recent studies, the results are sometimes contrasting and clearly dependent on the methods of analysis and on the frequency bands in which the EEG rhythms can be subdivided. The result variability can be therefore explained by the urgent need to share a methodological uniformity in the computation of graph construction and parameters' computation.

For AD, several studies have reported a disruptive reorganization of the brain networks, suggesting the SW index as biomarker for AD and for the conversion from MCI to AD. Other evidence has reported significant correlations between graph parameters and other biomarkers, such as the neuropsychological test, hippocampal volume, and genetic tests.

Moreover, it is strongly suggested that graph parameters can be awarded the role of diagnostic and evaluation biomarkers of AD and PD stages and rehabilitation treatments, and that it is possible to monitor the temporal development of PD through a follow up over the various cognitive impairment stages.

In conclusion, the graph theory could represent a promising tool for the identification, diagnosis, prognosis and even the identification of rehabilitation treatment for two of the main neurodegenerative diseases, such as AD and PD, to define the effects of the disease, increasing the information provided by traditional topographic mapping.

However, one of the major challenges of the application of graph theory to electrophysiological data is still to identify the measure that better describes the physiological mechanism under examination based on the data and the experiment. The aim of the current review was to describe the results of the application of graph theory analysis of electromagnetic brain signals in the pathological mechanisms in AD and PD. Several measures were presented, and each of them is more or less significant in the description of different aspects of the brain mechanism, and some of them show a significant correlation to each other. Further studies could be focused on the deeper understanding of the correlations between one measure and another and between the different brain mechanisms involved in the processes of neurodegeneration. This could solve the lack of standardization in methods, thus allowing to successively apply the more significantly measures to specific data and experiments, and to confirm the more promising results on larger populations. Finally, further reviews of the literature should explore the contribution of electrophysiological techniques in other neurodegenerative disorders (i.e., amyotrophic lateral sclerosis, fronto-temporal, vascular, Lewy body types of dementia, etc.).

Author Contributions: F.M. contributed to the conceptualization, methodology, and writing—original draft preparation. C.P. and L.N. were involved in the methodology data curation, and writing—reviewing and editing F.V., M.C., E.J., F.F. and P.M.R. helped in the supervision and writing—reviewing and editing the review. All authors have read and agreed to the published version of the manuscript.

Funding: This work was partially supported by the Italian Ministry of Health for Institutional Research (Ricerca corrente) and for the project "Prediction of conversion from Mild Cognitive Impairment to Alzheimer's disease based on TMS-EEG biomarkers" (GR-2016-02361802). The authors are also grateful to the Merck Sharp & Dohme (MSD) for the sponsorship.

Institutional Review Board Statement: Not applicable.

Informed Consent Statement: Not applicable.

Conflicts of Interest: The authors declare that there are not any conflict of interest.

References

1. Sporns, O.; Chialvo, D.R.; Kaiser, M.; Hilgetag, C.C. Organization, development and function of complex brain networks. *Trends Cogn. Sci.* **2004**, *8*, 418–425. [CrossRef] [PubMed]
2. Sporns, O. The human connectome: A complex network. *Ann. N. Y. Acad. Sci.* **2011**, *1224*, 109–125. [CrossRef] [PubMed]
3. Rossini, P.M.; Ferilli, M.A.; Rossini, L.; Ferreri, F. Clinical neurophysiology of brain plasticity in aging brain. *Curr. Pharm. Des.* **2013**, *19*, 6426–6439. [CrossRef]
4. Sporns, O.; Tononi, G.; Kötter, R. The human connectome: A structural description of the human brain. *PLoS Comput. Biol.* **2005**, *1*, e42. [CrossRef] [PubMed]
5. Rossini, P.M.; Di Iorio, R.; Bentivoglio, M.; Bertini, G.; Ferreri, F.; Gerloff, C.; Ilmoniemi, R.J.; Miraglia, F.; Nitsche, M.A.; Pestilli, F.; et al. Methods for analysis of brain connectivity: An IFCN-sponsored review. *Clin. Neurophysiol.* **2019**, *130*, 1833–1858. [CrossRef]
6. Tononi, G.; Sporns, O.; Edelman, G.M. A measure for brain complexity: Relating functional segregation and integration in the nervous system. *Proc. Natl. Acad. Sci. USA* **1994**, *91*, 5033–5037. [CrossRef]
7. Arendt, T. Synaptic degeneration in Alzheimer's disease. *Acta Neuropathol.* **2009**, *118*, 167–179. [CrossRef]
8. Blennow, K.; Cowburn, R.F. The neurochemistry of Alzheimer's disease. *Acta Neurol. Scand. Suppl.* **1996**, *168*, 77–86. [CrossRef]
9. Delbeuck, X.; Van der Linden, M.; Collette, F. Alzheimer's disease as a disconnection syndrome? *Neuropsychol. Rev.* **2003**, *13*, 79–92. [CrossRef]
10. Rossini, P.M.; Di Iorio, R.; Vecchio, F.; Anfossi, M.; Babiloni, C.; Bozzali, M.; Bruni, A.C.; Cappa, S.F.; Escudero, J.; Fraga, F.J.; et al. Early diagnosis of Alzheimer's disease: The role of biomarkers including advanced EEG signal analysis. Report from the IFCN-sponsored panel of experts. *Clin. Neurophysiol.* **2020**, *131*, 1287–1310. [CrossRef]
11. Rossini, P.M. Aging and brain connectivity via electroencephalographic recordings. *Neuroscience* **2019**, *422*, 228–229. [CrossRef] [PubMed]
12. Takahashi, R.H.; Capetillo-Zarate, E.; Lin, M.T.; Milner, T.A.; Gouras, G.K. Co-occurrence of Alzheimer's disease ß-amyloid and τ pathologies at synapses. *Neurobiol. Aging* **2010**, *31*, 1145–1152. [CrossRef] [PubMed]
13. Stoffers, D.; Bosboom, J.L.; Deijen, J.B.; Wolters, E.C.h.; Stam, C.J.; Berendse, H.W. Increased cortico-cortical functional connectivity in early-stage Parkinson's disease: An MEG study. *Neuroimage* **2008**, *41*, 212–222. [CrossRef]
14. Tessitore, A.; Esposito, F.; Vitale, C.; Santangelo, G.; Amboni, M.; Russo, A.; Corbo, D.; Cirillo, G.; Barone, P.; Tedeschi, G. Default-mode network connectivity in cognitively unimpaired patients with Parkinson disease. *Neurology* **2012**, *79*, 2226–2232. [CrossRef] [PubMed]
15. Ponsen, M.M.; Stam, C.J.; Bosboom, J.L.; Berendse, H.W.; Hillebrand, A. A three dimensional anatomical view of oscillatory resting-state activity and functional connectivity in Parkinson's disease related dementia: An MEG study using atlas-based beamforming. *Neuroimage Clin.* **2012**, *2*, 95–102. [CrossRef]
16. Chaudhuri, K.R.; Healy, D.G.; Schapira, A.H.; Excellence, N.I.f.C. Non-motor symptoms of Parkinson's disease: Diagnosis and management. *Lancet Neurol.* **2006**, *5*, 235–245. [CrossRef]
17. Sporns, O. Graph theory methods: Applications in brain networks. *Dialogues Clin. Neurosci.* **2018**, *20*, 111–121.
18. Friston, K.J.; Tononi, G.; Reeke, G.N.; Sporns, O.; Edelman, G.M. Value-dependent selection in the brain: Simulation in a synthetic neural model. *Neuroscience* **1994**, *59*, 229–243. [CrossRef]
19. Rubinov, M.; Sporns, O. Complex network measures of brain connectivity: Uses and interpretations. *Neuroimage* **2010**, *52*, 1059–1069. [CrossRef]
20. Li, Z.; Liu, C.; Wang, Q.; Liang, K.; Han, C.; Qiao, H.; Zhang, J.; Meng, F. Abnormal Functional Brain Network in Parkinson's Disease and the Effect of Acute Deep Brain Stimulation. *Front. Neurol.* **2021**, *12*, 715455. [CrossRef]
21. Vecchio, F.; Pappalettera, C.; Miraglia, F.; Alù, F.; Orticoni, A.; Judica, E.; Cotelli, M.; Pistoia, F.; Rossini, P.M. Graph Theory on Brain Cortical Sources in Parkinson's Disease: The Analysis of 'Small World' Organization from EEG. *Sensors* **2021**, *21*, 7266. [CrossRef] [PubMed]
22. Utianski, R.L.; Caviness, J.N.; van Straaten, E.C.; Beach, T.G.; Dugger, B.N.; Shill, H.A.; Driver-Dunckley, E.D.; Sabbagh, M.N.; Mehta, S.; Adler, C.H.; et al. Graph theory network function in Parkinson's disease assessed with electroencephalography. *Clin. Neurophysiol.* **2016**, *127*, 2228–2236. [CrossRef] [PubMed]
23. Revankar, G.S.; Kajiyama, Y.; Hattori, N.; Shimokawa, T.; Nakano, T.; Mihara, M.; Mori, E.; Mochizuki, H. Prestimulus Low-Alpha Frontal Networks Are Associated with Pareidolias in Parkinson's Disease. *Brain Connect.* **2021**, *11*, 772–782. [CrossRef] [PubMed]
24. Afshari, S.; Jalili, M. Directed Functional Networks in Alzheimer's Disease: Disruption of Global and Local Connectivity Measures. *IEEE J. Biomed. Health Inform.* **2017**, *21*, 949–955. [CrossRef] [PubMed]
25. Wang, R.; Wang, J.; Yu, H.; Wei, X.; Yang, C.; Deng, B. Decreased coherence and functional connectivity of electroencephalograph in Alzheimer's disease. *Chaos* **2014**, *24*, 033136. [CrossRef] [PubMed]

26. Hassan, M.; Chaton, L.; Benquet, P.; Delval, A.; Leroy, C.; Plomhause, L.; Moonen, A.J.; Duits, A.A.; Leentjens, A.F.; van Kranen-Mastenbroek, V.; et al. Functional connectivity disruptions correlate with cognitive phenotypes in Parkinson's disease. *Neuroimage Clin.* **2017**, *14*, 591–601. [CrossRef]
27. Franciotti, R.; Falasca, N.W.; Arnaldi, D.; Famà, F.; Babiloni, C.; Onofrj, M.; Nobili, F.M.; Bonanni, L. Cortical Network Topology in Prodromal and Mild Dementia Due to Alzheimer's Disease: Graph Theory Applied to Resting State EEG. *Brain Topogr.* **2019**, *32*, 127–141. [CrossRef]
28. McKhann, G.M.; Knopman, D.S.; Chertkow, H.; Hyman, B.T.; Jack, C.R., Jr.; Kawas, C.H.; Klunk, W.E.; Koroshetz, W.J.; Manly, J.J.; Mayeux, R.; et al. The diagnosis of dementia due to Alzheimer's disease: Recommendations from the National Institute on Aging-Alzheimer's Association workgroups on diagnostic guidelines for Alzheimer's disease. *Alzheimers Dement.* **2011**, *7*, 263–269. [CrossRef]
29. Stam, C.J.; Reijneveld, J.C. Graph theoretical analysis of complex networks in the brain. *Nonlinear Biomed. Phys.* **2007**, *1*, 3. [CrossRef]
30. Xie, T.; He, Y. Mapping the Alzheimer's brain with connectomics. *Front. Psychiatry* **2011**, *2*, 77. [CrossRef]
31. Tijms, B.M.; Wink, A.M.; de Haan, W.; van der Flier, W.M.; Stam, C.J.; Scheltens, P.; Barkhof, F. Alzheimer's disease: Connecting findings from graph theoretical studies of brain networks. *Neurobiol. Aging* **2013**, *34*, 2023–2036. [CrossRef] [PubMed]
32. Minati, L.; Varotto, G.; D'Incerti, L.; Panzica, F.; Chan, D. From brain topography to brain topology: Relevance of graph theory to functional neuroscience. *Neuroreport* **2013**, *24*, 536–543. [CrossRef] [PubMed]
33. Miraglia, F.; Vecchio, F.; Rossini, P.M. Searching for signs of aging and dementia in EEG through network analysis. *Behav. Brain Res.* **2017**, *317*, 292–300. [CrossRef] [PubMed]
34. Vecchio, F.; Miraglia, F.; Maria Rossini, P. Connectome: Graph theory application in functional brain network architecture. *Clin. Neurophysiol. Pract.* **2017**, *2*, 206–213. [CrossRef]
35. delEtoile, J.; Adeli, H. Graph Theory and Brain Connectivity in Alzheimer's Disease. *Neuroscientist* **2017**, *23*, 616–626. [CrossRef]
36. Rossini, P.; Miraglia, F.; Alù, F.; Cotelli, M.; Ferreri, F.; Di Iorio, R.; Iodice, F.; Vecchio, F. Neurophysiological Hallmarks of Neurodegenerative Cognitive Decline: The Study of Brain Connectivity as A Biomarker of Early Dementia. *J. Pers. Med.* **2020**, *10*, 34. [CrossRef]
37. Hallett, M.; de Haan, W.; Deco, G.; Dengler, R.; Di Iorio, R.; Gallea, C.; Gerloff, C.; Grefkes, C.; Helmich, R.C.; Kringelbach, M.L.; et al. Human brain connectivity: Clinical applications for clinical neurophysiology. *Clin. Neurophysiol.* **2020**, *131*, 1621–1651. [CrossRef]
38. de Haan, W.; Pijnenburg, Y.A.; Strijers, R.L.; van der Made, Y.; van der Flier, W.M.; Scheltens, P.; Stam, C.J. Functional neural network analysis in frontotemporal dementia and Alzheimer's disease using EEG and graph theory. *BMC Neurosci.* **2009**, *10*, 101. [CrossRef]
39. Stam, C.J.; de Haan, W.; Daffertshofer, A.; Jones, B.F.; Manshanden, I.; van Cappellen van Walsum, A.M.; Montez, T.; Verbunt, J.P.; de Munck, J.C.; van Dijk, B.W.; et al. Graph theoretical analysis of magnetoencephalographic functional connectivity in Alzheimer's disease. *Brain* **2009**, *132*, 213–224. [CrossRef]
40. Vecchio, F.; Miraglia, F.; Marra, C.; Quaranta, D.; Vita, M.G.; Bramanti, P.; Rossini, P.M. Human brain networks in cognitive decline: A graph theoretical analysis of cortical connectivity from EEG data. *J. Alzheimers Dis.* **2014**, *41*, 113–127. [CrossRef]
41. Poza, J.; Garcia, M.; Gomez, C.; Bachiller, A.; Carreres, A.; Hornero, R. Characterization of the spontaneous electroencephalographic activity in Alzheimer's disease using disequilibria and graph theory. *Annu. Int. Conf. IEEE Eng. Med. Biol. Soc.* **2013**, *2013*, 5990–5993. [CrossRef] [PubMed]
42. Vecchio, F.; Miraglia, F.; Quaranta, D.; Granata, G.; Romanello, R.; Marra, C.; Bramanti, P.; Rossini, P.M. Cortical connectivity and memory performance in cognitive decline: A study via graph theory from EEG data. *Neuroscience* **2016**, *316*, 143–150. [CrossRef] [PubMed]
43. Vecchio, F.; Miraglia, F.; Piludu, F.; Granata, G.; Romanello, R.; Caulo, M.; Onofrj, V.; Bramanti, P.; Colosimo, C.; Rossini, P.M. "Small World" architecture in brain connectivity and hippocampal volume in Alzheimer's disease: A study via graph theory from EEG data. *Brain Imaging Behav.* **2017**, *11*, 473–485. [CrossRef] [PubMed]
44. Frantzidis, C.A.; Vivas, A.B.; Tsolaki, A.; Klados, M.A.; Tsolaki, M.; Bamidis, P.D. Functional disorganization of small-world brain networks in mild Alzheimer's Disease and amnestic Mild Cognitive Impairment: An EEG study using Relative Wavelet Entropy (RWE). *Front. Aging Neurosci.* **2014**, *6*, 224. [CrossRef]
45. Vecchio, F.; Miraglia, F.; Iberite, F.; Lacidogna, G.; Guglielmi, V.; Marra, C.; Pasqualetti, P.; Tiziano, F.D.; Rossini, P.M. Sustainable method for Alzheimer dementia prediction in mild cognitive impairment: Electroencephalographic connectivity and graph theory combined with apolipoprotein E. *Ann. Neurol.* **2018**, *84*, 302–314. [CrossRef]
46. Miraglia, F.; Vecchio, F.; Marra, C.; Quaranta, D.; Alù, F.; Peroni, B.; Granata, G.; Judica, E.; Cotelli, M.; Rossini, P.M. Small World Index in Default Mode Network Predicts Progression from Mild Cognitive Impairment to Dementia. *Int. J. Neural. Syst.* **2020**, *30*, 2050004. [CrossRef]
47. Cecchetti, G.; Agosta, F.; Basaia, S.; Cividini, C.; Cursi, M.; Santangelo, R.; Caso, F.; Minicucci, F.; Magnani, G.; Filippi, M. Resting-state electroencephalographic biomarkers of Alzheimer's disease. *Neuroimage Clin.* **2021**, *31*, 102711. [CrossRef]
48. Vecchio, F.; Miraglia, F.; Gorgoni, M.; Ferrara, M.; Iberite, F.; Bramanti, P.; De Gennaro, L.; Rossini, P.M. Cortical connectivity modulation during sleep onset: A study via graph theory on EEG data. *Hum. Brain Mapp.* **2017**, *38*, 5456–5464. [CrossRef]

49. Abazid, M.; Houmani, N.; Boudy, J.; Dorizzi, B.; Mariani, J.; Kinugawa, K. A Comparative Study of Functional Connectivity Measures for Brain Network Analysis in the Context of AD Detection with EEG. *Entropy* **2021**, *23*, 1553. [CrossRef]
50. Kocagoncu, E.; Quinn, A.; Firouzian, A.; Cooper, E.; Greve, A.; Gunn, R.; Green, G.; Woolrich, M.W.; Henson, R.N.; Lovestone, S.; et al. Tau pathology in early Alzheimer's disease is linked to selective disruptions in neurophysiological network dynamics. *Neurobiol. Aging* **2020**, *92*, 141–152. [CrossRef]
51. Tait, L.; Stothart, G.; Coulthard, E.; Brown, J.T.; Kazanina, N.; Goodfellow, M. Network substrates of cognitive impairment in Alzheimer's Disease. *Clin. Neurophysiol.* **2019**, *130*, 1581–1595. [CrossRef]
52. Vecchio, F.; Quaranta, D.; Miraglia, F.; Pappalettera, C.; Di Iorio, R.; L'Abbate, F.; Cotelli, M.; Marra, C.; Rossini, P.M. Neuronavigated Magnetic Stimulation combined with cognitive training for Alzheimer's patients: An EEG graph study. *Geroscience* **2021**, *44*, 159–172. [CrossRef]
53. Vecchio, F.; Miraglia, F.; Alú, F.; Orticoni, A.; Judica, E.; Cotelli, M.; Rossini, P.M. Contribution of Graph Theory Applied to EEG Data Analysis for Alzheimer's Disease versus Vascular Dementia Diagnosis. *J. Alzheimers Dis.* **2021**, *82*, 871–879. [CrossRef]
54. Li, R.; Nguyen, T.; Potter, T.; Zhang, Y. Dynamic cortical connectivity alterations associated with Alzheimer's disease: An EEG and fNIRS integration study. *Neuroimage Clin.* **2019**, *21*, 101622. [CrossRef]
55. Vecchio, F.; Miraglia, F.; Alù, F.; Menna, M.; Judica, E.; Cotelli, M.; Rossini, P.M. Classification of Alzheimer's Disease with Innovative EEG Biomarkers in a Machine Learning Implementation. *J. Alzheimers Dis.* **2020**, *75*, 1253–1261. [CrossRef]
56. Bočková, M.; Rektor, I. Impairment of brain functions in Parkinson's disease reflected by alterations in neural connectivity in EEG studies: A viewpoint. *Clin. Neurophysiol.* **2019**, *130*, 239–247. [CrossRef]
57. Olde Dubbelink, K.T.; Hillebrand, A.; Stoffers, D.; Deijen, J.B.; Twisk, J.W.; Stam, C.J.; Berendse, H.W. Disrupted brain network topology in Parkinson's disease: A longitudinal magnetoencephalography study. *Brain* **2014**, *137*, 197–207. [CrossRef]
58. Peláez Suárez, A.A.; Berrillo Batista, S.; Pedroso Ibáñez, I.; Casabona Fernández, E.; Fuentes Campos, M.; Chacón, L.M. EEG-Derived Functional Connectivity Patterns Associated with Mild Cognitive Impairment in Parkinson's Disease. *Behav. Sci.* **2021**, *11*, 40. [CrossRef]
59. Mehraram, R.; Kaiser, M.; Cromarty, R.; Graziadio, S.; O'Brien, J.T.; Killen, A.; Taylor, J.P.; Peraza, L.R. Weighted network measures reveal differences between dementia types: An EEG study. *Hum. Brain Mapp.* **2020**, *41*, 1573–1590. [CrossRef]
60. Fogelson, N.; Li, L.; Li, Y.; Fernandez-Del-Olmo, M.; Santos-Garcia, D.; Peled, A. Functional connectivity abnormalities during contextual processing in schizophrenia and in Parkinson's disease. *Brain Cogn.* **2013**, *82*, 243–253. [CrossRef]
61. Bočková, M.; Výtvarová, E.; Lamoš, M.; Klimeš, P.; Jurák, P.; Halámek, J.; Goldemundová, S.; Baláž, M.; Rektor, I. Cortical network organization reflects clinical response to subthalamic nucleus deep brain stimulation in Parkinson's disease. *Hum. Brain Mapp.* **2021**, *42*, 5626–5635. [CrossRef]
62. Smith, S.M.; Miller, K.L.; Salimi-Khorshidi, G.; Webster, M.; Beckmann, C.F.; Nichols, T.E.; Ramsey, J.D.; Woolrich, M.W. Network modelling methods for FMRI. *Neuroimage* **2011**, *54*, 875–891. [CrossRef]

Review

Is Aducanumab for LMICs? Promises and Challenges

Illangage P. C. Gunawardena [1], Thaarvena Retinasamy [2] and Mohd. Farooq Shaikh [2,*]

1 Clinical School Johor Bahru, Jeffrey Cheah School of Medicine and Health Sciences, Monash University Malaysia, Johor Bahru 80100, Johor, Malaysia; igun0005@student.monash.edu
2 Neuropharmacology Research Strength, Jeffrey Cheah School of Medicine and Health Sciences, Monash University Malaysia, Bandar Sunway 47500, Selangor, Malaysia; thaarvena.retinasamy@monash.edu
* Correspondence: farooq.shaikh@monash.edu; Tel.: +60-3-5514-4483

Abstract: Aducanumab, a human monoclonal antibody, was approved in June of 2021 as the first disease-modifying treatment for Alzheimer's disease by the United States Food and Drug Administration (U.S. FDA). A substantial proportion of patients with Alzheimer's disease live in low- and middle-income countries (LMICs), and the debilitating effects of this disease exerts burdens on patients and caregivers in addition to the significant economic strains many nations bear. While the advantages of a disease-modifying therapy are clear in delaying the progression of disease to improve patient outcomes, aducanumab's approval by the U.S. FDA was met with controversy for a plethora of reasons. This paper will provide precursory insights into aducanumab's role, appropriateness, and cost-effectiveness in low- and middle-income countries. We extend some of the controversies associated with aducanumab, including the contradicting evidence from the two trials (EMERGE and ENGAGE) and the resources required to deliver the treatment safely and effectively to patients, among other key considerations.

Keywords: Alzheimer's disease; aducanumab; LMICs; APOE; burden of disease; treatment cost

Citation: Gunawardena, I.P.C.; Retinasamy, T.; Shaikh, M.F. Is Aducanumab for LMICs? Promises and Challenges. *Brain Sci.* **2021**, *11*, 1547. https://doi.org/10.3390/brainsci11111547

Academic Editors: Anke Sambeth and Andrew Clarkson

Received: 29 September 2021
Accepted: 20 November 2021
Published: 22 November 2021

Publisher's Note: MDPI stays neutral with regard to jurisdictional claims in published maps and institutional affiliations.

Copyright: © 2021 by the authors. Licensee MDPI, Basel, Switzerland. This article is an open access article distributed under the terms and conditions of the Creative Commons Attribution (CC BY) license (https://creativecommons.org/licenses/by/4.0/).

1. Introduction

Alzheimer's disease (AD) is an irreversible neurodegenerative disorder that is characterised by the progressive deterioration of certain parts of the brain that are essential for learning and memory. The disease progresses in stages over months or years, gradually affecting memory, reasoning, judgement, language, and eventually even simple tasks [1]. AD was first described by a German psychiatrist, Alois Alzheimer, in 1906, while performing an autopsy on a woman with memory and language impairment, where abnormal structures called senile plaques and neurofibrillary tangles were found throughout the cerebral cortex of her brain [2]. The clinical expression of AD is believed to begin decades before the onset of the disease, which is observed via the formation of specific AD pathology, amyloid-beta (Aβ) plaques between neurons, and the accumulation of intracellular neurofibrillary tangles composed of tau. These AD pathological hallmarks then trigger neuronal dysfunction, neurotoxicity, and inflammation, leading to cognitive dysfunction [3]. During the initial stages of the disease, neuronal and synaptic impairment occurs within the para-hippocampal regions, which are the regions of the brain responsible for the formation of new memories. However, as the disease progresses, the neuropathology continues to spread, triggering cortical atrophy and ventricular enlargement, which causes the total brain mass to reduce by up to 35% [4].

Among causes of dementia, such as cerebrovascular disease, Lewy body disease, and frontotemporal lobar degeneration (FTLD), AD is the most common cause of dementia worldwide [5]. AD accounts for 60–70% of all dementia cases, affecting at least 27 million people globally [6]. The incidence of AD increases exponentially with age [7]. Evidence suggests that dementias, including AD, are more prevalent among women than men [8,9]. It is often challenging to clinically distinguish AD from other types of dementia. The disease is usually difficult to diagnose until post-mortem, where Aβ plaques and tau neurofibrillary tangles within the brain are identified. Thus, a diagnosis is only feasible based on the symptoms and cognitive assessments, thereby making diagnosing, treating, and managing AD extremely demanding and, consequently, making analysing the burden of disease by dementia subtype challenging. This may also explain the scarcity of epidemiological data in low- and middle-income countries (LMICs).

Aducanumab is a human monoclonal antibody that works to reduce Aβ load in the brain; it is the first disease-modifying therapy to be approved for AD treatment [10]. On 7 June 2021, the United States Food and Drug Administration (U.S. FDA) approved aducanumab via an accelerated approval pathway [11]. Current AD treatment is centered on supportive care to manage the debilitating symptoms of dementia, and pharmacotherapy goals of mainstay classes of drugs, such as cholinesterase inhibitors (ChEIs) and N-methyl-D-aspartate (NMDA) receptor antagonists, do not modify the course of the disease. For this reason, there was much enthusiasm by those impacted by AD, including patients, caregivers, clinicians, and the broader community.

Nonetheless, this was overshadowed by controversy surrounding aducanumab's efficacy and performance in the two trials, namely EMERGE and ENGAGE, as well as questions of adverse events, the need for constant monitoring, and its high cost. Although aducanumab is currently new to the market, it is vital to gauge the suitability of this drug in the context of LMICs for the treatment of AD by taking a multifactorial approach. It is estimated that approximately half of the patient population with dementia (including all subtypes) live in LMICs [12]. LMICs have an increasingly ageing population, characterised by a demographic transition from a high to low shift in mortality and fertility rates. As a result, many LMICs face an increasing burden due to chronic, non-communicable diseases (NCDs) proportional to this epidemiological shift [13]; an increase in the number of AD cases in these countries is likely to be observed. This necessitates the need for a disease-modifying therapy, such as aducanumab.

2. Management of Alzheimer's Disease

Before the approval of aducanumab, a disease-modifying therapy, the mainstream treatment goals were improving cognitive and functional deficits, reducing behavioral disturbances, and ultimately improving patients' quality of life [14]. This was achieved using ChEIs, NMDA receptor antagonists, and SSRIs (selective serotonin-reuptake inhibitors) or SNRIs (serotonin-noradrenaline reuptake inhibitors). Most drugs currently in development at various stages of clinical trials are disease-modifying therapies; in fact, out of 126 agents in development in 2021, 104 (82.5% of all agents in trial) are disease-modifying agents [15]. A further 16 of these drugs target amyloid pathology specifically. Other disease-modifying therapies are centered on mechanisms based on synaptic plasticity and neuroprotection, tau pathology, oxidative stress, vasculature, and inflammation, to name some. The other non-disease modifying agents aim for the treatment of neuropsychiatric and behavioral symptoms associated with AD. For example, Sigma-1 receptor agonists, NMDA receptor antagonists, and Alpha-1 antagonists are being studied in various trials [15]. The following discussion will focus on drugs which are currently available on the market for the treatment of AD.

2.1. ChEIs and NMDA Receptor Antagonists

Two major classes of drugs, namely ChEIs and NMDA receptor antagonists, are currently used for improving symptoms and patient outcomes. ChEIs (donepezil, galantamine, or rivastigmine) enhance cortical cholinergic transmission and function by inhibiting synaptic cleft cholinesterase [16]. The benefit of symptomatic relief is modest; oral rivastigmine and oral galantamine are used in mild to moderate AD, while oral donepezil and transdermal rivastigmine are used in mild to severe AD [16,17]. NMDA receptor antagonists (memantine) work by blocking excessive, pathological NMDA receptor stimulation, and it is thought to be neuroprotective [18]. While memantine appears to have modest benefits in patients with moderate, severe, and advanced disease, especially when combined with a ChEI, there is limited evidence for its use in mild AD (although it is widely used off-label) [18].

2.2. SSRIs and SNRIs

Behavioral and psychological symptoms of AD are wide-ranging, from insomnia, anxiety, agitation, and apathy to depression, delusions, and hallucinations [19]. SSRIs, like sertraline, citalopram, and escitalopram, are preferred in depressed patients with AD, while atypical antidepressants and SNRIs may also be used [20]. Other pharmacological agents include low-dose trazodone or orexin antagonists for improved sleep. Trazodone and SSRIs may be considered for agitated behaviors, while risperidone, an atypical antipsychotic, may reduce general neurobehavioral symptoms in AD patients [14].

2.3. Aducanumab—Antibody-Based Immunotherapy

Aducanumab, which is marketed as ADUHELM, is AD's first targeted therapy, and its mechanism of action is based on the amyloid hypothesis. The amyloid hypothesis has been the mainstream concept underlying AD research for many years. It has been postulated that the misfolding and aggregation of Aβ peptides as senile plaques in the brain causes neurodegeneration [21]. Additionally, Aβ aggregation has also been said to be involved in triggering other closely related pathophysiological pathways, like tau hyperphosphorylation, inflammation, oxidative stress, and generation of reactive oxygen species (ROS), among others, which ultimately lead to neuronal toxicity and cell death [22]. Aβ peptides are obtained from the amyloid precursor protein (APP) via the action of secretases, which are protease enzymes. In normal physiological conditions, APP is usually cleaved by both α-secretase and γ-secretase, which produce non-toxic fragments that are cleared from the brain. However, under extreme pathophysiological conditions, the APP are cleaved by β-secretase and γ-secretase instead, producing Aβ peptide fragments, particularly $A\beta_{42}$, which are highly amyloidogenic. These fragments then proceed to misfold, forming toxic oligomers, protofibrils, fibrils, and senile plaques in extracellular regions of the brain (Figure 1) [22]. The resultant Aβ oligomers interfere with the signalling cascade in the synaptic cleft, ultimately causing synaptic dysfunction and neuronal death.

Aducanumab uses an antibody-based immunotherapeutic approach by choosing human B-cell clones-activated Aβ aggregates epitopes [23]. When various human memory B cells were selected for testing reactivity against aggregated Aβ, aducanumab, a human monoclonal antibody, was found to be selectively reactive with Aβ aggregates, soluble oligomers, as well as insoluble fibrils; hence, it was selected as the lead drug candidate [24]. Aducanumab was found to cross the blood-brain barrier in preclinical studies, which further reduced brain amyloid burden [23]. Preclinical studies using Tg2576 mice reported that intraperitoneally injected aducanumab bound and aided in the clearance of parenchymal plaques, preventing micro hemorrhages [25]. Additionally, the accumulation of brain macrophages around the remaining plaques was also reported, thus implying the possibility of phagocytosis being used to remove the Aβ plaques, potentially slowing neurodegeneration and disease progression [26]. Aducanumab has since been deemed as the first approved treatment to treat AD's root cause instead of just treating the symptoms.

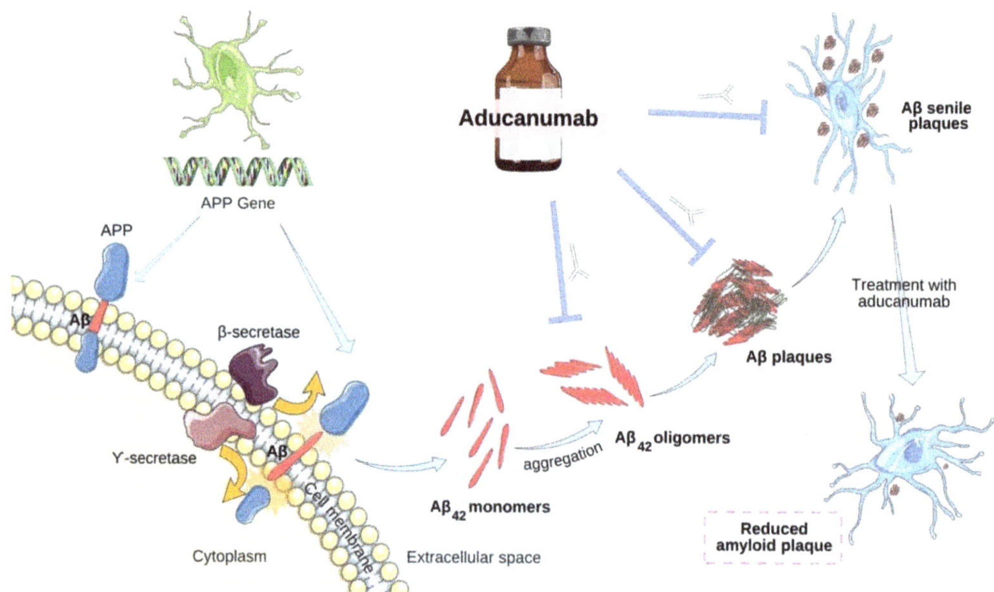

Figure 1. Aβ aggregation pathway and mechanism of action of aducanumab. APP is cleaved by β-secretase and γ-secretase to form $A\beta_{42}$ monomers, oligomers, and eventually senile plaques (note that protofibrils and fibrils are not illustrated in this figure); these are targets for clearance by aducanumab resulting in reduced amyloid burden. Parts of the figure were illustrated using images from Servier Medical Art by Servier, licensed under a Creative Commons Attribution 3.0 Unported License (https://creativecommons.org/licenses/by/3.0/ (accessed on 10 November 2021)).

3. Aducanumab in Low- and Middle-Income Countries

AD places burdens on an individual level, affecting patients, caregivers, and families while also burdening and straining healthcare systems at a societal level. In terms of years of life lived with disability (YLD) due to non-communicable disease, dementia (including AD) accounts for 11.9% of the total number of years [7]. In 2019, AD and all dementias accounted for 5.6% of all global disability-adjusted life years (DALYs) and was the fourth-highest cause of DALYs in patients aged 75 and older [27]. There is a demand for disease-modifying AD therapies, as an increase in the number of dementia and AD cases, especially in LMICs, would exert an unwarranted burden on patients and caregivers. At the same time, resource limitations would negatively impact healthcare systems in these countries. There are currently no biosimilars or generic equivalents of aducanumab. It is the only disease-modifying AD therapy currently in the market, targeting the underlying pathophysiology of the AD disease process [28–30]. The economic cost of dementia (including AD) increased by 35.4% from U.S. $604 billion in 2010 to U.S. $818 billion in 2015, which is 1.09% of the world's GDP [12,31]. An estimated U.S. $715.1 billion or 86% of these costs were from high-income countries (HICs), while LMICs accounted for a sum of U.S. $102.8 billion. A disease-modifying therapy could potentially decrease the burden of AD in terms of mortality and morbidity and improve health outcomes by generating an enduring clinical effect.

Comparing the efficacy of aducanumab directly with more conventional drug classes, such as ChEIs and NMDA receptor antagonists, is challenging, as the pharmacotherapy goals are inherently different. Aducanumab is a recombinant monoclonal antibody based on the principles of passive immunotherapy. It works by selectively binding Aβ fibrils and soluble oligomers, reducing amyloid-beta dose- and time-dependently [32]. ChEIs and NMDA receptor antagonists, on the other hand, aim for symptomatic alleviation. Evidence shows that donepezil, rivastigmine, and galantamine yields modest improvements in cognitive and clinical function in patients with mild to moderate AD in the short and long term [33]. Table 1 represents a crude comparison of the various characteristics of aducanumab, ChEIs, and NMDA receptor antagonists, including pharmacotherapy goals, mechanism of action, and efficacy, among others. The efficacy data are based on the most common primary outcome measures of cognitive function—with aducanumab's data based on the randomised-control trial (EMERGE), while meta-analyses were used for the remaining drug classes. Care should be taken when interpreting and comparing efficacy between the drugs, as it is subject to variabilities in study design, including patient characteristics and indications. This is due to a lack of head-to-head clinical trials between aducanumab and other drugs currently available on the market; hence, the data in Table 1 that directly compares the efficacy of different drug classes serves an exploratory purpose. Commenting on the robustness of the data presented, the quality of evidence was moderate in studies used in the meta-analyses for donepezil [34] and rivastigmine [35]. There was considerable heterogeneity in some outcome measures in the galantamine [36] and memantine [37] meta-analyses. All meta-analyses presented were subject to publication bias, and most studies included in the analyses were industry-funded.

Table 1. Direct comparison between aducanumab and mainstay pharmacotherapy of AD.

	Monoclonal Antibodies	Cholinesterase Inhibitors (ChEIs)	N-methyl-D-aspartate (NMDA) Receptor Antagonists
Drug	Aducanumab	Donepezil, rivastigmine, and galantamine	Memantine
Pharmacotherapy goal(s)	Disease-modifying treatment to reduce cognitive decline.	Symptomatic management of cognition and global functioning.	Symptomatic management of cognition and global functioning.
Mechanism of action	Selectively targets and clears Aβ aggregates, Aβ fibrils, and soluble oligomers. A reduction in Aβ build up in the brain is expected to demonstrate a reduction in cognitive decline in patients [23].	Increases cholinergic transmission by inhibiting cholinesterase at the synaptic cleft, thereby improving cortical cholinergic function [33].	Exerts neuroprotective effects by blocking pathological stimulation of glutamate NMDA receptors, thereby decreasing excitotoxicity [33].
Indication	Mild cognitive impairment (MCI), mild AD [10]	Mild to moderate AD, advanced disease [35]	Moderate to severe AD, mild AD (off-label) [18]
Route of administration	Intravenous infusion	Oral, transdermal patch (rivastigmine only)	Oral
Efficacy in terms of cognitive function	A statistically significant improvement in various scales of cognitive function was observed in the high-dose arm of EMERGE [29]. Difference vs. placebo: −0.39 (95% CI 0.69 to −0.09) [29] Score: CDR-SB * Difference vs. placebo: −1.400 ($p = 0.00967$) [29] Scale: ADAS-Cog ** Difference vs. placebo: 0.6 ($p = 0.05$) [29] Score: MMSE ‡	A meta-analysis of ChEIs revealed modest improvements [38]. **Donepezil** MD −2.67, (95% CI −3.31 to −2.02) [34] Scale: ADAS-Cog † MD 1.05, (95% CI 0.73 to 1.37) [34] Score: MMSE ‡ **Galantamine** MD −2.95, (95% CI −3.32, −2.57) [36] Scale: ADAS-Cog † MD 2.50, (95% CI 0.86 to 4.15) [36] Score: MMSE ‡ **Rivastigmine** MD −1.79, (95% CI −2.21 to −1.37) [35] Scale: ADAS-Cog † MD 0.74, (95% CI 0.52 to 0.97) [35] Score: MMSE ‡	A reduction in deterioration on different scales of clinical efficacy compared to placebo was observed in patients with moderate to severe AD [18]. MD −1.02, (95% CI −1.66 to −0.39, $p = 0.002$) [37] Scale: ADAS-Cog †

AD, Alzheimer's disease; Aβ, amyloid-beta protein; ADAS-Cog, Alzheimer's Disease Assessment Scale-Cognitive Subscale; CDR-SB, Clinical Dementia Rating-Sum of Boxes; CI, confidence interval; MD, mean difference; MMSE, Mini-Mental State Exam; SMD, standard mean difference. * CDR-SB scores range from 0 to 18; higher scores mean greater disease severity. † ADAS-Cog scores range from 0 to 70; higher scores mean greater cognitive impairment. ** The aducanumab trials utilised the ADAS-Cog-13 scale; scores range from 0 to 85; higher scores mean greater cognitive impairment. ‡ MMSE scores range from 0 to 30; higher scores mean less cognitive impairment.

3.1. Accelerated Approval and the Efficacy of Aducanumab

The first aducanumab trials started in 2011 with a phase I study (study 101) after pre-clinical trials with transgenic mice showed reduced amyloid burden in the brain [25,39]. The clinical trials that are important in assessing the efficacy of aducanumab are studies 103 (phase Ib) and the two identical phase III trials: 301 (ENGAGE) and 302 (EMERGE) [40]. Figure 2 is a timeline that summarizes the key events leading up to the accelerated approval of aducanumab.

Of the two phase III clinical trials, only the high-dose arm of one trial, EMERGE, met its primary endpoint by demonstrating improvements in the Clinical Dementia-Sum of Boxes (CDR-SB) score in addition to showing benefits in other secondary outcomes, such as the MMSE score, ADAS-Cog-13, and ADCS-ADL-MCI scores [40]. However, the low-dose arm did not reveal any benefit of aducanumab compared to the placebo, and no benefits were observed in either arm of the ENGAGE trial [29]. In fact, in the ENGAGE trial, it was noted that the CDR-SB score change in the high-dose arm was quantitatively worse than placebo at 78 weeks [40]. Prior to the current analysis by the manufacturer, the two trials were halted in March 2019 after a planned interim analysis met the criteria for futility [41]. On that account, confidence in the efficacy of aducanumab would need to be tempered due to the contradicting evidence presented from the two trials.

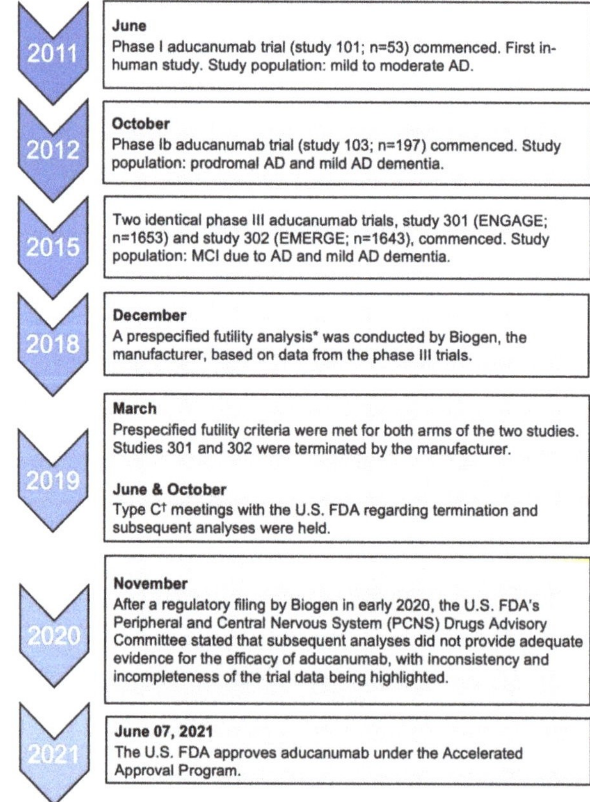

Figure 2. Timeline of clinical trials of aducanumab and key regulatory decisions. AD, Alzheimer's disease; MCI, mild cognitive impairment. * This was defined as the conditional power being less than 20% for both dose arms of the trials to meet the primary endpoint [40]. † Type C meetings are those that do not fall into type A meetings (due to stalled product development) or type B meetings (for example, after a trial's end-of-phase) [42].

Furthermore, the efficacy of aducanumab was determined in research settings, and therefore, clinical practice may vary. The placebo-controlled EMERGE and ENGAGE trial's population included individuals with early AD (i.e., those with mild cognitive impairment due to AD or those with mild AD) [29]. Moderate to severe AD patients make up approximately 50% of the total number of individuals living with AD [43]. The suitability and efficacy of aducanumab were not assessed in these patient groups, effectively limiting access to half of all patients living with AD, which will continue to contribute significantly to the burden of disease. Additionally, there is an intense debate surrounding the clinical significance observed in the EMERGE trial, as the hypothesis that clearance of Aβ protein equates to clinical improvement is inconclusive and is yet to be demonstrated [14]. As the trials have utilized a surrogate endpoint that facilitated the U.S. FDA's accelerated approval, the data can only predict a clinical outcome for the treatment of AD. As a result, full approval depends on a phase IV confirmatory trial, which aims to measure the clinical benefit [44]. Consequently, aducanumab's applicability to a limited subset of AD patients and its currently contended effectiveness would negatively affect the suitability of this drug in many countries, including in LMICs.

3.2. Treatment Challenges

An issue that would impact the suitability of aducanumab in many LMICs is the complexity of the treatment regimen and the need for robust healthcare systems to deliver therapies to patients. Aducanumab is administered by intravenous infusion every four weeks, with increasing titrations every two weeks for the first 32 weeks, followed by a constant high dose beyond week 36, with each infusion lasting 1 h [40]. Additionally, ascertaining the amyloid burden in patients before initiating treatment is crucial to guide clinical diagnosis and to assess the suitability of aducanumab. This is achieved by positron emission tomography (PET) for a visual read or through an invasive cerebrospinal fluid (CSF) quantitative analysis [10]. Amyloid-PET scans should be interpreted cautiously by trained radiologists and nuclear medicine specialists [10]. In addition to resource limitations and healthcare availability and accessibility, this complex treatment regimen highlights the importance of continuity and coordination in healthcare, which is associated with improved health outcomes [45]; however, such measures are lacking in many LMICs [46]. Safety and competency among healthcare workers are essential in delivering novel therapies effectively; evidence suggests that healthcare service competence and safety are deficient in LMICs [45]. The number of physicians per 1000 people in high-income countries was 3.1, while it was 1.3 in low- and middle-income countries [47], further emphasising resource limitations. On the contrary, a prospective advantage of aducanumab's approval is that it could lay the foundation for future therapies for AD in terms of advancing and improving treatment delivery to patients, which can improve health outcomes. For instance, long-term data from follow-up trials would be crucial in determining efficacy as well as providing an opportunity for head-to-head trials with existing therapies to be conducted. Due to the nature of AD and its multifaceted pathophysiology, combination therapy involving multiple targets may be necessary [48]. Therefore, aducanumab could serve as a catalyst towards better AD treatment in the future. Nonetheless, considering the importance of care continuity, follow-ups, and the general complexities associated with dosing intervals, which are all augmented by a general absence of high-quality healthcare coverage, introducing aducanumab to LMICs would be challenging.

3.3. Adverse Effects

Adequate follow-ups during the treatment phase are crucial for monitoring severe adverse events. ARIA (amyloid-related imaging abnormalities) due to oedema (ARIA-E) and brain microhemorrhage or localised superficial siderosis (ARIA-H) were frequently seen in the treatment groups [40]. In the high-dose treatment arm of the EMERGE and ENGAGE phase III trials, 41.3% of individuals experienced ARIAs compared to 10.3% in the placebo arm [40]. In addition to a pre-treatment MRI, frequent scans were performed to monitor the ARIAs of aducanumab. Individuals were scheduled to have five MRI scans of the brain in the first year of treatment alone, followed by two more scans in the last six months of treatment [40]. In clinical practice, this would mean a pre-treatment MRI, followed by two more brain MRI scans before the seventh and twelfth doses, which is in addition to more scans if patients experience symptoms related to ARIAs. To put it into perspective, in high-income countries, the number of MRI scanners per million inhabitants is 27.3, which contrasts with 3.4 scanners and 0.4 scanners per million inhabitants in upper-middle-income and lower-middle-income countries, respectively, and 0.2 scanners per million inhabitants in low-income countries [49]. If such monitoring practices cannot be implemented effectively in these countries, patients with severe ARIAs would not be identified and managed early, compounding the burden on patients, caregivers, healthcare systems, and economies.

3.4. Apolipoprotein E and Interethnic Differences

The synthesis, clearance, and accumulation of Aβ are influenced by a variety of factors. In the less common familial AD, mutations in the APP gene or PSEN1 gene may lead to increased Aβ accumulation [50]. Age-related processes, including neuronal stress, microglia-related inflammation, and a negative impact on protein homeostasis, may affect Aβ aggregation [51–53]. Other factors, such as insulin-like growth factor (IGF) resistance and diabetes, traumatic brain injuries, and the human microbiota, have also been studied [50]. However, for sporadic AD, which is more common, the presence of the ε4 allele of apolipoprotein E (APOEε4) is implicated, while BIN1 (bridging integrator-1) and TREM2 (Triggering Receptor Expressed On Myeloid Cells 2) also play a role [50]. Apolipoproteins usually aid in the transport of lipids in the body; however, APOE4 can also form stable complexes with Aβ, impacting its clearance from the brain. Therefore, APOEε4 genotype carriers have a higher amyloid load than non-carriers, and amyloid positivity is associated with greater cognitive impairment [54,55].

In the EMERGE and ENGAGE trials, participants underwent genetic screening for the presence of the APOEε4 allele [30]. It was noted that 65–75% of patients with AD carry the APOEε4 allele [56]. For instance, Mattsson and colleagues' [57] meta-analysis found that the frequency of APOEε4 carriers in patients with AD was 68.9% in Northern Europe and 52.1% in Asian populations. For patients with MCI, this was 52.5% vs. 33.3% in Northern Europe vs. Asia. The proportion of APOEε4 carriers with MCI who were Aβ positive in both Asian and European populations were significant compared to patients who were Aβ negative (Figure 3a,b) [57]. This is important, as the population assessed in the aducanumab trials were patients with confirmed amyloid pathology [10], which relates to the mechanism of action of the therapeutic. While more studies need to be conducted to confirm differences in APOEε4 allele frequencies across different populations, it may indicate the relative prevalence of amyloid pathology across different regions and ethnicities and, by extension, may guide cost-effectiveness assessments in nations including in LMICs.

Most notably, APOEε4 status was associated with an increased incidence of ARIA-E. As a result, dosing was adjusted in the initial stages of the trials to allow for lower doses in participants who carried the gene [29]. Figure 4 represents the number of participants who experienced ARIA-E in EMERGE and ENGAGE, stratified by APOEε4 carrier status compared to placebo. Furthermore, homozygous carriers may be predisposed to more frequent and severe ARIA than those who carry only one copy of the allele [58]. Although genotype testing is currently not indicated as part of the pre-treatment procedure, clinicians must carefully assess the risk-benefit balance. Additionally, if such risk-stratification measures are required as part of a full regulatory approval in the future, it would only serve to compound resource and economic burdens when delivering this complex therapy to patients.

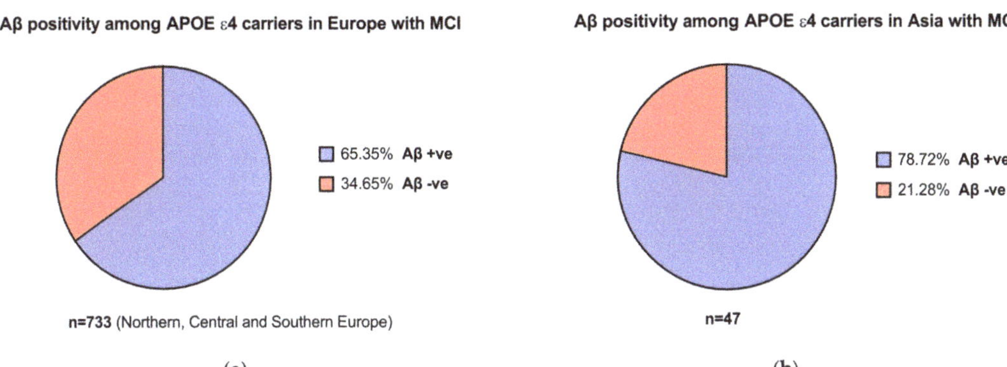

Figure 3. (**a**) Aβ positivity among APOEε4 carriers with mild cognitive impairment (MCI) in Europe (**b**) and in Asia. Data obtained from Mattson et al. [57].

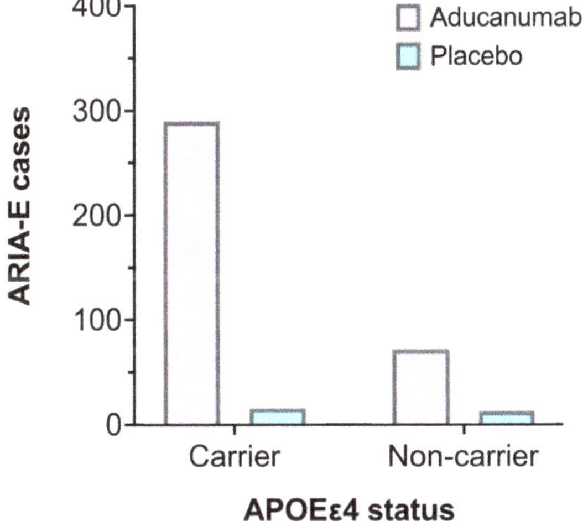

Figure 4. The number of ARIA-E cases by APOEε4 carrier status in EMERGE and ENGAGE (aducanumab 10 mg/kg, n = 362 vs. placebo, n = 29) [44].

The ethno-racial differences in AD pathology could impact the cost-effectiveness of aducanumab in LMICs. In diagnostics, P-tau biomarkers (p-tau181 and p-tau217) were associated with Aβ pathology on PET [59]. Brickman et al. [59] found that concentrations of p-tau biomarkers did not differ across Non-Hispanic Whites, Black people, and Hispanics. However, another study found that Black people had lower CSF phosphorylated-tau (p-tau181) and total tau (t-tau) levels when compared to Caucasians, independent of cognition [60]. Although there is no clear association between the prevalence of amyloid pathology and socioeconomic status, and due to limitations in the literature concerning the interethnic impact of both allele frequencies and biomarker levels on amyloid pathology, aducanumab's use in these populations must be assessed to further clarify its cost-effectiveness, especially in the context of LMICs.

3.5. The Economic Burden

As announced by the manufacturer, aducanumab's wholesale cost is U.S. $56,000 annually per patient [30]. The Institute for Clinical and Economic Review's (ICER) value-based analysis suggests a significantly lower price between U.S. $2500 and U.S. $8300 per annum, which is comparable to existing classes of drugs [30] (Table 2). The prevalence of AD is significant; however, its relatively low incidence compared to other non-communicable diseases could mean lower total public health expenditure. Nevertheless, ICER's report estimated the annual United States budget impact to be U.S. $819 million for aducanumab [30]. There is a pronounced incongruity between pharmaceutical spending in high-income countries and LMICs—while 84% of the global population reside in LMICs, it only accounted for 21.5% of the total global pharmaceutical expenditure [61]. Since aducanumab targets the early AD stage to prolong disease progression [40], healthcare systems would need to factor in the costs of managing patients over a longer period. Training individuals regarding safe administration, monitoring, follow-ups, and best practices would incur high costs. Moreover, the costs associated with pre-and post-treatment follow-ups, such as the need for multiple MRI scans of the brain, as described, are likely to burden national healthcare systems and economies while serving as a barrier by making healthcare inaccessible to patients. Conversely, the epidemiological shifts seen in LMICs towards lower mortality rates [13] may reflect improving economies, which could indicate a better level of economic sustainability for drugs such as aducanumab. Nevertheless, determining and comparing the cost-effectiveness of aducanumab between countries is challenging due to the different architectures of healthcare systems, inconsistent funding, utilisation of resources, and the individual nature of policy and legislation development.

Table 2. Additional differences between aducanumab and conventional drugs for the treatment of AD.

	Monoclonal Antibodies	Cholinesterase Inhibitors (ChEIs)	N-methyl-D-aspartate (NMDA) Receptor Antagonists
Drug	Aducanumab	Donepezil, rivastigmine, and galantamine	Memantine
Functional outcomes	MD 1.70 (95% CI 0.71 to 2.69) (high-dose arm of EMERGE) [29] Scale: ADCS-AD [a]	Donepezil: SMD 0.22 (95% CI 0.12–0.33) [34] Galantamine: SMD 0.19 (95% CI 0.01–0.37) [33] Rivastigmine: MD 1.80 (95% CI 0.20 to 3.40) [62] Scale: ADCS-AD [b]	MD 0.95 (95% CI 0.22 to 1.76) [18] Scale: ADCS-AD [c]
Entry to institutional or nursing care	Not assessed	No significant benefit in terms of delay of entry to institutional care [63].	No effect on the rate of nursing home placement [64].
Adverse effects	ARIA-E and ARIA-H Symptoms: headaches, confusion, dizziness, falls, vision changes, and nausea [40].	Donepezil: gastrointestinal symptoms (upset stomach, nausea, diarrhoea, and anorexia) Galantamine: gastrointestinal symptoms Rivastigmine (patch): nausea, vomiting, anorexia, headaches, dizziness General vagotonic symptoms: bradycardia and hypotension [38].	Dizziness, confusion, weight gain, hallucinations [18].
APOE genotyping	Not a requirement. However, genetic screening may help ascertain ARIA risk.	Not required	Not required
Pre-treatment amyloid status	Amyloid-PET or CSF analysis may be conducted [10].	Not applicable	Not applicable
Baseline MRI scan of the brain	Required (a recent scan within one year prior to initiating therapy) [10].	For clinical diagnosis	For clinical diagnosis
Follow-up MRI scans of the brain	Two scans prior to the seventh and twelfth doses. Additional monitoring of ARIAs with MRI if symptomatic [10].	Not required	Not required
Average annual cost (U.S. $)	56,000 *	2796 †	4096 ‡
Generics or biosimilars	None	Available	Available

AD, Alzheimer's disease; Aβ, amyloid-beta protein; ADCS-ADL, Alzheimer's Disease Cooperative Study-Activities of Daily Living Inventory; amyloid-PET, amyloid-positron emission tomography; ARIA-E, amyloid-related imaging abnormalities due to oedema; ARIA-H,amyloid-related imaging abnormalities due to brain microhemorrhage or localised superficial siderosis; CDR-SB, Clinical Dementia Rating-Sum of Boxes; MMSE, Mini-Mental State Exam; MD, mean difference; SMD, standard mean difference. [a] ADCS-ADL-MCI (aducanumab), adapted for patients with mild cognitive impairment (MCI), scored from 0 to 53; higher scores represent greater cognitive impairment. [b] ADCS-ADL (donepezil, rivastigmine and galantamine), scored from 0 to 78; higher scores represent greater cognitive impairment. [c] $ADCS\text{-}ADL_{19}$ (memantine), adapted for patients with severe AD, scored from 0 to 54; higher scores represent greater cognitive impairment. * Cost of the drug only, as set by the manufacturer [30]. Excludes other treatment-related costs. † Average retail price in the United States of America for one ChEI based on typical dosing in 2012 [65]. ‡ Average retail price in the United States of America based on typical dosing in 2012 [61].

4. Conclusions

In summary, disease-modifying treatment for Alzheimer's disease, such as aducanumab, has the potential to ease the burden at both an individual and societal level, and such an intervention would be highly beneficial in many countries. However, we have identified areas of concern with regards to the suitability of aducanumab in LMICs, including reservations about its clinical efficacy, the complexity associated with the safe delivery of aducanumab to patients, challenges with follow-ups and monitoring, the relative prevalence of apolipoprotein E in different regions, and most prominently, its high cost. These factors need to be considered when evaluating the cost-effectiveness of aducanumab in LMICs. By extension, these concerns could be applicable to future therapeutics, especially antibody-based immunotherapies, and highlights the need for more accessible options in the context of LMICs to reduce the burden of disease in these nations.

Author Contributions: M.F.S. conceptualised the idea; I.P.C.G. performed the literature review and wrote the manuscript with support from T.R. All authors have read and agreed to the published version of the manuscript.

Funding: The project is funded by the Ministry of Higher Education, Fundamental Research Grant Scheme (FRGS/1/2020/SKK0/MUSM/02/6).

Institutional Review Board Statement: Not applicable.

Data Availability Statement: Not applicable.

Conflicts of Interest: The authors declare no conflict of interest. The funders had no role in the design of the study; in the collection, analyses, or interpretation of data; in the writing of the manuscript, or in the decision to publish the results.

References

1. Gleerup, H.S.; Hasselbalch, S.G.; Simonsen, A.H. Biomarkers for Alzheimer's Disease in Saliva: A Systematic Review. *Dis. Markers* **2019**, *2019*, 4761054. [CrossRef] [PubMed]
2. Hardy, J. A Hundred Years of Alzheimer's Disease Research. *Neuron* **2006**, *52*, 3–13. [CrossRef]
3. Rajan, K.B.; Wilson, R.S.; Weuve, J.; Barnes, L.L.; Evans, D.A. Cognitive impairment 18 years before clinical diagnosis of Alzheimer disease dementia. *Neurology* **2015**, *85*, 898–904. [CrossRef] [PubMed]
4. Alves, L.; Correia, A.S.A.; Miguel, R.; Alegria, P.; Bugalho, P. Alzheimer's disease: A clinical practice-oriented review. *Front. Neurol.* **2012**, *3*, 63. [CrossRef]
5. Alzheimer's Association. Alzheimer's disease facts and figures. *Alzheimer's Dement.* **2020**, *16*, 391–460. [CrossRef]
6. Silva, M.V.; Loures, C.D.; Alves, L.C.; de Souza, L.C.; Borges, K.B.; das Graças Carvalho, M. Alzheimer's disease: Risk factors and potentially protective measures. *J. Biomed. Sci.* **2019**, *26*, 1. [CrossRef] [PubMed]
7. Prince, M.; Guerchet, M.; Prina, M. *WHO Thematic Briefing: The Epidemiology and Impact of Dementia—Current State and Future Trends*; World Health Organization: Geneva, Switzerland, 2015.
8. World Health Organization. *Global Action Plan on the Public Health Response to Dementia 2017–2025*; World Health Organization: Geneva, Switzerland, 2017; ISBN 9789241513487.
9. Nichols, E.; Szoeke, C.E.; Vollset, S.E.; Abbasi, N.; Abd-Allah, F.; Abdela, J.; Aichour, M.T.; Akinyemi, R.O.; Alahdab, F.; Asgedom, S.W.; et al. Global, regional, and national burden of Alzheimer's disease and other dementias, 1990–2016: A systematic analysis for the Global Burden of Disease Study 2016. *Lancet Neurol.* **2019**, *18*, 88–106. [CrossRef]
10. Cummings, J.; Aisen, P.; Apostolova, L.G.; Atri, A.; Salloway, S.; Weiner, M. Aducanumab: Appropriate Use Recommendations. *J. Prev. Alzheimer's Dis.* **2021**, *8*, 1–13. [CrossRef]
11. U.S Food & Drug Administration. Drugs (FDA) FDA-Approved Drugs. Available online: https://www.accessdata.fda.gov/scripts/cder/daf/index.cfm?event=overview.process&ApplNo=761178 (accessed on 3 August 2021).
12. Prince, M.; Wimo, A.; Guerchet, M.; Ali, G.-C.; Wu, Y.-T.; Prina, M. World Alzheimer Report 2015—The Global Impact of Dementia: An Analysis of Prevalence, Incidence, Cost and Trends. *Alzheimer's Dis. Int.* **2015**.
13. Guerchet, M.; Mayston, R.; Lloyd-Sherlock, P.; Prince, M.; Aboderin, I.; Akinyemi, R.; Paddick, S.-M.; Wimo, A.; Amoakoh-Coleman, M.; Uwakwe, R.; et al. *Dementia in Sub-Saharan Africa: Challenges and Opportunities*; Alzheimer's Disease International: London, UK, 2017.
14. Anand, A.; Patience, A.A.; Sharma, N.; Khurana, N. The present and future of pharmacotherapy of Alzheimer's disease: A comprehensive review. *Eur. J. Pharmacol.* **2017**, *815*, 364–375. [CrossRef]
15. Cummings, J.; Lee, G.; Zhong, K.; Fonseca, J.; Taghva, K. Alzheimer's disease drug development pipeline: 2021. *Alzheimer's Dement. Transl. Res. Clin. Interv.* **2021**, *7*, e12179. [CrossRef]

16. Blanco-Silvente, L.; Castells, X.; Saez, M.; Barceló, M.A.; Garre-Olmo, J.; Vilalta-Franch, J.; Capellà, D. Discontinuation, Efficacy, and Safety of Cholinesterase Inhibitors for Alzheimer's Disease: A Meta-Analysis and Meta-Regression of 43 Randomized Clinical Trials Enrolling 16 106 Patients. *Int. J. Neuropsychopharmacol.* **2017**, *20*, 519–528. [CrossRef] [PubMed]
17. Howard, R.; McShane, R.; Lindesay, J.; Ritchie, C.; Baldwin, A.; Barber, R.; Burns, A.; Dening, T.; Findlay, D.; Holmes, C.; et al. Donepezil and Memantine for Moderate-to-Severe Alzheimer's Disease. *N. Engl. J. Med.* **2012**, *366*, 893–903. [CrossRef]
18. McShane, R.; Westby, M.J.; Roberts, E.; Minakaran, N.; Schneider, L.; Farrimond, L.E.; Maayan, N.; Ware, J.; Debarros, J. Memantine for dementia. *Cochrane Database Syst. Rev.* **2019**, *3*, CD003154. [CrossRef] [PubMed]
19. Cerejeira, J.; Lagarto, L.; Mukaetova-Ladinska, E.B. Behavioral and psychological symptoms of dementia. *Front. Neurol.* **2012**, *3*, 73. [CrossRef]
20. Tible, O.P.; Riese, F.; Savaskan, E.; von Gunten, A. Best practice in the management of behavioural and psychological symptoms of dementia. *Ther. Adv. Neurol. Disord.* **2017**, *10*, 297–309. [CrossRef]
21. Selkoe, D.J.; Hardy, J. The amyloid hypothesis of Alzheimer's disease at 25 years. *EMBO Mol. Med.* **2016**, *8*, 595–608. [CrossRef] [PubMed]
22. Rajasekhar, K.; Govindaraju, T. Current progress, challenges and future prospects of diagnostic and therapeutic interventions in Alzheimer's disease. *RSC Adv.* **2018**, *8*, 23780–23804. [CrossRef]
23. Sevigny, J.; Chiao, P.; Bussière, T.; Weinreb, P.H.; Williams, L.; Maier, M.; Dunstan, R.; Salloway, S.; Chen, T.; Ling, Y.; et al. The antibody aducanumab reduces Aβ plaques in Alzheimer's disease. *Nature* **2016**, *537*, 50–56. [CrossRef] [PubMed]
24. Ferrero, J.; Williams, L.; Stella, H.; Leitermann, K.; Mikulskis, A.; O'Gorman, J.; Sevigny, J. First-in-human, double-blind, placebo-controlled, single-dose escalation study of aducanumab (BIIB037) in mild-to-moderate Alzheimer's disease. *Alzheimer's Dement.* **2016**, *2*, 169–176. [CrossRef]
25. Dunstan, R.; Bussiere, T.; Fahrer, D.; Quigley, C.; Zhang, X.; Themeles, M.; Engber, T.; Rhodes, K.; Arastu, M.; Li, M. Quantitation of beta-amyloid in transgenic mice using whole slide digital imaging and image analysis software. *Alzheimer's Dement.* **2011**, *7*, S700. [CrossRef]
26. Crehan, H.; Lemere, C.A. Chapter 7–Anti-Amyloid-β Immunotherapy for Alzheimer's Disease. In *Developing Therapeutics for Alzheimer's Disease*; Wolfe, M.S., Ed.; Academic Press: Boston, MA, USA, 2016; pp. 193–226.
27. Vos, T.; Lim, S.S.; Abbafati, C.; Abbas, K.M.; Abbasi, M.; Abbasifard, M.; Abbasi-Kangevari, M.; Abbastabar, H.; Abd-Allah, F.; Abdelalim, A.; et al. Global burden of 369 diseases and injuries in 204 countries and territories, 1990–2019: A systematic analysis for the Global Burden of Disease Study 2019. *Lancet* **2020**, *396*, 1204–1222. [CrossRef]
28. Alexander, G.C.; Karlawish, J. The Problem of Aducanumab for the Treatment of Alzheimer Disease. *Ann. Intern. Med.* **2021**. [CrossRef] [PubMed]
29. Haeberlein, S.B.; von Hehn, C.; Tian, Y.; Chalkias, S.; Muralidharan, K.K.; Chen, T.; Wu, S.; Skordos, L.; Nisenbaum, L.; Rajagovindan, R.; et al. Emerge and Engage topline results: Phase 3 studies of aducanumab in early Alzheimer's disease. *Alzheimer's Dement.* **2020**, *16*, e047259. [CrossRef]
30. Lin, G.A.; Whittington, M.D.; Synnott, P.G.; McKenna, A.; Campbell, J.; Pearson, S.D.; Rind, D.M. Aducanumab for Alzheimer's Disease: Effectiveness and Value; Final Evidence Report and Meeting Summary. Institute for Clinical and Economic Review, August 5, 2021. Available online: https://icer.org/assessment/alzheimers-disease-2021/ (accessed on 5 August 2021).
31. Wimo, A.; Guerchet, M.; Ali, G.-C.; Wu, Y.-T.; Prina, A.M.; Winblad, B.; Jönsson, L.; Liu, Z.; Prince, M. The worldwide costs of dementia 2015 and comparisons with 2010. *Alzheimer's Dement.* **2017**, *13*, 1–7. [CrossRef]
32. Hung, S.-Y.; Fu, W.-M. Drug candidates in clinical trials for Alzheimer's disease. *J. Biomed. Sci.* **2017**, *24*, 1–12. [CrossRef]
33. Li, D.-D.; Zhang, Y.-H.; Zhang, W.; Zhao, P. Meta-Analysis of Randomized Controlled Trials on the Efficacy and Safety of Donepezil, Galantamine, Rivastigmine, and Memantine for the Treatment of Alzheimer's Disease. *Front. Neurosci.* **2019**, *13*, 472. [CrossRef]
34. Birks, J.S.; Harvey, R.J. Donepezil for dementia due to Alzheimer's disease. *Cochrane Database Syst. Rev.* **2018**, *10*, 14651858. [CrossRef] [PubMed]
35. Birks, J.S.; Chong, L.Y.; Grimley Evans, J. Rivastigmine for Alzheimer's disease. *Cochrane Database Syst. Rev.* **2015**, *16*, 295–315. [CrossRef]
36. Jiang, D.; Yang, X.; Li, M.; Wang, Y.; Wang, Y. Efficacy and safety of galantamine treatment for patients with Alzheimer's disease: A meta-analysis of randomized controlled trials. *J. Neural Transm.* **2015**, *122*, 1157–1166. [CrossRef] [PubMed]
37. Kishi, T.; Matsunaga, S.; Oya, K.; Nomura, I.; Ikuta, T.; Iwata, N. Memantine for Alzheimer's Disease: An Updated Systematic Review and Meta-analysis. *J. Alzheimer's Dis.* **2017**, *60*, 401–425. [CrossRef]
38. Birks, J.S. Cholinesterase inhibitors for Alzheimer's disease. *Cochrane Database Syst. Rev.* **2006**, *1*, CD005593. [CrossRef]
39. Liu, K.Y.; Howard, R. Can we learn lessons from the FDA's approval of aducanumab? *Nat. Rev. Neurol.* **2021**, *17*, 715–722. [CrossRef]
40. U.S. Food & Drug Administration. *Briefing Document: Combined FDA and Applicant PCNS Drugs Advisory Committee.*. Available online: https://www.fda.gov/media/143502/download (accessed on 27 June 2021).
41. Retinasamy, T.; Shaikh, M.F. Aducanumab for Alzheimer's disease: An update. *Neurosci. Res. Notes* **2021**, *4*, 17–20. [CrossRef]
42. U.S. Food & Drug Administration. *Guidance for Industry: Formal Meetings between the FDA and Sponsors or Applicants.*. Available online: https://www.fda.gov/media/72253/download (accessed on 11 November 2021).

43. Yuan, J.; Maserejian, N.; Liu, Y.; Devine, S.; Gillis, C.; Massaro, J.; Au, R.; Bondi, M. Severity Distribution of Alzheimer's Disease Dementia and Mild Cognitive Impairment in the Framingham Heart Study. *J. Alzheimer's Dis.* **2021**, *79*, 807–817. [CrossRef] [PubMed]
44. U.S. Food & Drug Administration. *PCNS Drugs Advisory Committee: Aducanumab for the Treatment of Alzheimer's Disease.*. Available online: https://www.fda.gov/media/143506/download (accessed on 2 August 2021).
45. Bitton, A.; Fifield, J.; Ratcliffe, H.; Karlage, A.; Wang, H.; Veillard, J.H.; Schwarz, D.; Hirschhorn, L.R. Primary healthcare system performance in low-income and middle-income countries: A scoping review of the evidence from 2010 to 2017. *BMJ Glob. Health* **2019**, *4*, e001551. [CrossRef] [PubMed]
46. Schwarz, D.; Hirschhorn, L.R.; Kim, J.-H.; Ratcliffe, H.L.; Bitton, A. Continuity in primary care: A critical but neglected component for achieving high-quality universal health coverage. *BMJ Glob. Health* **2019**, *4*, e001435. [CrossRef] [PubMed]
47. World Bank Data. Physicians (per 1000 People). Available online: https://data.worldbank.org/indicator/SH.MED.PHYS.ZS (accessed on 27 June 2021).
48. Cummings, J.L.; Tong, G.; Ballard, C. Treatment Combinations for Alzheimer's Disease: Current and Future Pharmacotherapy Options. *J. Alzheimer's Dis.* **2019**, *67*, 779–794. [CrossRef]
49. International Atomic Energy Agency (IAEA). IMAGINE—IAEA Medical Imaging and Nuclear Medicine Global Resources Database. Available online: https://humanhealth.iaea.org/HHW/DBStatistics/IMAGINEMaps3.html (accessed on 27 June 2021).
50. Zhang, X.; Fu, Z.; Meng, L.; He, M.; Zhang, Z. The Early Events That Initiate β-Amyloid Aggregation in Alzheimer's Disease. *Front. Aging Neurosci.* **2018**, *10*, 359. [CrossRef] [PubMed]
51. Cai, Z.; Zhou, Y.; Xiao, M.; Yan, L.-J.; He, W. Activation of mTOR: A culprit of Alzheimer's disease? *Neuropsychiatr. Dis. Treat.* **2015**, *2015*, 1015–1030. [CrossRef]
52. Sarlus, H.; Heneka, M.T. Microglia in Alzheimer's disease. *J. Clin. Investig.* **2017**, *127*, 3240–3249. [CrossRef] [PubMed]
53. Arimon, M.; Takeda, S.; Post, K.L.; Svirsky, S.; Hyman, B.T.; Berezovska, O. Oxidative stress and lipid peroxidation are upstream of amyloid pathology. *Neurobiol. Dis.* **2015**, *84*, 109–119. [CrossRef] [PubMed]
54. Jansen, W.J.; Ossenkoppele, R.; Knol, D.L.; Tijms, B.M.; Scheltens, P.; Verhey, F.R.; Visser, P.J.; Aalten, P.; Aarsland, D.; Alcolea, D.; et al. Prevalence of cerebral amyloid pathology in persons without dementia: A meta-analysis. *JAMA* **2015**, *313*, 1924–1938. [CrossRef]
55. Li, C.; Loewenstein, D.A.; Duara, R.; Cabrerizo, M.; Barker, W.; Adjouadi, M.; Alzheimer's Disease Neuroimaging, I. The Relationship of Brain Amyloid Load and APOE Status to Regional Cortical Thinning and Cognition in the ADNI Cohort. *J. Alzheimer's Dis.* **2017**, *59*, 1269–1282. [CrossRef] [PubMed]
56. Kern, S.; Mehlig, K.; Kern, J.; Zetterberg, H.; Thelle, D.; Skoog, I.; Lissner, L.; Blennow, K.; Börjesson-Hanson, A. The Distribution of Apolipoprotein E Genotype Over the Adult Lifespan and in Relation to Country of Birth. *Am. J. Epidemiol.* **2015**, *181*, 214–217. [CrossRef] [PubMed]
57. Mattsson, N.; Groot, C.; Jansen, W.J.; Landau, S.M.; Villemagne, V.L.; Engelborghs, S.; Mintun, M.M.; Lleo, A.; Molinuevo, J.L.; Jagust, W.J.; et al. Prevalence of the apolipoprotein E ε4 allele in amyloid β positive subjects across the spectrum of Alzheimer's disease. *Alzheimer's Dement* **2018**, *14*, 913–924. [CrossRef] [PubMed]
58. VandeVrede, L.; Gibbs, D.M.; Koestler, M.; La Joie, R.; Ljubenkov, P.A.; Provost, K.; Soleimani-Meigooni, D.; Strom, A.; Tsoy, E.; Rabinovici, G.D.; et al. Symptomatic amyloid-related imaging abnormalities in an APOE ε4/ε4 patient treated with aducanumab. *Alzheimer's Dement. Diagn. Assess. Dis. Monit.* **2020**, *12*, e12101. [CrossRef] [PubMed]
59. Brickman, A.M.; Manly, J.J.; Honig, L.S.; Sanchez, D.; Reyes-Dumeyer, D.; Lantigua, R.A.; Lao, P.J.; Stern, Y.; Vonsattel, J.P.; Teich, A.F.; et al. Plasma p-tau181, p-tau217, and other blood-based Alzheimer's disease biomarkers in a multi-ethnic, community study. *Alzheimer's Dement.* **2021**, *17*, 1353–1364. [CrossRef]
60. Howell, J.C.; Watts, K.D.; Parker, M.W.; Wu, J.; Kollhoff, A.; Wingo, T.S.; Dorbin, C.D.; Qiu, D.; Hu, W.T. Race modifies the relationship between cognition and Alzheimer's disease cerebrospinal fluid biomarkers. *Alzheimer's Res. Ther.* **2017**, *9*, 88. [CrossRef] [PubMed]
61. Nguyen, T.A.; Knight, R.; Roughead, E.E.; Brooks, G.; Mant, A. Policy options for pharmaceutical pricing and purchasing: Issues for low- and middle-income countries. *Health Policy Plan.* **2015**, *30*, 267–280. [CrossRef]
62. Winblad, B.; Grossberg, G.; Frölich, L.; Farlow, M.; Zechner, S.; Nagel, J.; Lane, R. IDEAL: A 6-month, double-blind, placebo-controlled study of the first skin patch for Alzheimer disease. *Neurology* **2007**, *69*, S14–S22. [CrossRef] [PubMed]
63. Courtney, C.; Farrell, D.; Gray, R.; Hills, R.; Lynch, L.; Sellwood, E.; Edwards, S.; Hardyman, W.; Raftery, J.; Crome, P.; et al. Long-term donepezil treatment in 565 patients with Alzheimer's disease (AD2000): Randomised double-blind trial. *Lancet* **2004**, *363*, 2105–2115. [CrossRef] [PubMed]
64. Howard, R.; McShane, R.; Lindesay, J.; Ritchie, C.; Baldwin, A.; Barber, R.; Burns, A.; Dening, T.; Findlay, D.; Holmes, C.; et al. Nursing home placement in the Donepezil and Memantine in Moderate to Severe Alzheimer's Disease (DOMINO-AD) trial: Secondary and post-hoc analyses. *Lancet Neurol.* **2015**, *14*, 1171–1181. [CrossRef]
65. Consumers Union of U.S. Consumer Reports Best Buy Drugs. Evaluating Prescription Drugs Used to Treat: Alzheimer's Disease. Available online: https://article.images.consumerreports.org/prod/content/dam/cro/news_articles/health/PDFs/Alzheimer\T1\textquoterightsDisease_fullreport.pdf (accessed on 1 August 2021).

MDPI

Review

New Insights into the Molecular Interplay between Human Herpesviruses and Alzheimer's Disease—A Narrative Review

Evita Athanasiou [1,†], Antonios N. Gargalionis [2,*,†], Cleo Anastassopoulou [3], Athanassios Tsakris [3] and Fotini Boufidou [4,*]

1 Department of Biopathology, Primary Healthfund, Bank of Greece, 105 64 Athens, Greece; evitathan@yahoo.gr
2 Department of Biopathology, Eginition Hospital, Medical School, National and Kapodistrian University of Athens, 115 28 Athens, Greece
3 Department of Microbiology, Medical School, National and Kapodistrian University of Athens, 115 27 Athens, Greece; cleoa@med.uoa.gr (C.A.); atsakris@med.uoa.gr (A.T.)
4 Neurochemistry and Biological Markers Unit, 1st Department of Neurology, Eginition Hospital, Medical School, National and Kapodistrian University of Athens, 115 28 Athens, Greece
* Correspondence: agargal@med.uoa.gr (A.N.G.); fboufidou@med.uoa.gr (F.B.); Tel.: +30-2107289157 or +30-6977019252 (A.N.G.); +30-2107289125 or +30-6944284166 (F.B.)
† These authors contributed equally to this work.

Abstract: Human herpesviruses (HHVs) have been implicated as possible risk factors in Alzheimer's disease (AD) pathogenesis. Persistent lifelong HHVs infections may directly or indirectly contribute to the generation of AD hallmarks: amyloid beta (Aβ) plaques, neurofibrillary tangles composed of hyperphosphorylated tau proteins, and synaptic loss. The present review focuses on summarizing current knowledge on the molecular mechanistic links between HHVs and AD that include processes involved in Aβ accumulation, tau protein hyperphosphorylation, autophagy, oxidative stress, and neuroinflammation. A PubMed search was performed to collect all the available research data regarding the above mentioned mechanistic links between HHVs and AD pathology. The vast majority of research articles referred to the different pathways exploited by Herpes Simplex Virus 1 that could lead to AD pathology, while a few studies highlighted the emerging role of HHV 6, cytomegalovirus, and Epstein–Barr Virus. The elucidation of such potential links may guide the development of novel diagnostics and therapeutics to counter this devastating neurological disorder that until now remains incurable.

Keywords: Alzheimer's disease viral hypothesis; human herpesviruses; amyloid beta (Aβ); tau protein; neuroinflammation

Citation: Athanasiou, E.; Gargalionis, A.N.; Anastassopoulou, C.; Tsakris, A.; Boufidou, F. New Insights into the Molecular Interplay between Human Herpesviruses and Alzheimer's Disease—A Narrative Review. *Brain Sci.* **2022**, *12*, 1010. https://doi.org/10.3390/brainsci12081010

Academic Editor: Henry Mak

Received: 1 July 2022
Accepted: 28 July 2022
Published: 30 July 2022

Publisher's Note: MDPI stays neutral with regard to jurisdictional claims in published maps and institutional affiliations.

Copyright: © 2022 by the authors. Licensee MDPI, Basel, Switzerland. This article is an open access article distributed under the terms and conditions of the Creative Commons Attribution (CC BY) license (https://creativecommons.org/licenses/by/4.0/).

1. Introduction

Alzheimer's disease (AD) is an irreversible progressive neurodegenerative disorder and the most prevalent cause of dementia worldwide, accounting for an estimated 60% to 80% of cases [1]. According to the World Alzheimer Report 2018, about 50 million people are suffering from AD globally, and this number is predicted to reach 152 million by mid-century with a considerable socioeconomic impact. Clinically, AD is often diagnosed on the basis of memory dysfunction about recent events that evolve to involve impairment in cognitive, behavioral, and functional aspects of a patient's life. However, AD is considered to start decades earlier before clinical symptoms occur; thus, a better understanding of modifiable (e.g., cerebrovascular diseases, hypertension, diabetes, obesity, dyslipidemia, smoking) and non-modifiable (age, genetics, family history) risk factors might delay or prevent AD onset, as well as introduction of new therapeutic targets [2–5]. Age 65 is the threshold age for the differential diagnosis of early- from late-onset AD. Late-onset AD is characterized as sporadic, with high prevalence and morbidity, reflecting complex interactions between environmental and genetic factors [6]. Among them, the apolipoprotein E

(APOE) ε4 allele has been identified as a major susceptibility gene and the most ordinary genetic factor for late-onset AD [7].

The disease is named after the German psychiatrist and neuropathologist Alois Alzheimer, who first reported a severe medical condition of the cerebral cortex in the postmortem brain of his first patient that suffered from memory loss and behavioral changes, characterized by the histological features that are today associated with AD: a massive loss of neurons and the presence of amyloid neuritic plaques and neurofibrillary tangles [8]. Over a century later, the extracellular senile plaques formed by amyloid beta (Aβ), the intracellular neurofibrillary tangles composed of abnormally hyperphosphorylated tau proteins, and the synaptic loss remain the pathological hallmarks of AD [9,10]. Although the mechanisms related to the disease pathology have been under substantial investigation and several hypotheses have been proposed, there is no accepted theory at the moment for explaining AD pathogenesis [2,9]. Emerging evidence in recent years supports that a wide spectrum of infectious agents (viruses, bacteria, fungi, and protozoa) can cross the blood–brain barrier (BBB) and might play a triggering role in AD development [11–13]. Among them, herpesviruses represent the most studied family of neurotropic pathogens [14].

Association between HHVs and AD

Herpesviruses are double stranded-DNA viruses with a unique four-layered structure: a core containing the linear genome enclosed in an icosahedral capsid that is surrounded by an amorphous protein coat, the tegument, and by a lipid envelope bearing membrane-associated proteins [15]. Human herpesviruses (HHVs) are distributed worldwide, and more than 90% of adults are infected by one or multiple HHVs [16]. Based on their biological properties, HHVs are classified into three sub-families: the alpha (α), beta (β), and gamma (γ). Herpes simplex virus (HSV) 1, 2 and varicella zoster virus (VZV) belong to the α sub-family; cytomegalovirus (CMV) and HHV 6 and 7 are β-herpesviruses, whereas Epstein–Barr virus (EBV) and Kaposi sarcoma-associated herpesvirus (KSHV) are members of the γ sub-family. HHVs enter the nervous system by hematogenous or neuronal spread depending on the mechanism and location of exposure and can cause a neurologic disease during primary infection or following reactivation from a state of lifelong latency [17].

Almost three decades ago, different research groups demonstrated the presence of HSV 1 DNA in the postmortem brain of both AD patients and elderly people without AD [18–20]. In 1997, Itzhaki and colleagues first provided strong evidence for a linkage between HSV 1 infection and AD, supporting that the reactivation of HSV 1 in the central nervous system is more harmful in AD patients who carry the APOE-ε4 allele and that the combination of these two factors increases the likelihood of disease [21]. APOE-ε4 allele has been reported to promote brain infiltration by HSV 1, to enhance the attachment and the entry of HSV 1 into the host cells, as well as to increase the viral load in the brain [22,23]. The latter appears to have a detrimental effect on Aβ accumulation, further influencing the Aβ fibril formation and oligomeric Aβ stabilization that is essential for accelerating the early seeding of Aβ pathology [22,24]. It is also noteworthy that APOE facilitates lipid antigen presentation by CD1+ antigen presenting cells to naïve natural killer T (NKT) cells [25,26]. NKT cells have a critical role in the modulation of inflammatory responses that could lead to neuronal damage, and research has been recently conducted to shed light to NKT cells' involvement in AD [27]. On the other hand, NKT cells are recognized as a mechanism of the innate immunity against viral infections. However, it is revealed that HSV 1 downregulates CD1 from infected human primary dendritic cells suppressing NKT cell activation as an immune evasion strategy [28].

In 2005, Wozniak and colleagues found a greater localization of HSV-DNA in Aβ brains compared with the aged-matched non-affected controls, supporting the hypothesis that virus directly impacts the pathological hallmarks of AD [29]. Readhead et al. recently published massive molecular and bioinformatics analyses of postmortem brain samples from four independent cohorts of AD patients, reporting the increased presence of herpesviruses in AD brains versus matched controls [27]. Although HSV 1 is the most thoroughly studied microbe in the context of AD pathogenesis, other herpesviruses have also been under investigation for possible implication, including CMV [30–32], EBV [33–35], HHV 6, and HHV 7 [33,36]. However, it should be mentioned that most of these viruses were found in the brain in a relatively low proportion of AD patients and elderly controls, compared to HSV 1, apart from HHV 6, which was detected in 70% of AD patients and 40% of age-matched controls, respectively [37]. The considerable overlap of HHV 6 and HSV 1 in the brain suggests that the two HHVs might act together in the development of AD. It is well established that HSV 1 is responsible for the induction of the major features of AD, while the contribution of the other HVVs is suspected, but it remains to be confirmed [37,38]. Since no effective vaccine has yet been developed to prevent HSV infection in humans, HSV 1 therapy is based on antiviral drugs such as acyclovir, penciclovir, foscarnet, and BAY 57-1293 which target virus replication [39]. Several studies have demonstrated that the above drugs and other anti-herpetic compounds greatly reduce tau phosphorylation and Aβ formation driven by HSV 1 [40–43]. Thus, anti-herpetic molecules could be used as potent treatment agents to inhibit or moderate AD progression, especially in patients with a documented history of recurrent HSV 1 infection [44,45]. Consequently, different clinical trials are ongoing towards this direction [44], and these attempts are expected to add considerable information regarding the linkage between HHVs and AD.

Apart from HHVs, other neurotropic viruses are reported to have an impact on cognitive decline within the context of AD, including the human immunodeficiency virus (HIV), the human T cell leukemia virus type I (HTLV-1), the influenza virus, and the severe acute respiratory syndrome coronavirus 2 (SARS-CoV-2) [11,12,46]. Moreover, current findings support that patients with chronic hepatitis C virus (HCV) infection appear to exhibit cognitive impairment and that it may increase the risk for dementia [47–49]. However, the mechanisms by which HCV infection increases the risk of dementia remain to be clarified [50].

Different molecular mechanisms have been proposed to explain how HHVs promote AD development, such as mechanisms involved in extracellular Aβ deposition, tau protein hyperphosphorylation, autophagy, neuroinflammation, oxidative stress, and apoptosis. This review focuses on these molecular mechanistic links between HHVs and AD that could introduce potential new targets for preventing or treating the disease that until now remains incurable.

2. Methodology

A PubMed search was conducted using specific keywords in order to identify, at first, the articles that point out which HHVs are related to AD development and then the articles that describe the mechanisms by which HHV infections may factor into AD pathology. Emphasis was mainly given on recent experimental studies that illustrated molecular mechanisms involved in Aβ deposition, tau protein hyperphosphorylation, autophagy, oxidative stress, apoptosis, and neuroinflammation. The keywords used for the literature search in different combinations are mentioned below: dementia, AD, HHVs, HSV 1, HSV 2, EBV, CMV, HHV 6, HHV 7, Aβ, tau protein, autophagy, oxidative stress, apoptosis, inflammation, synaptic loss, neuronal death. The retrieved relevant articles were systematically reviewed and critically analyzed.

3. Aβ Accumulation and Tau Hyperphosphorylation in AD

Aβ peptides are formed within the amyloidogenic pathway through the successive cleavage of a neuronal trans-membrane glycoprotein, the amyloid precursor protein (APP), by membrane-bound proteolytic enzymes [2]. Firstly, the extracellular domain of APP is cleaved by β-secretase, resulting in the production of the soluble N-terminal fragment and the intramembranous C-terminal fragment. The latter is subsequently subjected to cleavage by γ-secretase to release the Aβ peptide and a cytoplasmic polypeptide named APP intracellular domain (AICD). At low concentrations, Aβ is considered to serve a number of physiological functions, including neurogenesis, neuronal growth and survival, protection against oxidative stress, regulation of synaptic activity and plasticity, maintenance of BBB structural integrity, and surveillance against infections [51,52]. On the other hand, the imbalance between Aβ production and clearance leads to Aβ monomer accumulation, which due to its sequence has a high tendency to aggregate-forming toxic oligomers [2].

Under physiological conditions, the microtubule-associated protein tau is basically expressed in neurons in the brain and peripheral nerves where it is mainly located in the cytoplasm of axons, with much lower amounts in somatodendritic compartments [53]. It is essential for the stability, assembly, and dynamics of neuronal microtubules, and the microtubules are important for cytoskeleton structure and activity to serve the trafficking of different vesicles and organelles [54]. In addition, tau contributes to several physiological processes including neurogenesis and synaptogenesis, synaptic functions in terms of learning and memory, neuronal excitability, myelination, motor function, iron homeostasis, and DNA protection and chromosomal stability [55]. In the brain, tau is subjected to various posttranslational modifications, and the role of these modifications in protein function remains unknown [56]. Among them, phosphorylation is the most studied. The hyperphosphorylation of tau protein leads to conformational changes and then to the formation of aggregates (paired helical filaments and/or neurofibrillary tangles), which are associated with microtubule destabilization, synaptic loss, and neurodegeneration [57].

HHVs Infections' Association with Aβ Accumulation and Tau Hyperphosphorylation

Mounting evidence demonstrates that HHVs can drive Aβ deposition and tau phosphorylation directly via interactions with the viral surface or specific viral proteins or indirectly by affecting upstream molecular pathways (Figure 1).

The term amyloid is generally used to describe a conformational state, where proteins are transformed into insoluble aggregates of fibrillar morphology, and several biological or non-biological surfaces have been proven that are able to lower the energy barrier to nucleation promoting amyloid aggregation via a catalytic pathway named heterogeneous nucleation mechanism [58]. This mechanism is currently considered a critical factor regarding amyloid depositions induced by viruses. Aβ was found to act directly on HSV 1 by binding to its surface glycoproteins in a cell-free system preventing viral entry into fibroblast, epithelial, and neuronal cells, and thus inhibiting HSV 1 replication [59]. Eimer and colleagues also revealed that the Aβ attachment to HHV surface glycoproteins accelerates Aβ deposition and results in protective viral entrapment activity in 5XFAD mouse and 3D human neural cell culture infection models against HSV 1, HHV 6A, and HHV 6B [60]. Likewise, Ezzat and colleagues found that HSV 1 catalyzes the amyloid formation of Aβ in vitro, and a higher tendency of virus-mediated amyloid catalysis is observed for the more amyloidogenic Aβ1-42 peptide compared with the shorter, less amyloidogenic Aβ1-40 peptide [61]. There is a structural, dynamical difference between the two isoforms of Aβ. The two extra residues of Aβ1-42 confer a β-sheet character in its C-terminus that could explain the greater aggregation propensity of this isoform [62–64]. It is well established that β-sheets have been implicated in the formation of the fibrils and protein aggregates observed in amyloidosis [65]. In the same study, a significant increase in Aβ1-42 accumulation was noticed in the hippocampi and cortices of 5XFAD mice intracranially injected with HSV 1 in comparison with the non-infected animals. Taken together, the above findings indicate that Aβ peptides could be considered a novel class of antimicrobial agent against neurotropic infections caused by HHVs, but their overproduction may contribute to amyloid plaque formation and AD progression.

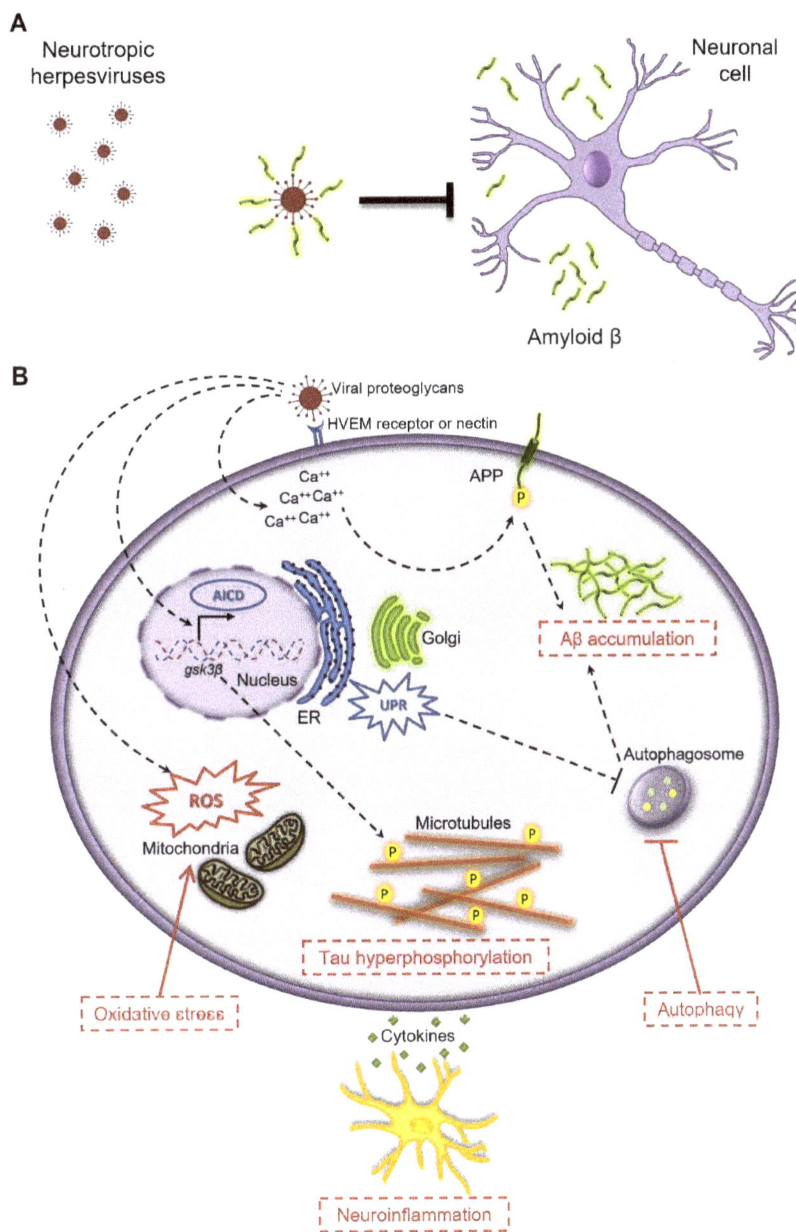

Figure 1. Neurotropic human herpesviruses drive the generation of AD hallmarks directly via interactions with the viral surface (**A**), or indirectly by affecting different molecular mechanisms (**B**), such as mechanisms involved in Aβ deposition, tau protein hyperphosphorylation, autophagy, oxidative stress, and neuroinflammation. HVEM: Herpesvirus Entry Mediator, AICD: APP Intracellular Domain, *gsk3β*: glycogen synthase kinase 3β, UPR: Unfolded Protein Response, APP: Amyloid Precursor Protein, Aβ: Amyloid β, ER: Endoplasmic Reticulum, ROS: Reactive Oxygen Species.

Furthermore, it was shown that HSV 1 infection promotes a striking increase in the intracellular levels of Aβ in cultured neuronal and glial cells [66–69]. Piacentini and colleagues demonstrated that HSV 1 binding to the neuronal cell membrane provokes a complex of electrophysiological responses that lead to significantly elevated intracellular calcium levels [67]. These signals trigger the phosphorylation of threonine at position 668 of APP, and thus, APP is subjected to multiple cleavages by secretases and caspases, resulting in Aβ accumulation. Moreover, HSV 1 infections seem to alter APP metabolism. HSV 1 activates the double-stranded (ds) RNA-activated protein kinase (PKR) that catalyzes the phosphorylation of the eukaryotic initiation factor 2-alpha (eIF2-α), a GTP-binding protein that in turn activates β-secretase translation, leading to Aβ production [70]. Research data also revealed that intracellular HSV 1 particles undergo frequent dynamic interplay with APP in a manner that facilitates viral transport and alters APP subcellular distribution and transport, upregulating its amyloidogenic processing [71]. In more details, HSV 1 capsids and APP-containing membranes co-localize and travel together through the cytoplasm at fast-anterograde transport rates. As a result, APP is mis-localized in HSV 1 infected cells, and then, it may be subjected to phosphorylation and proteolysis [69,71]. During HSV 1 infection, AICD accumulates within the nucleus of infected neuronal cells and binds to the promoter region of nep gene that encodes neprilysin (NEP), increasing its transcription [72]. NEP is a zinc metallopeptidase that has the capability of degrading monomeric and oligomeric forms of Aβ and plays a pivotal role in the maintenance of Aβ homeostasis in the brain [73]. NEP function as an Aβ-degrading enzyme introduces a protective mechanism against amyloid plaque formation in AD pathology. However, in a recent study, intranasal delivery of NEP in a transgenic AD mouse model rapidly eliminated Aβ plaques, but later on, this elimination caused a dramatic compensation of plaques in the cortex [74]. The authors proposed that NEP degrades the large Aβ plaques in the brain, which then are seen as smaller Aβ plaques, or that the transgenic mice activate a compensatory process to produce more Aβ plaques. MiR-H1, a miRNA encoded by HSV 1 and mainly expressed in productive infection, was found to target ubiquitin protein ligase E3 component n-recognin 1 (Ubr1), a RING-type E3 ubiquitin ligase that causes the degradation of proteins bearing "destabilizing" N-terminal residues, such as Aβ [75]. Ubr1 silencing mediated by miR-H1 reveals a novel molecular mechanism by which HSV 1 boosts Aβ gathering. The HHV 6A infection of microglial cells also promoted Aβ accumulation, as well as tau phosphorylation [76]. Recently, HHV 6A U4 protein was found to interact with cullin-RING E3 ubiquitin-protein ligases [77], a highly polymorphic E3 collection composed of a cullin backbone onto which carriers of activated ubiquitin and a diverse assortment of adaptors are bound to recruit appropriate substrates for ubiquitylation [78]. The above research findings led to the hypothesis that HHV 6A U4 protein competes with APP for binding to E3 ubiquitin ligase, and thus it prevents APP proteasomal degradation leading to increased APP expression and Aβ deposition.

In 2009, Wozniak and colleagues elucidated the mechanism involved in tau phosphorylation caused by HSV 1 infection, demonstrating that HSV 1 induces the activation of two enzymes, the glycogen synthase kinase 3β (GSK3β) and the protein kinase A (PKA) [79]. In this study, GSK3β proved to be responsible for tau phosphorylation at serine 202, threonine 212, serine 396, and serine 404, while tau phosphorylation at serine 214 occurred via the activity of PKA. Previous research data revealed that the viral US3 protein kinase interacts with and activates host PKA, further mimicking its functions [80]. In accordance with the above findings, AICD accumulated during HSV 1 infection within the nucleus of neuronal cells binds to the promoter region of gsk3β gene, enhancing its transcription [72]. Furthermore, neuroblastoma cells infected by HSV 1 generate increased levels of phosphorylated tau proteins that show nuclear localization at sites of viral DNA replication, whereas cyclin-dependent kinase (cdk) inhibitors reverse the above effect speculating that HSV 1 could contribute to tau phosphorylation through the deregulation of cdk activity [81]. Like HSV 1, the infection of human neuroblastoma cells with HSV 2 induces the main neurodegeneration markers associated with AD, including tau hyperphosphorylation, abnormalities in

APP proteolytic processing resulting in intracellular Aβ accumulation, and impairment of autophagy [82].

4. Autophagy Impairment Caused by HHVs Infections

Autophagy is a catabolic process through which cells remove toxic components such as protein aggregates, misfolded proteins, and damaged organelles ensuring the maintenance of homeostasis and the adaptation to stressful conditions. This mechanism is highly important for the survival and homeostasis of post-mitotic cells, such as neurons that cannot dilute the unwanted cellular components by cell division. Autophagy is involved in the metabolism of Aβ [83] and tau proteins [84]. This fact, together with the observation that HHVs may dysregulate or impair the autophagic pathways of the host in order to survive, suggests another mechanistic link between infection by HHVs and AD pathology (Figure 1).

A recent study demonstrated for the first time that HHV 6A infection reduces autophagy in astrocytoma cells and primary neurons and activates unfolded protein response (UPR), a highly specific signaling pathway in the endoplasmic reticulum (ER) that recognizes the accumulation of unfolded or misfolded proteins [85]. According to the research results, autophagy reduction could be attributed to the altered lysosomal acidification caused by HHV 6A infection and could be considered a mechanism through which the virus increases Aβ production. The interplay between autophagy and UPR activation is of high importance, since the latter may promote the autophagic process in order to alleviate cells from ER stress. In the same study, UPR activation results in an increase of tau protein phosphorylation at multiple serine residues (such as 202, 396, and 404), which may be induced by the activation of the ER stress sensor protein PERK (Protein Kinase RNA-like ER Kinase) that also acts by activating GSK3β [86,87].

HSV 1 infection also appears to exert a modifying influence on the autophagic process [88,89]. Beclin 1 is an adaptor multi-domain protein with a large interactome that is subjected to different post-translational modifications and plays a key role in the synthesis and maturation of autophagosomes, whereas it may regulate autophagy in either a positive or a negative manner [90]. The HSV 1 ICP34.5 (Infected Cell Protein 34.5) neurovirulence protein was found to directly interact with Beclin 1 in mammalian cells, antagonizing and then impeding the host autophagy response [91]. In the same study, it was demonstrated that a mutant HSV 1 lacking the ICP34.5 domain that binds to Beclin 1 fails to impair autophagy in neurons and to cause lethal encephalitis in mice. Like ICP34.5, HSV 1 Us11 protein also displays anti-autophagic activity via its direct interaction with PKR probably by blocking its activity [92]. Human neuroblastoma cells infected by HSV 1 exhibit an increased formation of microtubule-associated protein 1 light chain 3-II (LC3-II)-positive autophagic vesicles in which intracellular Aβ is localized, as well as an ineffective fusion between autophagosomes and lysosomes, indicating that this infection cause functional impairments in the late stages of autophagy [68,93].

Autophagy modulation has recently emerged as an effective strategy in the fight of neurodegenerative disorders, including AD [7,94–96]. Two different approaches have been under investigation towards this direction: the use of small molecule therapeutics (such as berberine, sirolimus, or trehalose) and the genetic intervention (i.e., gene therapy targeting TFEB or BECN1). Both approaches exhibited promising outcomes in both in vitro and in vivo models of AD [97–101], whereas drug repositioning studies are ongoing to reveal new autophagy modulators [94]. Several potential therapeutic strategies have also been proposed that target the ER stress signaling to combat AD pathology [102]. The use of pharmacological modulators of UPR, such as inhibitors of the integrated stress response, chemical chaperones, and anti-inflammatory drugs demonstrated promising results ameliorating cognitive deficits in AD mouse models [103–106].

It is worthwhile to mention that autophagy also occurs in oxidative stress [107], a process that reflects the significant imbalance in the equilibrium between the antioxidant mechanisms and the generation of reactive oxygen species (ROS) and reactive nitrogen species (RNS). Numerous studies support that oxidative stress contributes to AD pathogenesis and progression [108,109], whereas different viruses are commonly associated with the presence of oxidative stress in infected cells [110,111]. HSV 1 infection was found to increase ROS levels in mouse P19N neural cells as a prerequisite for viral replication [112]. ROS production was also induced in mouse microglial cells infected by HSV 1 via the viral stimulation of toll-like receptor (TLR) 2, resulting in lipid peroxidation and neurotoxicity [113]. Santana and colleagues revealed that oxidative stress enhances the autophagic flux impairment that is provoked by HSV 1 infection in neuronal cell models [114]. In line with the above findings, a recent study combining experimental data and a functional whole genome analysis based on microarrays demonstrated that oxidative stress and HSV 1 infection enhance the lysosomal content, inhibit the activity of lysosomal enzymes, and impair the endocytosis-mediated degradation of the epidermal growth factor receptor (EGFR) in a cell model of neurodegeneration [115]. In more details, it was shown that the activity of cathepsins B, K, S, and D/E was significantly reduced in this model. Cathepsins are proteases with serine, cysteine, or aspartic acid residues, which are activated in the acidic environment of lysosomes and are vital for a wide range of physiological functions including autophagy [116]. Lysosome alkalization caused by oxidative stress as a response to HSV 1 infection may impair cathepsins activity. The inactivation of cysteine cathepsins B and L was found to affect the proper lysosomal function (as it was assessed by a decreased EGFR degradation and the accumulation of several lysosomal proteins), and then to get involved in the generation of amyloidogenic products (Aβ, C-terminal fragment, and β-secretase) that are derived from the APP cleavage [117]. Currently, combined biochemical and redox proteomic approaches revealed that HSV 1 reactivations trigger oxidative modifications of proteins linked to AD pathogenesis in the cortex of infected mice [118]. In this context, the oxidation of the glucose-regulated protein 78 (GRP78) leads to its dysfunction and consequent aberrant UPR activation, while the oxidation of collapsin response-mediated protein 2 (CRMP2) was found preserved. CRMP2 is a multifunctional adaptor protein in the central nervous system able to bind and stabilize microtubules playing a key role in cytoskeletal dynamics, vesicle trafficking, and synaptic physiology [119]. Upon oxidation on Cys504, a disulphide-linked CRMP2 homodimer forms a transient complex with thirodoxin1, leading to the activation of GSK3β [120], which is involved in tau phosphorylation.

A recent systematic study based on the analysis of transcriptome datasets from seropositive or seronegative patients for CMV, EBV or HHV 6, and AD or Parkinson's disease patients revealed that infected by HHVs patients share common molecular signatures with AD patients. In more detail, host response against the abovementioned HHVs seems to have an impact in oxidative stress mechanisms involved in AD pathology through the activation of sirtuin-1 and peroxisome proliferator-activated receptor-gamma coactivator (PGC)-1alpha pathway [121]. Sirtuin-1 has been shown to mediate the deacetylation of PGC-1alpha, resulting in the upregulation of its transcriptional function, and the overexpression of PGC1α leads to reduced Aβ generation, particularly by regulating the expression of β-secretase [122].

However, oxidative stress does not represent a HHVs-specific phenomenon with an impact on AD pathology. Neuron cells are particularly vulnerable to oxidative stress due to their high polyunsaturated fatty acid content in membranes, high oxygen consumption, and weak antioxidant defense [123]. Environmental stress, nutritional factors (e.g., redox-active metals), and inflammation may increase the oxidative stress, leading to higher Aβ production [124–126]. Elderly individuals are more prone to oxidative stress, which partially accounts for AD susceptibility in aging populations [127,128].

5. Neuroinflammation Induced by HHVs Infections

Taking into account that the brain is no longer considered an immune-privileged organ and that patients with AD demonstrate elevated levels of inflammatory markers and AD risk genes related to innate immune functions, neuroinflammation has emerged as a major driving factor in neurodegeneration and AD pathology besides the classical hallmarks

(Figure 1) [129]. The inflammatory response in AD brain is characterized by the activation of astrocytes and microglial cells that are thought to be important elements of both early and late AD pathogenesis [130–132]. Although a number of studies have demonstrated that glial activation prevents AD progression by facilitating Aβ clearance in the brain, a growing body of evidence supports that the exacerbated or hindered activation of glial cells enhances the production of proinflammatory cytokines and Aβ in the brain, triggering a cascade of events that eventually lead to neurodegeneration and cognitive decline [130]. Thus, neuroinflammation could represent a promising drug target to avert AD progression.

Neurotropic viral infections may interplay with or exacerbate inflammatory processes in the central nervous system and then alter host cell immunity, contributing to AD development [133]. In this context, HSV 1 is again the most investigated virus among HHVs. It was shown that human trigeminal ganglia latently infected by HSV 1 exhibit persisting infiltration by lymphocytes and increased expression of cytokines that affect viral replication (interferon (IFN)-γ and tumor necrosis factor (TNF)-α), chemokines that attract immune cells (IFN-γ-inducible protein 10 (IP-10)), and the C-C motif chemokine ligand 5 (CCL5 or RANTES), [134]. These findings also highlighted the fact that HSV 1 latency is accompanied by a chronic inflammatory process but without any neuronal destruction. Interestingly, the infection of human primary neural cells with HSV 1 induced the upregulation of a brain-enriched miRNA (miRNA-146a) that is associated with proinflammatory signaling in stressed brain cells and AD, suggesting a possible role in viral evasion from the complement system and the activation of key elements of the arachidonic acid cascade known to contribute to Alzheimer-type neuropathological change [135,136].

Microglial cells represent the first line of defense against HSV 1 by releasing proinflammatory cytokines and chemokines [137], whereas they are considered the major cellular source of inducible nitric oxide synthase (iNOS) facilitating the production of nitric oxide (NO) [138]. Enhanced NO synthesis is an important contributor to oxidative stress-associated neurodegeneration, while NO is reported to create some sort of a cycle triggering Aβ deposition, which in turn activates resident glial cells that secrete NO; this continuous cycle is thought to have a detrimental impact on AD patients [139]. NO also aggravates the toxic effect of glutamate via an ionotropic receptor in microglia, the N-methyl-D-aspartate (NMDA) receptor [140]. Moreover, HSV 1 infection was found to prompt murine microglia to generate ROS through Toll-like receptor (TLR) 2, which exacerbated inflammation leading to neural oxidative damage [113,141]. In response to toxicity, damaged neurons can generate damage associated molecular patterns (DAMPs) and other microglia activators, including matrix metalloproteinase-3 (MMP3), a-synuclein and melanin, and further interfere with AD pathogenesis [137]. Following HSV 1 infection, microglia are the main producers of type I IFNs among cells in the central nervous system via the cyclic GMP-AMP synthetase (cGAS) and stimulator of IFN genes (STING) pathways [97,142], and it is well known that type I IFN response drives neuroinflammation and synapse loss in AD [143,144].

It is worthy to mention that, as an AD pathological feature, neuroinflammation is not restricted to HHVs. Other neurotropic viruses provoke inflammatory responses that may have an impact on AD development. SARS-CoV-2 is a highly neuroinvasive neurotropic virus that invades cells through angiotensin-converting enzyme 2 (ACE2) receptor-driven pathway. Upon the activation of microglia by SARS-CoV-2, cytokines such as IL-1 and TNF activate astrocytes, which in turn can produce inflammatory molecules, including TNF-α, ROS, and NO [145,146]. Furthermore, neurotropic RNA viruses are able to induce inflammatory events after viral entry that may prime the central nervous system to develop neurodegenerative disorders [147].

There is accumulating evidence that myelin damage is an important part of the pathological changes observed in AD, and may even precede Aβ and tau pathologies [148]. Although oligodendrocytes are the main cells that orchestrate myelination, astrocytes and microglia also contribute to this process and could play either beneficial or detrimental roles by promoting or impairing the endogenous capability of oligodendrocyte progenitor cells to promote spontaneous re-myelination after myelin loss [149]. The myelin basic protein is the major structural component of myelin and has been proven to bind Aβ and inhibit Aβ fibril formation, possibly regulating Aβ1-42 deposition and then amyloid plaques formation in the brains of AD patients [150,151]. On the other hand, Aβ induces oligodendrocytes death by activating the sphingomyelinase–ceramide pathway and hinders myelin formation [152]. Thus, myelin loss and decreased levels of myelin basic protein may accelerate Aβ deposition and Aβ plaque formation in AD patients [148]. In a recent study, it was demonstrated that HSV 1 infection generates brain inflammation and multifocal demyelination in the cotton rat *Sigmodon hispidus*, while the process of re-myelination that rapidly follows demyelination leads to the formation of partially re-myelinated plaques [153].

Although a few studies have been conducted regarding the role of EBV in AD pathogenesis compared to HSV 1, it is well established that during its latency and reactivation phases EBV may generate a systemic immune response to stress, which promotes inflammation associated with cognitive decline during aging [33,35]. Under normal circumstances, peripheral lymphocytes cross the BBB to patrol the brain parenchyma. EBV may infect peripheral blood mononuclear cells and use them to enter the brain via the "Trojan horse" mechanism [154]. In a recent study, B lymphocytes isolated from an AD patient were immortalized upon infection with EBV and then co-cultured with neuronal cells in order to construct a cellular model to mimic the normal conditions of the bloodstream [155]. B lymphocytes were able to produce high levels of the inflammatory cytokine TNF-α that in turn promoted Aβ accumulation and tau protein hyperphosphorylation [155,156]. Gate and colleagues suggested that the adaptive immunity might also play a significant role in disease progression based on the discovery of CD8+ T effector memory CD45RA+ (TEMRA) cells with proinflammatory and cytotoxic functions in AD patients and the identification of two EBV antigens that triggered this immune response, the Epstein–Barr nuclear antigen 3 (EBNA3A) and the trans-activator protein BZLF1 [34,157,158]. In another attempt to provide insights into the mechanistic role of EBV in AD, almost a hundred EBV proteins were analyzed for their aggregation proclivity using an in silico analysis [159]. BNLF-2a is the tail-anchored protein encoded by EBV, which inserts ER after translation and blocks the transporter protein (TAP) associated with antigen processing, thereby providing immune escape properties to infected cells [160]. BNLF-2a can further promote AD progression by inhibiting TAP and down-regulating major human histocompatibility complex (MHC)-I and II expression, leading to the accumulation of neuronal cells and viral polypeptides in the environment [159].

6. Concluding Remarks

During their long-lasting stay in the human brain, HHVs have several opportunities to interfere with practically every mechanism that has been proposed to underlie AD pathophysiology (Table 1). Strong evidence supports that infections by HHVs trigger molecular events associated with autophagy, oxidative stress, and neuroinflammation with a profound effect on AD pathology. Interestingly, there is an interplay between the aforementioned biological processes, since they may induce and/or enhance each other. HSV 1 and HHV 6A have been reported to dysregulate autophagy in neuronal cells with an impact on Aβ accumulation and tau protein metabolism. HSV 1 impairs autophagy through its neurovirulence proteins: ICP34.5 [91] and US11 [92], whereas HHV 6A dysregulates the autophagic process via the activation of UPR [85]. Accumulating evidence also suggests that HHVs infection results in the increased production of oxidant species, like they do infections by other neurotropic viruses such as HIV [161]. Oxidative stress induced by HSV 1 modifies lysosomes by increasing their content, decreasing the activity of the lysosomal enzymes, or impairing cathepsins that are involved in the generation of amyloidogenic products. CMV, EBV, and HHV 6 have been reported that they also have an impact on oxidative stress via the activation of the sirtuin-1 and PCG-1alpha pathway, which influences Aβ generation [121,122]. Furthermore, it is well established that the autophagy reduction and the increased production of oxidant species promote inflammation. The release of inflammatory agents stimulates Aβ production in astrocytes, while soluble tau oligomers may be secreted into the extracellular environment and contribute, independently or combined with Aβ, to synaptic dysfunction [162]. Glial cells play either a protective or restorative role in neurons, and thus impairment of their function by HHVs infection could contribute to AD progression. Although there is a great number of molecules derived from HSV 1, HHV 6A, EBV, and CMV that are identified to play a key role in AD pathology, there is a long way to pave in order to elucidate all the molecular mechanisms involved, to examine the possible etiology of other HHVs or to investigate the potent contribution of other factors such as gut microbiota. Shedding more light on the molecular pathways involved in the aforementioned biological processes could lead to new therapeutic targets in order to prevent or hinder the neurodegeneration that leads to AD advancement.

Table 1. HHVs proteins and their targets in neuronal cells within the context of AD pathology.

HHV Type	Viral Molecule	Host Cell Target	Biological Process	Reference
HSV 1	MiR-H1	Urb1	Aβ accumulation	[75]
	US3 protein kinase	PKA	Tau phosphorylation	[80]
	ICP34.5 protein	Beclin 1	Autophagy blockage	[91]
	Us11 protein	PKR	Autophagy blockage	[92]
		MiRNA-146a	Neuroinflammation	[135,136]
HHV 6A	U4 protein	APP	Aβ accumulation & tau phosphorylation	[77]
HHV 6 EBV CMV		Sirtuin-1	Oxidative stress	[121] [122]
EBV	EBNA3A, BZLF1	TEMRA cells	Neuroinflammation	[157]
	BNLF-2a	TAP	Immune evasion	[159]

Abbreviations: HHVs; Human Herpesviruses, AD; Alzheimer's Disease, HSV 1; Human Simplex Virus 1, EBV; Epstein–Barr Virus, CMV; Cytomegalovirus, MiR-H1; microRNA-H1, ICP34.5; Infected cell protein 34.5, EBNA3A; Epstein–Barr nuclear antigen 3, TEMRA; CD8+ T effector memory CD45RA+, TAP; Transporter protein associated with antigen processing, Urb1; ubiquitin protein ligase E3 component n-recognin 1, PKA; protein kinase A, PKR; double-stranded (ds) RNA-activated protein kinase, APP; Amyloid Precursor Protein, Aβ; amyloid β.

The aim of this review was to highlight potential mechanistic links between HHVs and AD and, thus, to help identify possible new treatment targets. Given the series of failure over the past two decades on disease-modifying treatments and the controversial approval of Aducanumab by the Food and Drug Administration (FDA) in 2021, in conjunction with the multifactorial nature of the disease, strategies aiming at the elimination of molecules that could serve as putative triggers, if not causative factors, for AD is a reasonable approach. This possibility is supported by an observational retrospective cohort study on electronic health databases in Taiwan that provided evidence for a lower incidence of dementia in HSV-infected patients who had received an antiviral drug compared to non-treated subjects [45]. Currently, there are two ongoing Phase II placebo–control clinical trials of valacyclovir in patients with mild AD and HSV seropositivity (NCT03282916) and in mild cognitive impairment patients exhibiting AD biomarkers (NCT04710030). These studies are expected to shed light on the HSV involvement in AD progression and pathogenesis.

Additionally, over the last five years, there is increasing evidence for an altered folding of several host encoded cellular proteins associated with their self-aggregation into disease specific pathological lesions within the brain [163]. Besides this, evidence for the transmissibility of unusual pathogenic Aβ and tau isoforms has given rise to the "prion hypothesis" for AD and the conceptualization of a double prion disorder [164]. Under this perspective, the augmentation of abnormal protein folding by HHVs and other environmental factors in AD and the specific molecular mechanistic links involved glares as an intriguing new horizon.

Author Contributions: A.N.G. and F.B. conceived the idea and structured the manuscript. E.A., A.N.G. and F.B. drafted the manuscript. E.A. designed the figure and the table. C.A. and A.T. edited and reviewed the manuscript. All authors have read and agreed to the published version of the manuscript.

Funding: This research received no external funding.

Institutional Review Board Statement: Not applicable.

Informed Consent Statement: Not applicable.

Data Availability Statement: Not applicable.

Conflicts of Interest: The authors declare no conflict of interest.

References

1. 2021 Alzheimer's disease facts and figures. *Alzheimers Dement.* **2021**, *17*, 327–406. [CrossRef] [PubMed]
2. Knopman, D.S.; Amieva, H.; Petersen, R.C.; Chetelat, G.; Holtzman, D.M.; Hyman, B.T.; Nixon, R.A.; Jones, D.T. Alzheimer disease. *Nat. Rev. Dis. Primers* **2021**, *7*, 33. [CrossRef] [PubMed]
3. Scheltens, P.; De Strooper, B.; Kivipelto, M.; Holstege, H.; Chetelat, G.; Teunissen, C.E.; Cummings, J.; van der Flier, W.M. Alzheimer's disease. *Lancet* **2021**, *397*, 1577–1590. [CrossRef]
4. Silva, M.V.F.; Loures, C.M.G.; Alves, L.C.V.; de Souza, L.C.; Borges, K.B.G.; Carvalho, M.D.G. Alzheimer's disease: Risk factors and potentially protective measures. *J. Biomed. Sci.* **2019**, *26*, 33. [CrossRef]
5. Zhang, X.X.; Tian, Y.; Wang, Z.T.; Ma, Y.H.; Tan, L.; Yu, J.T. The Epidemiology of Alzheimer's Disease Modifiable Risk Factors and Prevention. *J. Prev. Alzheimers Dis.* **2021**, *8*, 313–321. [CrossRef] [PubMed]
6. Hou, Y.; Dan, X.; Babbar, M.; Wei, Y.; Hasselbalch, S.G.; Croteau, D.L.; Bohr, V.A. Ageing as a risk factor for neurodegenerative disease. *Nat. Rev. Neurol.* **2019**, *15*, 565–581. [CrossRef] [PubMed]
7. Liu, C.C.; Kanekiyo, T.; Xu, H.; Bu, G. Apolipoprotein E and Alzheimer disease: Risk, mechanisms and therapy. *Nat. Rev. Neurol.* **2013**, *9*, 106–118. [CrossRef] [PubMed]
8. Cipriani, G.; Dolciotti, C.; Picchi, L.; Bonuccelli, U. Alzheimer and his disease: A brief history. *Neurol. Sci.* **2011**, *32*, 275–279. [CrossRef] [PubMed]
9. Breijyeh, Z.; Karaman, R. Comprehensive Review on Alzheimer's Disease: Causes and Treatment. *Molecules* **2020**, *25*, 5789. [CrossRef] [PubMed]
10. DeTure, M.A.; Dickson, D.W. The neuropathological diagnosis of Alzheimer's disease. *Mol. Neurodegener.* **2019**, *14*, 32. [CrossRef] [PubMed]
11. Lotz, S.K.; Blackhurst, B.M.; Reagin, K.L.; Funk, K.E. Microbial Infections Are a Risk Factor for Neurodegenerative Diseases. *Front. Cell Neurosci.* **2021**, *15*, 691136. [CrossRef]
12. Shi, M.; Li, C.; Tian, X.; Chu, F.; Zhu, J. Can Control Infections Slow Down the Progression of Alzheimer's Disease? Talking About the Role of Infections in Alzheimer's Disease. *Front. Aging Neurosci.* **2021**, *13*, 685863. [CrossRef]
13. Sochocka, M.; Zwolinska, K.; Leszek, J. The Infectious Etiology of Alzheimer's Disease. *Curr. Neuropharmacol.* **2017**, *15*, 996–1009. [CrossRef]
14. Seaks, C.E.; Wilcock, D.M. Infectious hypothesis of Alzheimer disease. *PLoS Pathog.* **2020**, *16*, e1008596. [CrossRef]
15. Whitley, R.J. Chapter 68. Herpesviruses. In *Medical Microbiology*, 4th ed.; University of Texas Medical Branch at Galveston: Galveston, TX, USA, 1996.
16. Lan, K.; Luo, M.H. Herpesviruses: Epidemiology, pathogenesis, and interventions. *Virol. Sin.* **2017**, *32*, 347–348. [CrossRef] [PubMed]
17. Baldwin, K.J.; Cummings, C.L. Herpesvirus Infections of the Nervous System. *Continuum* **2018**, *24*, 1349–1369. [CrossRef]
18. Baringer, J.R.; Pisani, P. Herpes simplex virus genomes in human nervous system tissue analyzed by polymerase chain reaction. *Ann. Neurol.* **1994**, *36*, 823–829. [CrossRef]
19. Jamieson, G.A.; Maitland, N.J.; Wilcock, G.K.; Craske, J.; Itzhaki, R.F. Latent herpes simplex virus type 1 in normal and Alzheimer's disease brains. *J. Med. Virol.* **1991**, *33*, 224–227. [CrossRef] [PubMed]
20. Jamieson, G.A.; Maitland, N.J.; Wilcock, G.K.; Yates, C.M.; Itzhaki, R.F. Herpes simplex virus type 1 DNA is present in specific regions of brain from aged people with and without senile dementia of the Alzheimer type. *J. Pathol.* **1992**, *167*, 365–368. [CrossRef]
21. Itzhaki, R.F.; Lin, W.R.; Shang, D.; Wilcock, G.K.; Faragher, B.; Jamieson, G.A. Herpes simplex virus type 1 in brain and risk of Alzheimer's disease. *Lancet* **1997**, *349*, 241–244. [CrossRef]
22. Khokale, R.; Kang, A.; Buchanan-Peart, K.R.; Nelson, M.L.; Awolumate, O.J.; Cancarevic, I. Alzheimer's Gone Viral: Could Herpes Simplex Virus Type-1 Be Stealing Your Memories? *Cureus* **2020**, *12*, e11726. [CrossRef]
23. Linard, M.; Letenneur, L.; Garrigue, I.; Doize, A.; Dartigues, J.F.; Helmer, C. Interaction between APOE4 and herpes simplex virus type 1 in Alzheimer's disease. *Alzheimers Dement.* **2020**, *16*, 200–208. [CrossRef]

24. Parhizkar, S.; Holtzman, D.M. APOE mediated neuroinflammation and neurodegeneration in Alzheimer's disease. *Semin. Immunol.* **2022**, 101594. [CrossRef]
25. van den Elzen, P.; Garg, S.; Leon, L.; Brigl, M.; Leadbetter, E.A.; Gumperz, J.E.; Dascher, C.C.; Cheng, T.Y.; Sacks, F.M.; Illarionov, P.A.; et al. Apolipoprotein-mediated pathways of lipid antigen presentation. *Nature* **2005**, *437*, 906–910. [CrossRef]
26. Zhang, H.L.; Wu, J.; Zhu, J. The immune-modulatory role of apolipoprotein E with emphasis on multiple sclerosis and experimental autoimmune encephalomyelitis. *Clin. Dev. Immunol.* **2010**, *2010*, 186813. [CrossRef] [PubMed]
27. Solana, C.; Tarazona, R.; Solana, R. Immunosenescence of Natural Killer Cells, Inflammation, and Alzheimer's Disease. *Int. J. Alzheimers Dis.* **2018**, *2018*, 3128758. [CrossRef]
28. Yuan, W.; Dasgupta, A.; Cresswell, P. Herpes simplex virus evades natural killer T cell recognition by suppressing CD1d recycling. *Nat. Immunol.* **2006**, *7*, 835–842. [CrossRef]
29. Wozniak, M.A.; Shipley, S.J.; Combrinck, M.; Wilcock, G.K.; Itzhaki, R.F. Productive herpes simplex virus in brain of elderly normal subjects and Alzheimer's disease patients. *J. Med. Virol.* **2005**, *75*, 300–306. [CrossRef]
30. Barnes, L.L.; Capuano, A.W.; Aiello, A.E.; Turner, A.D.; Yolken, R.H.; Torrey, E.F.; Bennett, D.A. Cytomegalovirus infection and risk of Alzheimer disease in older black and white individuals. *J. Infect. Dis.* **2015**, *211*, 230–237. [CrossRef] [PubMed]
31. Lee, K.H.; Kwon, D.E.; Do Han, K.; La, Y.; Han, S.H. Association between cytomegalovirus end-organ diseases and moderate-to-severe dementia: A population-based cohort study. *BMC Neurol.* **2020**, *20*, 216. [CrossRef]
32. Stebbins, R.C.; Noppert, G.A.; Yang, Y.C.; Dowd, J.B.; Simanek, A.; Aiello, A.E. Association Between Immune Response to Cytomegalovirus and Cognition in the Health and Retirement Study. *Am. J. Epidemiol.* **2021**, *190*, 786–797. [CrossRef] [PubMed]
33. Carbone, I.; Lazzarotto, T.; Ianni, M.; Porcellini, E.; Forti, P.; Masliah, E.; Gabrielli, L.; Licastro, F. Herpes virus in Alzheimer's disease: Relation to progression of the disease. *Neurobiol. Aging* **2014**, *35*, 122–129. [CrossRef]
34. Kang, J.S.; Liu, P.P. Human herpesvirus 4 and adaptive immunity in Alzheimer's disease. *Signal Transduct. Target. Ther.* **2020**, *5*, 48. [CrossRef]
35. Shim, S.M.; Cheon, H.S.; Jo, C.; Koh, Y.H.; Song, J.; Jeon, J.P. Elevated Epstein-Barr Virus Antibody Level is Associated with Cognitive Decline in the Korean Elderly. *J. Alzheimers Dis* **2017**, *55*, 293–301. [CrossRef] [PubMed]
36. Readhead, B.; Haure-Mirande, J.V.; Funk, C.C.; Richards, M.A.; Shannon, P.; Haroutunian, V.; Sano, M.; Liang, W.S.; Beckmann, N.D.; Price, N.D.; et al. Multiscale Analysis of Independent Alzheimer's Cohorts Finds Disruption of Molecular, Genetic, and Clinical Networks by Human Herpesvirus. *Neuron* **2018**, *99*, 64–82 e67. [CrossRef] [PubMed]
37. Itzhaki, R.F. Herpes and Alzheimer's Disease: Subversion in the Central Nervous System and How It Might Be Halted. *J. Alzheimers Dis.* **2016**, *54*, 1273–1281. [CrossRef] [PubMed]
38. Itzhaki, R.F. Herpes simplex virus type 1 and Alzheimer's disease: Possible mechanisms and signposts. *FASEB J.* **2017**, *31*, 3216–3226. [CrossRef]
39. James, S.H.; Prichard, M.N. Current and future therapies for herpes simplex virus infections: Mechanism of action and drug resistance. *Curr. Opin. Virol.* **2014**, *8*, 54–61. [CrossRef]
40. Hui, Z.; Zhijun, Y.; Yushan, Y.; Liping, C.; Yiying, Z.; Difan, Z.; Chunglit, C.T.; Wei, C. The combination of acyclovir and dexamethasone protects against Alzheimer's disease-related cognitive impairments in mice. *Psychopharmacology* **2020**, *237*, 1851–1860. [CrossRef]
41. Wozniak, M.; Bell, T.; Denes, A.; Falshaw, R.; Itzhaki, R. Anti-HSV1 activity of brown algal polysaccharides and possible relevance to the treatment of Alzheimer's disease. *Int. J. Biol. Macromol.* **2015**, *74*, 530–540. [CrossRef]
42. Wozniak, M.A.; Frost, A.L.; Itzhaki, R.F. The helicase-primase inhibitor BAY 57-1293 reduces the Alzheimer's disease-related molecules induced by herpes simplex virus type 1. *Antivir. Res.* **2013**, *99*, 401–404. [CrossRef] [PubMed]
43. Wozniak, M.A.; Frost, A.L.; Preston, C.M.; Itzhaki, R.F. Antivirals reduce the formation of key Alzheimer's disease molecules in cell cultures acutely infected with herpes simplex virus type 1. *PLoS ONE* **2011**, *6*, e25152. [CrossRef] [PubMed]
44. Protto, V.; Marcocci, M.E.; Miteva, M.T.; Piacentini, R.; Li Puma, D.D.; Grassi, C.; Palamara, A.T.; De Chiara, G. Role of HSV-1 in Alzheimer's disease pathogenesis: A challenge for novel preventive/therapeutic strategies. *Curr. Opin. Pharmacol.* **2022**, *63*, 102200. [CrossRef] [PubMed]
45. Tzeng, N.S.; Chung, C.H.; Lin, F.H.; Chiang, C.P.; Yeh, C.B.; Huang, S.Y.; Lu, R.B.; Chang, H.A.; Kao, Y.C.; Yeh, H.W.; et al. Anti-herpetic Medications and Reduced Risk of Dementia in Patients with Herpes Simplex Virus Infections-a Nationwide, Population-Based Cohort Study in Taiwan. *Neurotherapeutics* **2018**, *15*, 417–429. [CrossRef]
46. Filgueira, L.; Larionov, A.; Lannes, N. The Influence of Virus Infection on Microglia and Accelerated Brain Aging. *Cells* **2021**, *10*, 1836. [CrossRef]
47. Adinolfi, L.E.; Nevola, R.; Lus, G.; Restivo, L.; Guerrera, B.; Romano, C.; Zampino, R.; Rinaldi, L.; Sellitto, A.; Giordano, M.; et al. Chronic hepatitis C virus infection and neurological and psychiatric disorders: An overview. *World J. Gastroenterol.* **2015**, *21*, 2269–2280. [CrossRef]
48. Chiu, W.C.; Tsan, Y.T.; Tsai, S.L.; Chang, C.J.; Wang, J.D.; Chen, P.C.; Health Data Analysis in Taiwan (hDATa) Research Group. Hepatitis C viral infection and the risk of dementia. *Eur. J. Neurol.* **2014**, *21*, 1068-e59. [CrossRef] [PubMed]
49. Choi, H.G.; Soh, J.S.; Lim, J.S.; Sim, S.Y.; Lee, S.W. Association between dementia and hepatitis B and C virus infection. *Medicine* **2021**, *100*, e26476. [CrossRef]
50. Huang, L.; Wang, Y.; Tang, Y.; He, Y.; Han, Z. Lack of Causal Relationships Between Chronic Hepatitis C Virus Infection and Alzheimer's Disease. *Front. Genet.* **2022**, *13*, 828827. [CrossRef]

51. Brothers, H.M.; Gosztyla, M.L.; Robinson, S.R. The Physiological Roles of Amyloid-beta Peptide Hint at New Ways to Treat Alzheimer's Disease. *Front. Aging Neurosci.* **2018**, *10*, 118. [CrossRef]
52. Morley, J.E.; Farr, S.A.; Nguyen, A.D.; Xu, F. Editorial: What is the Physiological Function of Amyloid-Beta Protein? *J. Nutr. Health Aging* **2019**, *23*, 225–226. [CrossRef]
53. Guo, T.; Noble, W.; Hanger, D.P. Roles of tau protein in health and disease. *Acta Neuropathol.* **2017**, *133*, 665–704. [CrossRef]
54. Venkatramani, A.; Panda, D. Regulation of neuronal microtubule dynamics by tau: Implications for tauopathies. *Int. J. Biol. Macromol.* **2019**, *133*, 473–483. [CrossRef] [PubMed]
55. Kent, S.A.; Spires-Jones, T.L.; Durrant, C.S. The physiological roles of tau and Abeta: Implications for Alzheimer's disease pathology and therapeutics. *Acta Neuropathol.* **2020**, *140*, 417–447. [CrossRef] [PubMed]
56. Morris, M.; Knudsen, G.M.; Maeda, S.; Trinidad, J.C.; Ioanoviciu, A.; Burlingame, A.L.; Mucke, L. Tau post-translational modifications in wild-type and human amyloid precursor protein transgenic mice. *Nat. Neurosci.* **2015**, *18*, 1183–1189. [CrossRef] [PubMed]
57. Vanessa, J. De-Paula, M.R., Breno, S. Diniz & Orestes, V. Forlenza Alzheimer's Disease. In *Protein Aggregation and Fibrillogenesis in Cerebral and Systemic Amyloid Disease*; Springer: Dordrecht/Drechtsteden, The Netherlands, 2012; Volume 65.
58. Malmberg, M.; Malm, T.; Gustafsson, O.; Sturchio, A.; Graff, C.; Espay, A.J.; Wright, A.P.; El Andaloussi, S.; Linden, A.; Ezzat, K. Disentangling the Amyloid Pathways: A Mechanistic Approach to Etiology. *Front. Neurosci.* **2020**, *14*, 256. [CrossRef]
59. Bourgade, K.; Garneau, H.; Giroux, G.; Le Page, A.Y.; Bocti, C.; Dupuis, G.; Frost, E.H.; Fulop, T., Jr. beta-Amyloid peptides display protective activity against the human Alzheimer's disease-associated herpes simplex virus-1. *Biogerontology* **2015**, *16*, 85–98. [CrossRef]
60. Eimer, W.A.; Vijaya Kumar, D.K.; Navalpur Shanmugam, N.K.; Rodriguez, A.S.; Mitchell, T.; Washicosky, K.J.; Gyorgy, B.; Breakefield, X.O.; Tanzi, R.E.; Moir, R.D. Alzheimer's Disease-Associated beta-Amyloid Is Rapidly Seeded by Herpesviridae to Protect against Brain Infection. *Neuron* **2018**, *100*, 1527–1532. [CrossRef]
61. Ezzat, K.; Pernemalm, M.; Palsson, S.; Roberts, T.C.; Jarver, P.; Dondalska, A.; Bestas, B.; Sobkowiak, M.J.; Levanen, B.; Skold, M.; et al. The viral protein corona directs viral pathogenesis and amyloid aggregation. *Nat. Commun.* **2019**, *10*, 2331. [CrossRef] [PubMed]
62. Jarrett, J.T.; Berger, E.P.; Lansbury, P.T., Jr. The carboxy terminus of the beta amyloid protein is critical for the seeding of amyloid formation: Implications for the pathogenesis of Alzheimer's disease. *Biochemistry* **1993**, *32*, 4693–4697. [CrossRef]
63. Lim, K.H.; Collver, H.H.; Le, Y.T.; Nagchowdhuri, P.; Kenney, J.M. Characterizations of distinct amyloidogenic conformations of the Abeta (1-40) and (1-42) peptides. *Biochem. Biophys. Res. Commun.* **2007**, *353*, 443–449. [CrossRef]
64. Snyder, S.W.; Ladror, U.S.; Wade, W.S.; Wang, G.T.; Barrett, L.W.; Matayoshi, E.D.; Huffaker, H.J.; Krafft, G.A.; Holzman, T.F. Amyloid-beta aggregation: Selective inhibition of aggregation in mixtures of amyloid with different chain lengths. *Biophys. J.* **1994**, *67*, 1216–1228. [CrossRef]
65. Periole, X.; Huber, T.; Bonito-Oliva, A.; Aberg, K.C.; van der Wel, P.C.A.; Sakmar, T.P.; Marrink, S.J. Energetics Underlying Twist Polymorphisms in Amyloid Fibrils. *J. Phys. Chem. B* **2018**, *122*, 1081–1091. [CrossRef]
66. De Chiara, G.; Marcocci, M.E.; Civitelli, L.; Argnani, R.; Piacentini, R.; Ripoli, C.; Manservigi, R.; Grassi, C.; Garaci, E.; Palamara, A.T. APP processing induced by herpes simplex virus type 1 (HSV-1) yields several APP fragments in human and rat neuronal cells. *PLoS ONE* **2010**, *5*, e13989. [CrossRef] [PubMed]
67. Piacentini, R.; Civitelli, L.; Ripoli, C.; Marcocci, M.E.; De Chiara, G.; Garaci, E.; Azzena, G.B.; Palamara, A.T.; Grassi, C. HSV-1 promotes Ca^{2+}-mediated APP phosphorylation and Abeta accumulation in rat cortical neurons. *Neurobiol. Aging* **2011**, *32*, 2323 e13–2323 e26. [CrossRef] [PubMed]
68. Santana, S.; Recuero, M.; Bullido, M.J.; Valdivieso, F.; Aldudo, J. Herpes simplex virus type I induces the accumulation of intracellular beta-amyloid in autophagic compartments and the inhibition of the non-amyloidogenic pathway in human neuroblastoma cells. *Neurobiol. Aging* **2012**, *33*, 430.e19–430.e33. [CrossRef]
69. Wozniak, M.A.; Itzhaki, R.F.; Shipley, S.J.; Dobson, C.B. Herpes simplex virus infection causes cellular beta-amyloid accumulation and secretase upregulation. *Neurosci. Lett.* **2007**, *429*, 95–100. [CrossRef]
70. Ill-Raga, G.; Palomer, E.; Wozniak, M.A.; Ramos-Fernandez, E.; Bosch-Morato, M.; Tajes, M.; Guix, F.X.; Galan, J.J.; Clarimon, J.; Antunez, C.; et al. Activation of PKR causes amyloid ss-peptide accumulation via de-repression of BACE1 expression. *PLoS ONE* **2011**, *6*, e21456. [CrossRef] [PubMed]
71. Cheng, S.B.; Ferland, P.; Webster, P.; Bearer, E.L. Herpes simplex virus dances with amyloid precursor protein while exiting the cell. *PLoS ONE* **2011**, *6*, e17966. [CrossRef] [PubMed]
72. Civitelli, L.; Marcocci, M.E.; Celestino, I.; Piacentini, R.; Garaci, E.; Grassi, C.; De Chiara, G.; Palamara, A.T. Herpes simplex virus type 1 infection in neurons leads to production and nuclear localization of APP intracellular domain (AICD): Implications for Alzheimer's disease pathogenesis. *J. Neurovirol.* **2015**, *21*, 480–490. [CrossRef] [PubMed]
73. Wang, S.; Wang, R.; Chen, L.; Bennett, D.A.; Dickson, D.W.; Wang, D.S. Expression and functional profiling of neprilysin, insulin-degrading enzyme, and endothelin-converting enzyme in prospectively studied elderly and Alzheimer's brain. *J. Neurochem.* **2010**, *115*, 47–57. [CrossRef] [PubMed]
74. Humpel, C. Intranasal Delivery of Collagen-Loaded Neprilysin Clears Beta-Amyloid Plaques in a Transgenic Alzheimer Mouse Model. *Front. Aging Neurosci.* **2021**, *13*, 649646. [CrossRef] [PubMed]

75. Zheng, K.; Liu, Q.; Wang, S.; Ren, Z.; Kitazato, K.; Yang, D.; Wang, Y. HSV-1-encoded microRNA miR-H1 targets Ubr1 to promote accumulation of neurodegeneration-associated protein. *Virus Genes* **2018**, *54*, 343–350. [CrossRef] [PubMed]
76. Bortolotti, D.; Gentili, V.; Rotola, A.; Caselli, E.; Rizzo, R. HHV-6A infection induces amyloid-beta expression and activation of microglial cells. *Alzheimers Res. Ther.* **2019**, *11*, 104. [CrossRef]
77. Tang, T.; Jia, J.; Garbarino, E.; Chen, L.; Ma, J.; Li, P.; Chen, X.; Wang, L.; Wen, Y.; Wang, Y.; et al. Human Herpesvirus 6A U4 Inhibits Proteasomal Degradation of the Amyloid Precursor Protein. *J. Virol.* **2022**, *96*, e0168821. [CrossRef]
78. Hua, Z.; Vierstra, R.D. The cullin-RING ubiquitin-protein ligases. *Annu. Rev. Plant Biol.* **2011**, *62*, 299–334. [CrossRef] [PubMed]
79. Wozniak, M.A.; Frost, A.L.; Itzhaki, R.F. Alzheimer's disease-specific tau phosphorylation is induced by herpes simplex virus type 1. *J. Alzheimers Dis.* **2009**, *16*, 341–350. [CrossRef] [PubMed]
80. Benetti, L.; Roizman, B. Herpes simplex virus protein kinase US3 activates and functionally overlaps protein kinase A to block apoptosis. *Proc. Natl. Acad. Sci. USA* **2004**, *101*, 9411–9416. [CrossRef]
81. Alvarez, G.; Aldudo, J.; Alonso, M.; Santana, S.; Valdivieso, F. Herpes simplex virus type 1 induces nuclear accumulation of hyperphosphorylated tau in neuronal cells. *J. Neurosci. Res.* **2012**, *90*, 1020–1029. [CrossRef]
82. Kristen, H.; Santana, S.; Sastre, I.; Recuero, M.; Bullido, M.J.; Aldudo, J. Herpes simplex virus type 2 infection induces AD-like neurodegeneration markers in human neuroblastoma cells. *Neurobiol. Aging* **2015**, *36*, 2737–2747. [CrossRef] [PubMed]
83. Zhou, F.; van Laar, T.; Huang, H.; Zhang, L. APP and APLP1 are degraded through autophagy in response to proteasome inhibition in neuronal cells. *Protein Cell* **2011**, *2*, 377–383. [CrossRef] [PubMed]
84. Wang, Y.; Mandelkow, E. Degradation of tau protein by autophagy and proteasomal pathways. *Biochem. Soc. Trans.* **2012**, *40*, 644–652. [CrossRef] [PubMed]
85. Romeo, M.A.; Gilardini Montani, M.S.; Gaeta, A.; D'Orazi, G.; Faggioni, A.; Cirone, M. HHV-6A infection dysregulates autophagy/UPR interplay increasing beta amyloid production and tau phosphorylation in astrocytoma cells as well as in primary neurons, possible molecular mechanisms linking viral infection to Alzheimer's disease. *Biochim. Biophys. Acta Mol. Basis Dis.* **2020**, *1866*, 165647. [CrossRef] [PubMed]
86. Nijholt, D.A.; Nolle, A.; van Haastert, E.S.; Edelijn, H.; Toonen, R.F.; Hoozemans, J.J.; Scheper, W. Unfolded protein response activates glycogen synthase kinase-3 via selective lysosomal degradation. *Neurobiol. Aging* **2013**, *34*, 1759–1771. [CrossRef]
87. van der Harg, J.M.; Nolle, A.; Zwart, R.; Boerema, A.S.; van Haastert, E.S.; Strijkstra, A.M.; Hoozemans, J.J.; Scheper, W. The unfolded protein response mediates reversible tau phosphorylation induced by metabolic stress. *Cell Death Dis.* **2014**, *5*, e1393. [CrossRef] [PubMed]
88. Itzhaki, R.F.; Cosby, S.L.; Wozniak, M.A. Herpes simplex virus type 1 and Alzheimer's disease: The autophagy connection. *J. Neurovirol.* **2008**, *14*, 1–4. [CrossRef] [PubMed]
89. O'Connell, D.; Liang, C. Autophagy interaction with herpes simplex virus type-1 infection. *Autophagy* **2016**, *12*, 451–459. [CrossRef]
90. Hill, S.M.; Wrobel, L.; Rubinsztein, D.C. Post-translational modifications of Beclin 1 provide multiple strategies for autophagy regulation. *Cell Death Differ.* **2019**, *26*, 617–629. [CrossRef] [PubMed]
91. Orvedahl, A.; Alexander, D.; Talloczy, Z.; Sun, Q.; Wei, Y.; Zhang, W.; Burns, D.; Leib, D.A.; Levine, B. HSV-1 ICP34.5 confers neurovirulence by targeting the Beclin 1 autophagy protein. *Cell Host Microbe* **2007**, *1*, 23–35. [CrossRef] [PubMed]
92. Lussignol, M.; Queval, C.; Bernet-Camard, M.F.; Cotte-Laffitte, J.; Beau, I.; Codogno, P.; Esclatine, A. The herpes simplex virus 1 Us11 protein inhibits autophagy through its interaction with the protein kinase PKR. *J. Virol.* **2013**, *87*, 859–871. [CrossRef] [PubMed]
93. Santana, S.; Bullido, M.J.; Recuero, M.; Valdivieso, F.; Aldudo, J. Herpes simplex virus type I induces an incomplete autophagic response in human neuroblastoma cells. *J. Alzheimers Dis.* **2012**, *30*, 815–831. [CrossRef] [PubMed]
94. Eshraghi, M.; Ahmadi, M.; Afshar, S.; Lorzadeh, S.; Adlimoghaddam, A.; Rezvani Jalal, N.; West, R.; Dastghaib, S.; Igder, S.; Torshizi, S.R.N.; et al. Enhancing autophagy in Alzheimer's disease through drug repositioning. *Pharmacol. Ther.* **2022**, *237*, 108171. [CrossRef] [PubMed]
95. Heras-Sandoval, D.; Perez-Rojas, J.M.; Pedraza-Chaverri, J. Novel compounds for the modulation of mTOR and autophagy to treat neurodegenerative diseases. *Cell. Signal.* **2020**, *65*, 109442. [CrossRef] [PubMed]
96. Rahman, M.A.; Rahman, M.R.; Zaman, T.; Uddin, M.S.; Islam, R.; Abdel-Daim, M.M.; Rhim, H. Emerging Potential of Naturally Occurring Autophagy Modulators Against Neurodegeneration. *Curr. Pharm. Des.* **2020**, *26*, 772–779. [CrossRef]
97. Chen, Z.; Zhong, D.; Li, G. The role of microglia in viral encephalitis: A review. *J. Neuroinflamm.* **2019**, *16*, 76. [CrossRef]
98. Martini-Stoica, H.; Xu, Y.; Ballabio, A.; Zheng, H. The Autophagy-Lysosomal Pathway in Neurodegeneration: A TFEB Perspective. *Trends Neurosci.* **2016**, *39*, 221–234. [CrossRef]
99. Pupyshev, A.B.; Belichenko, V.M.; Tenditnik, M.V.; Bashirzade, A.A.; Dubrovina, N.I.; Ovsyukova, M.V.; Akopyan, A.A.; Fedoseeva, L.A.; Korolenko, T.A.; Amstislavskaya, T.G.; et al. Combined induction of mTOR-dependent and mTOR-independent pathways of autophagy activation as an experimental therapy for Alzheimer's disease-like pathology in a mouse model. *Pharmacol. Biochem. Behav.* **2022**, *217*, 173406. [CrossRef]
100. Zhang, W.; Wang, J.; Yang, C. Celastrol, a TFEB (transcription factor EB) agonist, is a promising drug candidate for Alzheimer disease. *Autophagy* **2022**, *18*, 1740–1742. [CrossRef] [PubMed]

101. Zhang, X.; Chen, S.; Song, L.; Tang, Y.; Shen, Y.; Jia, L.; Le, W. MTOR-independent, autophagic enhancer trehalose prolongs motor neuron survival and ameliorates the autophagic flux defect in a mouse model of amyotrophic lateral sclerosis. *Autophagy* **2014**, *10*, 588–602. [CrossRef]
102. Uddin, M.S.; Tewari, D.; Sharma, G.; Kabir, M.T.; Barreto, G.E.; Bin-Jumah, M.N.; Perveen, A.; Abdel-Daim, M.M.; Ashraf, G.M. Molecular Mechanisms of ER Stress and UPR in the Pathogenesis of Alzheimer's Disease. *Mol. Neurobiol.* **2020**, *57*, 2902–2919. [CrossRef]
103. Cuadrado-Tejedor, M.; Ricobaraza, A.L.; Torrijo, R.; Franco, R.; Garcia-Osta, A. Phenylbutyrate is a multifaceted drug that exerts neuroprotective effects and reverses the Alzheimer s disease-like phenotype of a commonly used mouse model. *Curr. Pharm. Des.* **2013**, *19*, 5076–5084. [CrossRef] [PubMed]
104. Hafycz, J.M.; Strus, E.; Naidoo, N. Reducing ER stress with chaperone therapy reverses sleep fragmentation and cognitive decline in aged mice. *Aging Cell* **2022**, *21*, e13598. [CrossRef]
105. Halliday, M.; Radford, H.; Sekine, Y.; Moreno, J.; Verity, N.; le Quesne, J.; Ortori, C.A.; Barrett, D.A.; Fromont, C.; Fischer, P.M.; et al. Partial restoration of protein synthesis rates by the small molecule ISRIB prevents neurodegeneration without pancreatic toxicity. *Cell Death Dis.* **2015**, *6*, e1672. [CrossRef] [PubMed]
106. Rivera-Krstulovic, C.; Duran-Aniotz, C. Unfolded protein response as a target in the treatment of Alzheimer's disease. *Rev. Med. Chil.* **2020**, *148*, 216–223. [CrossRef]
107. Yun, H.R.; Jo, Y.H.; Kim, J.; Shin, Y.; Kim, S.S.; Choi, T.G. Roles of Autophagy in Oxidative Stress. *Int. J. Mol. Sci.* **2020**, *21*, 3289. [CrossRef] [PubMed]
108. Butterfield, D.A.; Halliwell, B. Oxidative stress, dysfunctional glucose metabolism and Alzheimer disease. *Nat. Rev. Neurosci.* **2019**, *20*, 148–160. [CrossRef] [PubMed]
109. Martins, R.N.; Villemagne, V.; Sohrabi, H.R.; Chatterjee, P.; Shah, T.M.; Verdile, G.; Fraser, P.; Taddei, K.; Gupta, V.B.; Rainey-Smith, S.R.; et al. Alzheimer's Disease: A Journey from Amyloid Peptides and Oxidative Stress, to Biomarker Technologies and Disease Prevention Strategies-Gains from AIBL and DIAN Cohort Studies. *J. Alzheimers Dis.* **2018**, *62*, 965–992. [CrossRef] [PubMed]
110. Camini, F.C.; da Silva Caetano, C.C.; Almeida, L.T.; de Brito Magalhaes, C.L. Implications of oxidative stress on viral pathogenesis. *Arch. Virol.* **2017**, *162*, 907–917. [CrossRef] [PubMed]
111. Foo, J.; Bellot, G.; Pervaiz, S.; Alonso, S. Mitochondria-mediated oxidative stress during viral infection. *Trends Microbiol.* **2022**, *30*, 679–692. [CrossRef]
112. Kavouras, J.H.; Prandovszky, E.; Valyi-Nagy, K.; Kovacs, S.K.; Tiwari, V.; Kovacs, M.; Shukla, D.; Valyi-Nagy, T. Herpes simplex virus type 1 infection induces oxidative stress and the release of bioactive lipid peroxidation by-products in mouse P19N neural cell cultures. *J. Neurovirol.* **2007**, *13*, 416–425. [CrossRef]
113. Schachtele, S.J.; Hu, S.; Little, M.R.; Lokensgard, J.R. Herpes simplex virus induces neural oxidative damage via microglial cell Toll-like receptor-2. *J. Neuroinflamm.* **2010**, *7*, 35. [CrossRef]
114. Santana, S.; Sastre, I.; Recuero, M.; Bullido, M.J.; Aldudo, J. Oxidative stress enhances neurodegeneration markers induced by herpes simplex virus type 1 infection in human neuroblastoma cells. *PLoS ONE* **2013**, *8*, e75842. [CrossRef] [PubMed]
115. Kristen, H.; Sastre, I.; Munoz-Galdeano, T.; Recuero, M.; Aldudo, J.; Bullido, M.J. The lysosome system is severely impaired in a cellular model of neurodegeneration induced by HSV-1 and oxidative stress. *Neurobiol. Aging* **2018**, *68*, 5–17. [CrossRef] [PubMed]
116. Patel, S.; Homaei, A.; El-Seedi, H.R.; Akhtar, N. Cathepsins: Proteases that are vital for survival but can also be fatal. *Biomed. Pharm.* **2018**, *105*, 526–532. [CrossRef] [PubMed]
117. Cermak, S.; Kosicek, M.; Mladenovic-Djordjevic, A.; Smiljanic, K.; Kanazir, S.; Hecimovic, S. Loss of Cathepsin B and L Leads to Lysosomal Dysfunction, NPC-Like Cholesterol Sequestration and Accumulation of the Key Alzheimer's Proteins. *PLoS ONE* **2016**, *11*, e0167428. [CrossRef]
118. Protto, V.; Tramutola, A.; Fabiani, M.; Marcocci, M.E.; Napoletani, G.; Iavarone, F.; Vincenzoni, F.; Castagnola, M.; Perluigi, M.; Di Domenico, F.; et al. Multiple Herpes Simplex Virus-1 (HSV-1) Reactivations Induce Protein Oxidative Damage in Mouse Brain: Novel Mechanisms for Alzheimer's Disease Progression. *Microorganisms* **2020**, *8*, 972. [CrossRef]
119. Hensley, K.; Venkova, K.; Christov, A.; Gunning, W.; Park, J. Collapsin response mediator protein-2: An emerging pathologic feature and therapeutic target for neurodisease indications. *Mol. Neurobiol.* **2011**, *43*, 180–191. [CrossRef]
120. Morinaka, A.; Yamada, M.; Itofusa, R.; Funato, Y.; Yoshimura, Y.; Nakamura, F.; Yoshimura, T.; Kaibuchi, K.; Goshima, Y.; Hoshino, M.; et al. Thioredoxin mediates oxidation-dependent phosphorylation of CRMP2 and growth cone collapse. *Sci. Signal.* **2011**, *4*, ra26. [CrossRef] [PubMed]
121. Costa Sa, A.C.; Madsen, H.; Brown, J.R. Shared Molecular Signatures Across Neurodegenerative Diseases and Herpes Virus Infections Highlights Potential Mechanisms for Maladaptive Innate Immune Responses. *Sci. Rep.* **2019**, *9*, 8795. [CrossRef]
122. Mota, B.C.; Sastre, M. The Role of PGC1alpha in Alzheimer's Disease and Therapeutic Interventions. *Int. J. Mol. Sci.* **2021**, *22*, 5769. [CrossRef]
123. Liu, Z.; Zhou, T.; Ziegler, A.C.; Dimitrion, P.; Zuo, L. Oxidative Stress in Neurodegenerative Diseases: From Molecular Mechanisms to Clinical Applications. *Oxid. Med. Cell. Longev.* **2017**, *2017*, 2525967. [CrossRef] [PubMed]
124. Aseervatham, G.S.; Sivasudha, T.; Jeyadevi, R.; Arul Ananth, D. Environmental factors and unhealthy lifestyle influence oxidative stress in humans—An overview. *Environ. Sci. Pollut. Res. Int.* **2013**, *20*, 4356–4369. [CrossRef]

125. Moulton, P.V.; Yang, W. Air pollution, oxidative stress, and Alzheimer's disease. *J. Environ. Public Health* **2012**, *2012*, 472751. [CrossRef] [PubMed]
126. Smith, M.A.; Zhu, X.; Tabaton, M.; Liu, G.; McKeel, D.W., Jr.; Cohen, M.L.; Wang, X.; Siedlak, S.L.; Dwyer, B.E.; Hayashi, T.; et al. Increased iron and free radical generation in preclinical Alzheimer disease and mild cognitive impairment. *J. Alzheimers Dis.* **2010**, *19*, 363–372. [CrossRef]
127. Hamilton, A.; Holscher, C. The effect of ageing on neurogenesis and oxidative stress in the APP(swe)/PS1(deltaE9) mouse model of Alzheimer's disease. *Brain Res.* **2012**, *1449*, 83–93. [CrossRef] [PubMed]
128. Leyane, T.S.; Jere, S.W.; Houreld, N.N. Oxidative Stress in Ageing and Chronic Degenerative Pathologies: Molecular Mechanisms Involved in Counteracting Oxidative Stress and Chronic Inflammation. *Int. J. Mol. Sci.* **2022**, *23*, 7273. [CrossRef] [PubMed]
129. Leng, F.; Edison, P. Neuroinflammation and microglial activation in Alzheimer disease: Where do we go from here? *Nat. Rev. Neurol.* **2021**, *17*, 157–172. [CrossRef]
130. Fakhoury, M. Microglia and Astrocytes in Alzheimer's Disease: Implications for Therapy. *Curr. Neuropharmacol.* **2018**, *16*, 508–518. [CrossRef] [PubMed]
131. Giovannoni, F.; Quintana, F.J. The Role of Astrocytes in CNS Inflammation. *Trends Immunol.* **2020**, *41*, 805–819. [CrossRef] [PubMed]
132. Salter, M.W.; Stevens, B. Microglia emerge as central players in brain disease. *Nat. Med.* **2017**, *23*, 1018–1027. [CrossRef]
133. Duggan, M.R.; Torkzaban, B.; Ahooyi, T.M.; Khalili, K. Potential Role for Herpesviruses in Alzheimer's Disease. *J. Alzheimers Dis.* **2020**, *78*, 855–869. [CrossRef] [PubMed]
134. Theil, D.; Derfuss, T.; Paripovic, I.; Herberger, S.; Meinl, E.; Schueler, O.; Strupp, M.; Arbusow, V.; Brandt, T. Latent herpesvirus infection in human trigeminal ganglia causes chronic immune response. *Am. J. Pathol.* **2003**, *163*, 2179–2184. [CrossRef]
135. Cokaric Brdovcak, M.; Zubkovic, A.; Jurak, I. Herpes Simplex Virus 1 Deregulation of Host MicroRNAs. *Noncoding RNA* **2018**, *4*, 36. [CrossRef]
136. Hill, J.M.; Zhao, Y.; Clement, C.; Neumann, D.M.; Lukiw, W.J. HSV-1 infection of human brain cells induces miRNA-146a and Alzheimer-type inflammatory signaling. *Neuroreport* **2009**, *20*, 1500–1505. [CrossRef] [PubMed]
137. Wang, Y.; Jia, J.; Wang, Y.; Li, F.; Song, X.; Qin, S.; Wang, Z.; Kitazato, K.; Wang, Y. Roles of HSV-1 infection-induced microglial immune responses in CNS diseases: Friends or foes? *Crit. Rev. Microbiol.* **2019**, *45*, 581–594. [CrossRef] [PubMed]
138. Marques, C.P.; Cheeran, M.C.; Palmquist, J.M.; Hu, S.; Lokensgard, J.R. Microglia are the major cellular source of inducible nitric oxide synthase during experimental herpes encephalitis. *J. Neurovirol.* **2008**, *14*, 229–238. [CrossRef] [PubMed]
139. Asiimwe, N.; Yeo, S.G.; Kim, M.S.; Jung, J.; Jeong, N.Y. Nitric Oxide: Exploring the Contextual Link with Alzheimer's Disease. *Oxid. Med. Cell. Longev.* **2016**, *2016*, 7205747. [CrossRef] [PubMed]
140. Colonna, M.; Butovsky, O. Microglia Function in the Central Nervous System During Health and Neurodegeneration. *Annu. Rev. Immunol.* **2017**, *35*, 441–468. [CrossRef]
141. Hu, S.; Sheng, W.S.; Schachtele, S.J.; Lokensgard, J.R. Reactive oxygen species drive herpes simplex virus (HSV)-1-induced proinflammatory cytokine production by murine microglia. *J. Neuroinflamm.* **2011**, *8*, 123. [CrossRef]
142. Reinert, L.S.; Lopusna, K.; Winther, H.; Sun, C.; Thomsen, M.K.; Nandakumar, R.; Mogensen, T.H.; Meyer, M.; Vaegter, C.; Nyengaard, J.R.; et al. Sensing of HSV-1 by the cGAS-STING pathway in microglia orchestrates antiviral defence in the CNS. *Nat. Commun.* **2016**, *7*, 13348. [CrossRef]
143. Roy, E.R.; Wang, B.; Wan, Y.W.; Chiu, G.; Cole, A.; Yin, Z.; Propson, N.E.; Xu, Y.; Jankowsky, J.L.; Liu, Z.; et al. Type I interferon response drives neuroinflammation and synapse loss in Alzheimer disease. *J. Clin. Investig.* **2020**, *130*, 1912–1930. [CrossRef] [PubMed]
144. Taylor, J.M.; Moore, Z.; Minter, M.R.; Crack, P.J. Type-I interferon pathway in neuroinflammation and neurodegeneration: Focus on Alzheimer's disease. *J. Neural. Transm.* **2018**, *125*, 797–807. [CrossRef] [PubMed]
145. Almutairi, M.M.; Sivandzade, F.; Albekairi, T.H.; Alqahtani, F.; Cucullo, L. Neuroinflammation and Its Impact on the Pathogenesis of COVID-19. *Front. Med.* **2021**, *8*, 745789. [CrossRef] [PubMed]
146. Amruta, N.; Chastain, W.H.; Paz, M.; Solch, R.J.; Murray-Brown, I.C.; Befeler, J.B.; Gressett, T.E.; Longo, M.T.; Engler-Chiurazzi, E.B.; Bix, G. SARS-CoV-2 mediated neuroinflammation and the impact of COVID-19 in neurological disorders. *Cytokine Growth Factor Rev.* **2021**, *58*, 1–15. [CrossRef]
147. Klein, R.S.; Garber, C.; Funk, K.E.; Salimi, H.; Soung, A.; Kanmogne, M.; Manivasagam, S.; Agner, S.; Cain, M. Neuroinflammation During RNA Viral Infections. *Annu. Rev. Immunol.* **2019**, *37*, 73–95. [CrossRef]
148. Papuc, E.; Rejdak, K. The role of myelin damage in Alzheimer's disease pathology. *Arch. Med. Sci.* **2020**, *16*, 345–351. [CrossRef]
149. Traiffort, E.; Kassoussi, A.; Zahaf, A.; Laouarem, Y. Astrocytes and Microglia as Major Players of Myelin Production in Normal and Pathological Conditions. *Front. Cell Neurosci.* **2020**, *14*, 79. [CrossRef]
150. Hoos, M.D.; Ahmed, M.; Smith, S.O.; Van Nostrand, W.E. Myelin basic protein binds to and inhibits the fibrillar assembly of Abeta42 in vitro. *Biochemistry* **2009**, *48*, 4720–4727. [CrossRef]
151. Liao, M.C.; Ahmed, M.; Smith, S.O.; Van Nostrand, W.E. Degradation of amyloid beta protein by purified myelin basic protein. *J. Biol. Chem.* **2009**, *284*, 28917–28925. [CrossRef]
152. Lee, J.T.; Xu, J.; Lee, J.M.; Ku, G.; Han, X.; Yang, D.I.; Chen, S.; Hsu, C.Y. Amyloid-beta peptide induces oligodendrocyte death by activating the neutral sphingomyelinase-ceramide pathway. *J. Cell Biol.* **2004**, *164*, 123–131. [CrossRef]

153. Boukhvalova, M.S.; Mortensen, E.; Mbaye, A.; Lopez, D.; Kastrukoff, L.; Blanco, J.C.G. Herpes Simplex Virus 1 Induces Brain Inflammation and Multifocal Demyelination in the Cotton Rat Sigmodon hispidus. *J. Virol.* **2019**, *94*, e01161-19. [CrossRef] [PubMed]
154. Kristensson, K. Microbes' roadmap to neurons. *Nat. Rev. Neurosci.* **2011**, *12*, 345–357. [CrossRef]
155. Dezfulian, M. A new Alzheimer's disease cell model using B cells to induce beta amyloid plaque formation and increase TNF alpha expression. *Int. Immunopharmacol.* **2018**, *59*, 106–112. [CrossRef] [PubMed]
156. Ounanian, A.; Guilbert, B.; Seigneurin, J.M. Characteristics of Epstein-Barr virus transformed B cell lines from patients with Alzheimer's disease and age-matched controls. *Mech. Ageing Dev.* **1992**, *63*, 105–116. [CrossRef]
157. Gate, D.; Saligrama, N.; Leventhal, O.; Yang, A.C.; Unger, M.S.; Middeldorp, J.; Chen, K.; Lehallier, B.; Channappa, D.; De Los Santos, M.B.; et al. Clonally expanded CD8 T cells patrol the cerebrospinal fluid in Alzheimer's disease. *Nature* **2020**, *577*, 399–404. [CrossRef]
158. Hur, J.Y.; Frost, G.R.; Wu, X.; Crump, C.; Pan, S.J.; Wong, E.; Barros, M.; Li, T.; Nie, P.; Zhai, Y.; et al. The innate immunity protein IFITM3 modulates gamma-secretase in Alzheimer's disease. *Nature* **2020**, *586*, 735–740. [CrossRef]
159. Tiwari, D.; Singh, V.K.; Baral, B.; Pathak, D.K.; Jayabalan, J.; Kumar, R.; Tapryal, S.; Jha, H.C. Indication of Neurodegenerative Cascade Initiation by Amyloid-like Aggregate-Forming EBV Proteins and Peptide in Alzheimer's Disease. *ACS Chem. Neurosci.* **2021**, *12*, 3957–3967. [CrossRef]
160. Wycisk, A.I.; Lin, J.; Loch, S.; Hobohm, K.; Funke, J.; Wieneke, R.; Koch, J.; Skach, W.R.; Mayerhofer, P.U.; Tampe, R. Epstein-Barr viral BNLF2a protein hijacks the tail-anchored protein insertion machinery to block antigen processing by the transport complex TAP. *J. Biol. Chem.* **2011**, *286*, 41402–41412. [CrossRef] [PubMed]
161. Ivanov, A.V.; Valuev-Elliston, V.T.; Ivanova, O.N.; Kochetkov, S.N.; Starodubova, E.S.; Bartosch, B.; Isaguliants, M.G. Oxidative Stress during HIV Infection: Mechanisms and Consequences. *Oxid. Med. Cell. Longev.* **2016**, *2016*, 8910396. [CrossRef] [PubMed]
162. Mielcarska, M.B.; Skowronska, K.; Wyzewski, Z.; Toka, F.N. Disrupting Neurons and Glial Cells Oneness in the Brain-The Possible Causal Role of Herpes Simplex Virus Type 1 (HSV-1) in Alzheimer's Disease. *Int. J. Mol. Sci.* **2021**, *23*, 242. [CrossRef]
163. Lukiw, W.J. Recent Advances in Our Molecular and Mechanistic Understanding of Misfolded Cellular Proteins in Alzheimer's Disease (AD) and Prion Disease (PrD). *Biomolecules* **2022**, *12*, 166. [CrossRef] [PubMed]
164. Ayers, J.I.; Paras, N.A.; Prusiner, S.B. Expanding spectrum of prion diseases. *Emerg. Top. Life Sci.* **2020**, *4*, 155–167. [CrossRef] [PubMed]

Article

Changes in Tyrosine Hydroxylase Activity and Dopamine Synthesis in the Nigrostriatal System of Mice in an Acute Model of Parkinson's Disease as a Manifestation of Neurodegeneration and Neuroplasticity

Anna Kolacheva *, Leyla Alekperova, Ekaterina Pavlova, Alyona Bannikova and Michael V. Ugrumov

Koltzov Institute of Developmental Biology, Russian Academy of Sciences, 26 Vavilova Street, 119334 Moscow, Russia; al.alekperova@gmail.com (L.A.); guchia@gmail.com (E.P.); mukurokun354@gmail.com (A.B.); michael.ugrumov@mail.ru (M.V.U.)
* Correspondence: annakolacheva@gmail.com; Tel.: +7-499-135-8842

Abstract: The progressive degradation of the nigrostriatal system leads to the development of Parkinson's disease (PD). The synthesis of dopamine, the neurotransmitter of the nigrostriatal system, depends on the rate-limiting enzyme, tyrosine hydroxylase (TH). In this study, we evaluated the synthesis of dopamine during periods of neurodegradation and neuroplasticity in the nigrostriatal system on a model of the early clinical stage of PD. It was shown that the concentration of dopamine correlated with activity of TH, while TH activity did not depend on total protein content either in the SN or in the striatum. Both during the period of neurodegeneration and neuroplasticity, TH activity in SN was determined by the content of P19-TH, and in the striatum it was determined by P31-TH and P40-TH (to a lesser extent). The data obtained indicate a difference in the regulation of dopamine synthesis between DA-neuron bodies and their axons, which must be considered for the further development of symptomatic pharmacotherapy aimed at increasing TH activity.

Keywords: Parkinson's disease; MPTP model; mice; tyrosine hydroxylase; phosphorylation; neurodegeneration; neuroplasticity

Citation: Kolacheva, A.; Alekperova, L.; Pavlova, E.; Bannikova, A.; Ugrumov, M.V. Changes in Tyrosine Hydroxylase Activity and Dopamine Synthesis in the Nigrostriatal System of Mice in an Acute Model of Parkinson's Disease as a Manifestation of Neurodegeneration and Neuroplasticity. *Brain Sci.* **2022**, *12*, 779. https://doi.org/10.3390/brainsci12060779

Academic Editors: Christina Piperi, Chiara Villa and Yam Nath Paudel

Received: 14 May 2022
Accepted: 11 June 2022
Published: 14 June 2022

Publisher's Note: MDPI stays neutral with regard to jurisdictional claims in published maps and institutional affiliations.

Copyright: © 2022 by the authors. Licensee MDPI, Basel, Switzerland. This article is an open access article distributed under the terms and conditions of the Creative Commons Attribution (CC BY) license (https://creativecommons.org/licenses/by/4.0/).

1. Introduction

The nigrostriatal system is a part of the basal ganglia; it is involved in the regulation of motor function and motor memory. The main neurotransmitter in this system is dopamine (DA), which is synthesized from tyrosine by two enzymes: tyrosine hydroxylase (TH) and aromatic L-amino acid decarboxylase (AADC). It is believed that the rate-limiting stage of DA synthesis is the formation of L-DOPA from tyrosine by TH; BH4 and Fe act as co-factors [1–3]. DA synthesized in the cytosol is taken up into vesicles by the vesicular monoamine transporter type 2 (VMAT2). DA is released from the DA axon terminals via vesicle exocytosis in response to an action potential. After action on the receptors of target neurons, DA is degraded by catechol-O-methyl transferase (COMT) and monoamine oxidase (MAO) contained in the intercellular space, glial cells, and axon terminals [4–6] or recaptured into the synaptic terminal with a dopamine transporter (DAT) and then into vesicles with VMAT2 for reuse. In the normal striatum, DA reuptake prevails over its degradation [7].

Parkinson's disease (PD) develops in the process of degradation of the nigrostriatal system. PD is the second most common neurodegenerative disease that affects 1–3% of the world's population over 60 years old [8]. The appearance of motor symptoms typical for PD is preceded by a long-term latent degradation of the nigrostriatal system, up to the degeneration of 50–60% of DA neuron bodies and a decrease in DA to 20–30% in the striatum (threshold values) [9]. In this case, the depletion of neuroplasticity in the brain also occurs [10–13]. One of these compensatory mechanisms may be an increase in the functional activity of surviving DA neurons. Nagatsu and coworkers showed that in PD, despite a decrease in the total content of TH and its activity, homospecific activity (TH activity per enzyme amount) increases [14,15].

TH regulation can be divided into fast, which includes post-translational changes of the protein (phosphorylation, inhibition by catecholamines, etc.), and slow, which is transcription activation [16,17]. The key post-translational process that increases the activity of TH is its phosphorylation at serine at positions 8, 19, 31, and 40. It has been shown that phosphorylation at positions P31-TH and P40-TH increases TH activity, and phosphorylation at position P19-TH increases the availability site 40 for phosphorylation (P40-TH). The role of phosphorylation at position P8 (P8-TH) is not completely clear [16–22]. It has also been shown that the phosphorylated protein at positions P19-TH is a marker of its degradation [18,23].

The regulation of TH activity by phosphorylation has been studied in various PD models in vivo, e.g., using neurotoxic models with the administration of 1-methyl-4-phenyl-1,2,3,6-tetrahydropyridine (MPTP) [24–26] and 6-hydroxydopamine (6-OHDA) [27]. However, in previous studies, the evaluation of TH activity and/or the content of its phosphorylated forms was performed without taking into account the progressive degradation of the nigrostriatal system. It seems important to study the regulation of TH during the period of neurodegeneration considering the peculiarities of the pathogenesis of the disease, where death of DA neurons is a permanent process. This will also allow to find new approaches to the development of symptomatic pharmacotherapy aimed at replenishing the level of DA in the striatum in PD.

Earlier in our laboratory, we developed a mice model of PD. The model uses four-fold administration of MPTP, which leads to the development of an early clinical stage in key parameters: the presence of impaired motor function, a decrease in the concentration of DA in the striatum and number of DA neuron bodies in the substantia nigra (SN) to a strictly defined threshold value [28]. This model was used to estimate the period of degradation of the nigrostriatal system not only in neuron bodies in the SN, but also in their axon terminals in the striatum [29,30]. Based on these studies, we chose a period of neurodegradation on DA neurons and a period when neuroplasticity begins to develop in the surviving DA neurons.

The goal of this study was to evaluate the synthesis of DA in DA neurons during the period indicated above, with an emphasis on the molecular mechanisms responsible for the change in TH activity. We evaluated the activity of TH, the content of TH and its phosphorylated forms (P19-, P31-, and P40-TH) in the striatum and SN, as well as the concentration of DA in the striatum during the degradation of the nigrostriatal system and a few hours after its completion.

2. Materials and Methods

2.1. Animals and Experimental Procedures

Male mice C57BL/6 aging 8–12 weeks and weighing 22–26 g (n = 130) were used in this study. The animals were housed at 21–23 °C in a 12 h light–dark cycle with free access to food and water. The experimental procedures were carried out in accordance with the National Institutes of Health Guide for the Care and Use of Laboratory Animals (8th edition, 2011) and were approved by the Animal Care and Use Committee of the Koltzov Institute of Developmental Biology of the Russian Academy of Sciences (protocol No. 44 from 24 December 2020 and No. 55 from 9 December 2021).

2.2. Modeling of the Early Clinical Stage of Parkinson's Disease

An early clinical PD model was reproduced in mice with four subcutaneous injections of MPTP (Sigma Aldrich, USA) at a dose of 12 mg/kg of body weight (free base) in saline with 2 h intervals. The control animals received saline only [28,29].

2.3. Design of Experiments

The scheme of experiments is shown in Figure 1.

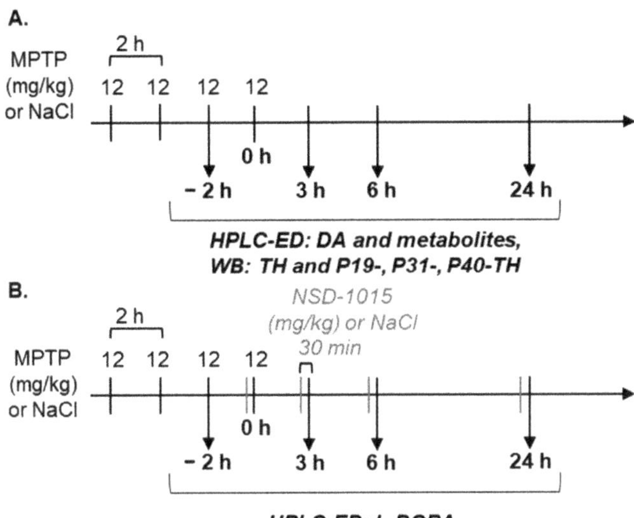

Figure 1. The scheme of the experiments for assessing: (**A**) the concentration of DA in the striatum as well as the content of TH and its phosphorylated forms at serine at positions P19-, P31-, and P40-TH and (**B**) the activity of TH during inhibition of DAA in the striatum and substantia nigra during the period of neurodegeneration (from −2 h to 6 h after MPTP injections) and neuroplasticity (24 h) in the nigrostriatal system in an early clinical PD model. DA—dopamine, HPLC-ED—high-performance liquid chromatography with electrochemical detection; L-DOPA—L-3,4-dihydroxyphenylalanine; MPTP—1-methyl-4-phenyl-1,2,3,6-tetrahydropyridine; NSD-1015—3-hydroxybenzylhydrazine; TH—tyrosine hydroxylase; SN—substantia nigra, WB—Western blot.

In the first series of experiments, the content of TH and its phosphorylated forms in the striatum and SN was determined 3, 6, and 24 h after the fourth injections of MPTP (n = 8–10 in each group). Taking into account that the degradation of the terminal DA axon begins earlier than that of the cell bodies of DA neurons, the striatum was also isolated 2 h after the second MPTP injections (shown as −2 h in the diagram). In addition, for all indicated time intervals, the concentration of DA in the striatum was assessed.

The animals were anesthetized with isoflurane and decapitated. The striatum and SN were excised from the brain according to previously described methods [28,31]. All samples were weighed, frozen in liquid nitrogen, and stored at −70 °C until the analysis. The striatum from one half of the brain was used to determine the concentration of catecholamines by high-performance liquid chromatography with electrochemical detection (HPLC-ED). The striatum from the remaining half of the brain and the SN from both halves of the brain were used to determine the content of TH and its phosphorylated forms by Western blotting (WB).

In the second series of experiments, TH activity was determined. Half an hour before the decapitation (2 h after 2 injections of MPTP and 3, 6, and 24 h after 4 injections of MPTP) mice of both experimental and control groups (n = 7–10 in each group) were i.p. administered with an AADC inhibitor 3-hydroxybenzylhydrazine (NSD-1015, Sigma Aldrich, USA) at a dose of 100 mg/kg of body weight [32]. After anesthesia and decapitation, the striatum and SN were excised from the brain, weighed, frozen in liquid nitrogen, and stored at −70 °C until analysis of the level of L-3,4-dihydroxyphenylalanine (L-DOPA) by HPLC-ED.

2.4. Sample Analysis

2.4.1. High-Performance Liquid Chromatography with Electrochemical Detection

In brain tissue samples, the concentration of DA and its metabolites (3,4-dihydroxyphenylacetic acid (DOPAC), 3-methoxytyramine (3MT), homovanillic acid (HVA)) or L-DOPA after DAA inhibition with NSD-1015 was determined by HPLC-ED. The samples were homogenized with a Labsonic M ultrasonic homogenizer (Sartorius, France) in 0.1 *n* $HClO_4$ (Sigma Aldrich, USA) with 250 pmol/mL internal standard 3,4-dihydroxybenzylamine hydrobromide (Sigma Aldrich, USA). After that, the solution was centrifuged for 20 min at 2000× *g*.

HPLC separation was carried out on a reversed-phase column ReproSil-Pur, ODS-3, 4 × 100 mm with a pore diameter of 3 μm (Dr. Majsch, Germany) at +30 °C and a mobile phase speed of 1 mL/min supported by a liquid chromatograph LC-20ADsp (Shimadzu, Japan). The mobile phase consisted of 0.1 M citrate-phosphate buffer, 0.3 mM sodium octanesulfonate, 0.1 mM EDTA, and 8% acetonitrile (all reagents from Sigma Aldrich, USA), pH 2.5. An electrochemical detector Decade II (Antec Leyden, The Netherlands) was equipped with a working glassy carbon electrode (+0.85 V) and an Ag/AgCl reference electrode. Peaks of interest and internal standard were identified by their release time in the standard solution. The monoamine concentrations were calculated by the internal standard method using a calibration curve with LabSolutions software (Shimadzu, Japan). The concentration of DA was calculated according to the following equation:

The concentration of DA = (the area of DA peak in the sample solution/the area of DA peak in the standard solution) × (the area of DHBA peak in the standard solution/the area of DHBA peak in the sample solution (250 pmol/mL for striatum or 50 pmol/mL for SN) × concentration standart solution (250 pmol/mL for striatum or 50 pmol/mL for SN) × V,

where V is the volume of sample solution: SN—0.120 mL, striatum—0.400 mL. Striatal samples were normalized to tissue weight. The concentration of DA metabolites was calculated in a similar way. Retention time of DA and its metabolites was: DHBA—3.0 min, DOPAC—3.8 min, DA—4.3 min, HVA—8.7 min, 3MT—9.7 min.

The DA turnover was defined as the ratio of DA metabolites (DOPAC, 3MT, or HVA) to DA.

2.4.2. Western Blot

Striatal and SN samples from each experimental and control group were used for the Western blot assay. Tissue samples were homogenized in a RIPA buffer with a ×2 protease inhibitor cocktail (Thermo Fischer, USA) and a ×1 phosphatase inhibitor cocktail (Cell Signaling, USA). To the striatum and SN were added 300 and 140 μL of the buffer, and the samples were centrifuged at 20,000× *g* for 20 min. Then, the protein concentration was determined using a Bicinchoninic Acid Solution (BCA) as a protein assay test [33]. The samples were denatured at 95 °C for 5 min in Laemmli sample buffer consisting of 2% SDS, 10% glycerol, 5%-mercaptoethanol, 62.5 mM Tris (pH 6.8), and 0.004% bromophenol blue. Protein extracts were separated by electrophoresis in a 12% polyacrylamide gel with SDS in a buffer and transfered overnight to a nitrocellulose membrane (Hybond-enhanced chemiluminescence, Amersham Biosciences, USA). Equal loading was reconfirmed by Ponceau-S staining of each Western blot lane on the membrane [34,35]. The Ponceau-S was then removed by washing with a TBS buffer. The blots were blocked in TBS with 0.05% Tween 20 and 5% bovine serum albumin (BSA) for 1 h at RT and incubated with mouse monoclonal antibodies to TH (1:1000) (Sigma Aldrich, USA), or rabbit polyclonal antibodies to P19-TH (1:1000) (Thermo Fisher, USA), or to P40-TH (1:1000) (BioNovus, USA) in TBST with 1% BSA, or to P31-TH (1:1000) (Cell Signaling, USA) in TBST with 5% BSA, overnight at 4 °C. Then, the blots were washed in TBST, incubated with secondary horseradish peroxidase-conjugated anti-mouse or anti-rabbit IgG antibodies (Jackson ImmunoResearch Laboratories, USA) in TBST for 2 h and washed in TBST. The conjugated antibodies were visualized using enhanced chemiluminescence in 0.1 M Tris-HCl with 12.5 mM luminol,

2 mM coumaric acid, and 0.03% H_2O_2 (pH 8.5). The intensity of TH and P19-, P31-, and P40-TH protein bands was measured by densitometry using ImageLab software (Biorad, USA) and then were normalized to the Ponceau signal which is more stable than β-actin in the case of the neurodegeneration [34–37]. The results were expressed as relative optical density in the experimental groups compared with those in the saline injected group taken as 100%. Western blot of TH, P31-TH, P40-TH and P19-TH immunoreactivity and Ponceau staining in the striatum and in the SN are represented in the Figures S1 and S2.

2.5. Statistical Analysis

Data are presented as the group mean ± standard error of mean. The correspondence of the data to the normal distribution was checked using the Shapiro–Wilk test. The results were statistically processed with the GraphPad Prism 6.0 software package (GraphPad Software, USA) using one-way ANOVA and parametric Student's *t*-test. $p \leq 0.05$ was used everywhere as the significance criterion.

3. Results

3.1. The Concentration of Dopamine and Its Metabolites in the Striatum

The average concentration of DA in the striatum in control animals was 102.6 ± 0.8 pmol/mg (Figure 2A). The DA level decreased to 65.0 ± 2.1 pmol/mg 2 h after two MPTP injections. It was 4.8 ± 0.6 and 6.0 ± 0.4 pmol/mg 3 and 6 h after four MPTP injections, respectively, and after 24 h it was 10.5 ± 0.5 pmol/mg.

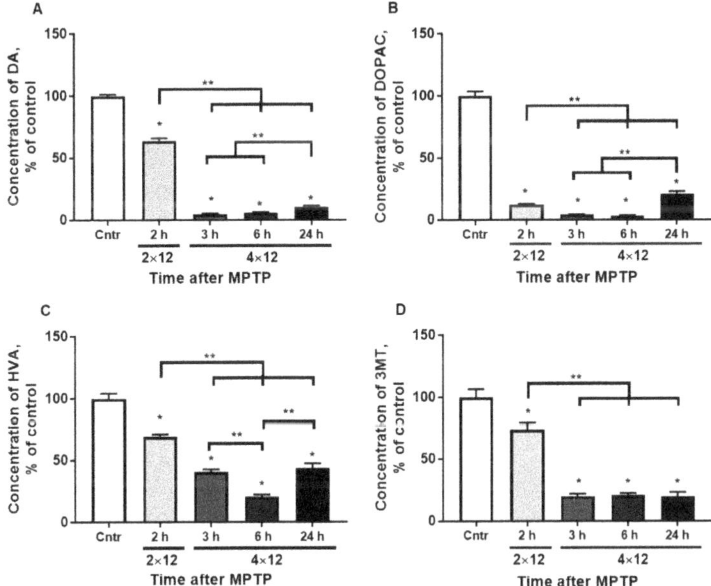

Figure 2. The concentration of DA (**A**), DOPAC (**B**), HVA (**C**), and 3MT (**D**) in the striatum of the control and 2 h after 2 × 12 mg/kg of MPTP and 3, 6, and 24 h after 4 × 12 mg/kg of MPTP. The results are presented as percentages of those of the control (100%). * $p < 0.05$ compared with the control (saline). ** $p < 0.05$ compared with selected MPTP groups.

A significant difference in the concentration of DA was observed between the experimental group with 2 h after the two MPTP injections and the groups with 3 and 6 h after the four MPTP injections.

The average concentration of DOPAC in the striatum in the control groups was 6.5 ± 0.2 pmol/mg (Figure 2B). The DOPAC concentration decreased to 0.7 ± 0.03 pmol/mg 2 h after two injections of MPTP. It was 0.3 ± 0.02 and 0.2 ± 0.01 pmol/mg 3 and 6 h after four MPTP injections, respectively, and after 24 h it was 1.3 ± 0.04 pmol/mg. A significant difference was observed between the experimental group with 2 h after two MPTP injections, and the groups with 3, 6, and 24 h after four MPTP injections as well as between the groups with 3, 6, and 24 h after the injections.

The average concentration of HVA in the striatum in the control groups was 7.7 ± 0.3 pmol/mg (Figure 2C). The concentration of HVA decreased to 4.0 ± 0.2 pmol/mg 2 h after two MPTP injections. It was 4.1 ± 0.6 pmol/mg 3 h after four MPTP injections, 1.8 ± 0.2 pmol/mg after 6 h, and 3.7 ± 0.1 pmol/ mg after 24 h. Significant differences were observed between the experimental group with 2 h after the two MPTP injections and the groups with 3, 6, and 24 h after four MPTP injections as well as between the groups with 3 and 6 h and with 6 and 24 h after injections.

The average concentration of 3MT in the striatum in the control groups was 2.1 ± 0.1 pmol/mg (Figure 2D). The 3MT level decreased to 1.4 ± 0.1 pmol/mg 2 h after two MPTP injections. The concentration of 3MT was 1.1 ± 0.6 and 0.5 ± 0.05 pmol/mg 3 and 6 h after four MPTP injections, respectively, and after 24 h it was 0.4 ± 0.02 pmol/mg. Significant differences were observed between the experimental group with 2 h after the two MPTP injections, and the groups with 3, 6, and 24 h after four MPTP injections.

3.2. Dopamine Turnover in the Striatum

The DOPAC/DA ratio decreased to 19% and 51% 2 h after two MPTP injections and 6 h after four MPTP injections, respectively. The DOPAC/DA ratio was at the control level after 3 h, and after 24 h it increased to 184% (Figure 3A). Significant differences between all experimental groups were observed.

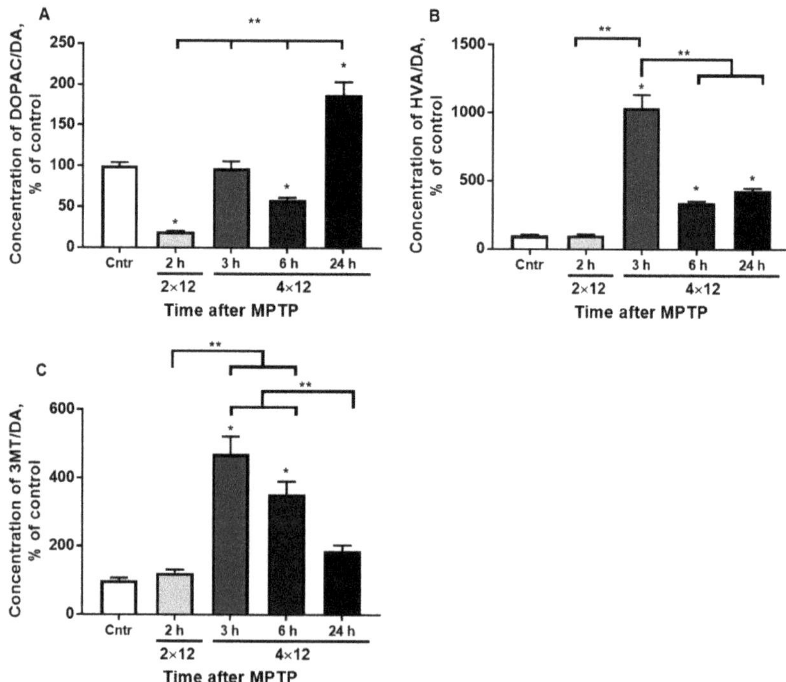

Figure 3. The DA turnover calculated by DOPAC/DA (**A**), HVA/DA (**B**), and 3MT/DA (**C**) in the striatum of the control and 2 h after 2 × 12 mg/kg of MPTP and 3, 6, and 24 h after 4 × 12 mg/kg of MPTP. The results are presented as percentages of those in the control (100%). * $p < 0.05$ compared with the control (saline). ** $p < 0.05$ compared with selected MPTP groups.

The HVA/DA ratio 2 h after two MPTP injections remained at the control level. The HVA/DA ratio was 1010% 3 h after four injections of MPTP, and it was about 400% after 6 and 24 h (Figure 3B). Significant differences were found between the experimental group collected 3 h after four MPTP injections and all the other experimental groups.

The 3MT/DA ratio remained at the control level 2 h after two MPTP injections. The 3MT/DA ratio was about 400% on average 3 and 6 h after four MPTP injections, and 184% 24 h after MPTP (Figure 3C). Significant differences were found between the experimental groups with 3 and 6 h after four MPTP injections and the groups with 2 h after two MPTP injections and 24 h after four MPTP injections.

3.3. The Concentration of L-DOPA in the Striatum and SN after Inhibition of AADC with NSD-1015

The average concentration of L-DOPA in the striatum in the control group 30 min after the administration of NSD-1015 at a dose of 100 mg/kg was 17.1 ± 1.4 pmol/mg (Figure 4A). The concentration of L-DOPA was 7.1 ± 0.3, 5.4 ± 0.8, and 7.2 ± 2.2% of that in the control 2 h after two MPTP injections, 3 and 6 h after four MPTP injections, respectively, and after 24 h it was 32.8 ± 3.4%. Significant differences were found between the experimental group with 24 h after the MPTP injections and all other experimental groups.

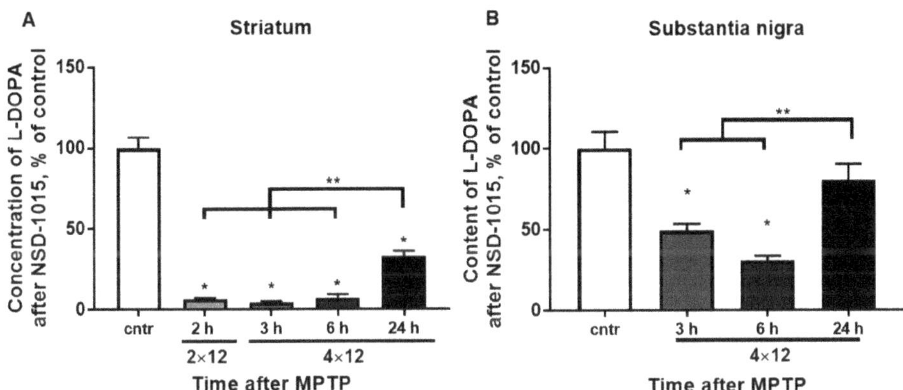

Figure 4. (**A**) The concentration of L-DOPA in the striatum upon inhibition of AADC by NSD-1015 (100 mg/kg 30 min before decapitation) of the control and 2 h after 2 × 12 mg/kg of MPTP and 3, 6, and 24 h after 4 × 12 mg/kg of MPTP. (**B**) The content of L-DOPA in the SN upon inhibition of AADC by NSD-1015 (100 mg/kg 30 min before decapitation) of the control and 3, 6, 24 h after 4 × 12 mg/kg of MPTP. The results are presented as percentages of those in the control (100%). * $p < 0.05$ compared with the control (saline). ** $p < 0.05$ compared with selected MPTP groups.

The average concentration of L-DOPA in the SN in the control group 30 min after the administration of NSD-1015 at a dose of 100 mg/kg was 11.5 ± 0.6 pmol (Figure 4B). The L-DOPA concentration was 49.3 ± 4.2 and 30.8 ± 2.8% of that in the control 3 and 6 h after four MPTP injections, respectively, and after 24 h, the L-DOPA concentration was at the control level. Significant differences were found between the experimental groups with 3 and 6 h and the group with 24 h after four MPTP injections.

3.4. The Content of TH and Its Phosphorylated Forms in the Striatum

The total TH content in the striatum (Figure 5A,B) remained at the control level during the first 6 h after four MPTP injections and decreased by 43% after 24 h. Significant differences were found between the experimental group with 24 h after four MPTP injections and the groups with 2 h after two MPTP injections and with 3 and 6 h after four MPTP injections.

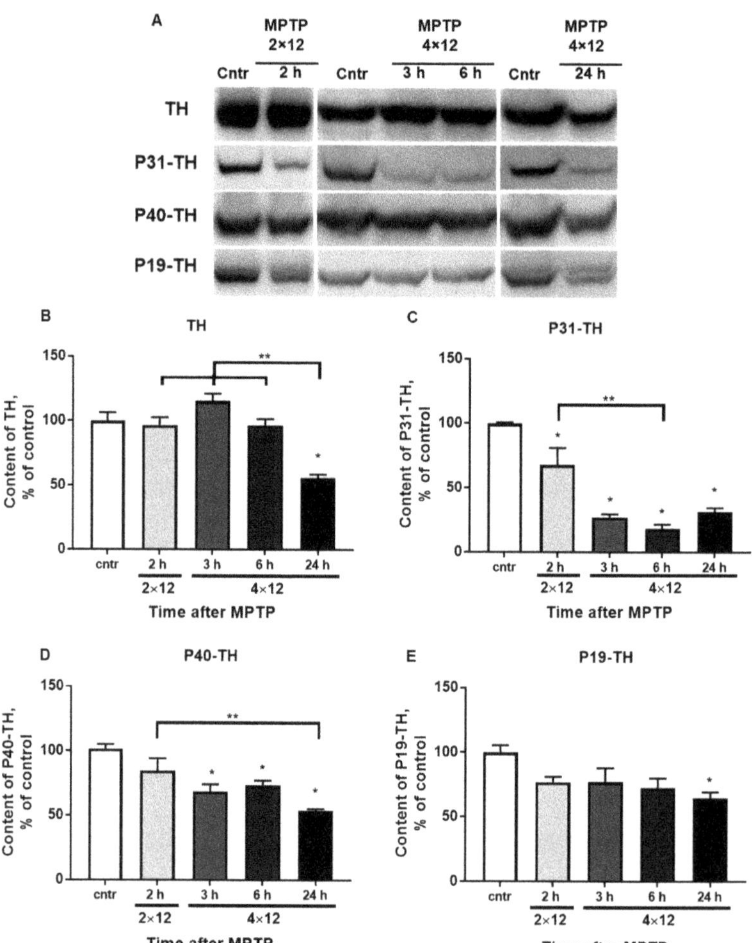

Figure 5. (**A**) Western blot representation of TH, P31-TH, P40-TH, and P19-TH immunoreactivity in the striatum of the control and 2 h after 2 × 12 mg/kg of MPTP and 3, 6, and 24 h after 4 × 12 mg/kg of MPTP. (**B–E**) Bar graph representation of TH (**B**), P31-TH (**C**), P40-TH (**D**), and P19-TH (**E**) in the striatum of the control (saline) and 2 h after 2 × 12 mg/kg of MPTP and 3, 6, and 24 h after 4 × 12 mg/kg of MPTP. The results are presented as percentages of those in the control (100%). * $p < 0.05$ compared with the control (saline). ** $p < 0.05$ compared with selected MPTP groups.

P31-TH content (Figure 5A,C) decreased by 50% compared with that of the control 2 h after two MPTP injections. The content of this phosphorylated form of TH was 25, 14 and 30% after 3, 6, and 24 h, respectively. Significant differences were found between the experimental groups collected 2 h after two MPTP injections and 6 h after four MPTP injections.

P40-TH content (Figure 5A,D) did not change compared with that of the control 2 h after two MPTP injections. It decreased by 30% 3 and 6 h after four MPTP injections, respectively, and by 45% after 24 h. Significant differences were found between the experimental group with 2 h after two MPTP injections and the group with 24 h after four MPTP injections.

P19-TH content (Figure 5A,E) did not change up to 6 h after four MPTP injections, and after 24 h it decreased to 65%.

3.5. The Content of TH and Its Phosphorylated Forms in the SN

The content of total TH in the SN (Figure 6A,B) remained at the control level during the first 6 h after four MPTP injections. It decreased by 13% 24 h after four MPTP injections. Significant differences were found between the experimental groups with 3 and 24 h after four MPTP injections.

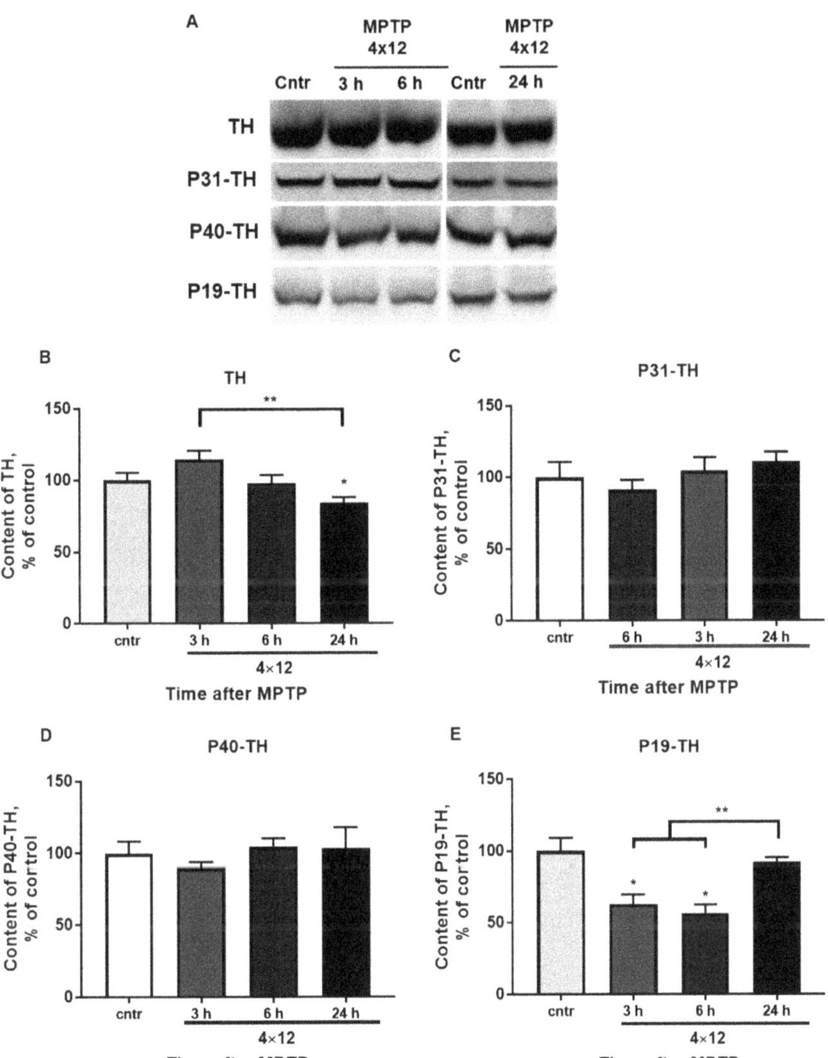

Figure 6. (**A**) Western blot representation of TH, P31-TH, P40-TH, and P19-TH immunoreactivity in the SN of the control and 3, 6, and 24 h after 4 × 12 mg/kg of MPTP. (**B–E**) Bar graph representation of TH (**B**), P31-TH (**C**), P40-TH (**D**), and P19-TH (**E**) in the SN of the control (saline) and 3, 6, and 24 h after 4 × 12 mg/kg of MPTP. The results are presented as percentages of those in the control (100%). * $p < 0.05$ compared with the control (saline). ** $p < 0.05$ compared with selected MPTP groups.

The content of P31-TH and P40-TH (Figure 6A,C,D) did not change in the studied intervals after MPTP administration.

The content of P19-TH (Figure 6E) decreased by 43 and 38% 3 and 6 h after four MPTP injections, respectively. The content of P19-TH was at the control level after 24 h. Significant differences were found between the experimental group with 24 h after four MPTP injections and the groups with 3 and 6 h after four MPTP injections.

4. Discussion

The prevailing number of studies use acute modeling of PD in mice, i.e., administration of MPTP within one day using one to four injections at doses ranging from 10 to 50 mg/kg [38]. Previously, an acute model of PD was developed in our laboratory with four-fold administration of 12 mg/kg of MPTP with a 2 h interval. For this model, threshold degradation of the nigrostriatal system with impaired motor behavior in mice was shown two weeks after neurotoxin administration [30]. This model, as well as similar ones, are commonly used to study neuroplasticity processes that develop in the presence of an emerging DA deficiency and loss of DA neurons [31,39,40].

We evaluated the period of degradation of the nigrostriatal system using the developed model [30] in order to study neurodegenerative processes. We demonstrated that the terminals of DA axons react earlier to the injection of a neurotoxin, and the period of their degradation lasts longer than the death of neuronal bodies in the SN. The degree of functional inhibition (decrease in DA concentration) in the striatum prevailed over the degree of functional inhibition in the SN. At the same time, in contrast to those in the striatum, surviving DA neurons in the SN demonstrated partial recovery of their functional state (Figure 7).

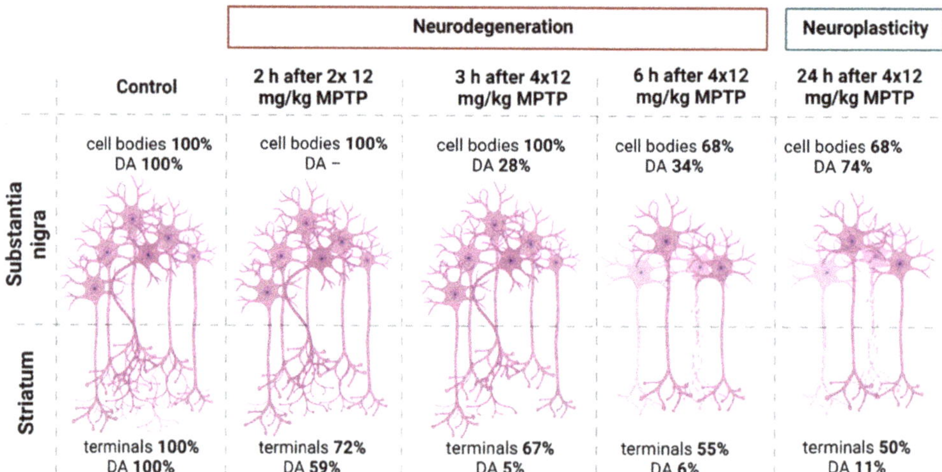

Figure 7. The quantity of DA neuron cell bodies and content of DA in the SN, the quantity of terminals of DA axons and DA concentration in the striatum in the control (saline) and during 24 h after 4 × 12 mg/kg of MPTP (the data adapted with permission from Ref. [29]). The figure was created using Biorender (www.biorender.com, the date of last access to the link is 14 June 2022).

In this study, we evaluated DA synthesis during the period of neurodegeneration in the nigrostriatal system and during the first hours after its completion—the period of neuroplasticity. The key indicators used were the total content and activity of TH, as well as the content of its phosphorylated serine forms at positions 19, 31, and 40 (P19-, P31-, and P40-TH).

4.1. Dopamine Synthesis in the Substantia Nigra

The content of TH in a neuron body is the result of three processes—synthesis, degradation, and transportation along processes (axons, dendrites). According to the data obtained, the content of TH remained at the control level during 6 h after the last (fourth) injection of MPTP. At the same time, after 3 h, the number of neuron bodies did not change, and after 6 h, it decreased by 32%, i.e., in the surviving DA neurons, the content of TH increased compared to that in the control. The number of neuron bodies did not change, while the level of TH decreased by 13%, 24 h after the last injection. The discovered changes are not associated with a decrease in the content of TH in the neuron bodies (by IHC) [30]. This is probably partly due to the reduced activity of the ubiquitin-dependent system of protein proteolysis one day after a similar regimen of MPTP administration [41,42]. Probably, the observed decrease in the TH content reflects the restoration of anterograde protein transport along the fibers [43,44].

The TH activity decreased by 50–60% 3 and 6 h after the fourth injection of MPTP compared to the control level. The content of DA decreased by approximately the same amount [30]. This confirms a direct relationship between enzyme activity and neurotransmitter levels in SN. After MPP+ enters DA neurons via DAT, in addition to inhibition of the Mitochondrial Respiratory Complex I and initiation of oxidative stress, pumping of MPP+ (MPTP metabolite which is a toxin for DA neurons) into synaptic vesicles via VMAT2 begins where MPP+ competitively replaces DA [45–50]. Meanwhile, an increased level of cytosolic DA has not only a toxic effect on the DA neuron [50,51], but can also bind to the TH enzyme, inhibiting the latter [52,53].

Interestingly, in the same period (3 and 6 h after four injections of MPTP), a decrease in the content of TH phosphorylated by serine at position 19 (P19-TH) was shown. Dephosphorylation of P19-TH in the SN and the striatum is performed by protein phosphatase 2A (PP2A) [54]. An increase in the expression of α-synuclein, mitochondrial dysfunction, induction of oxidative stress, and apoptosis leads to an increase in the activity and/or content of PP2A [54,55]. In chronic modeling of PD by administration of MPTP, mice show an increase in the PP2A content in SN [56], and monkeys (Cynomolgus monkeys) demonstrate a decrease in PP2A activity [57]. However, in the above-mentioned studies, the content and activity of PP2A were evaluated after the completion of neurodegeneration and the onset of the development of neuroplasticity. We assume that under our experimental design, the decrease in P19-TH 3 and 6 h after four injections of MPTP was provoked by an increase in the activity and/or content of PP2A, which occurred because of mitochondrial dysfunction and induction of oxidative stress by MPP+.

During the first day after MPTP administration, correlations between TH activity and P19-TH content were observed. Both criteria were lowered 3 and 6 h after four injections of MPTP, and by 24 h, they were restored to the control level. The relationship between the activity of TH and P19-TH was not obvious. Moreover, according to the available data, the phosphorylated form does not directly affect the activity of TH *in vitro* and *in vivo* [22,27]. However, on the MN9D line (hybrid neuroblastoma-immortalized DA mesencephalic neurons from C57Bl/6 mice embryos), it was shown that P19-TH binds to chaperone proteins of the 14-3-3 family, which can lead to an increase in TH activity [58–60]. In our case, the dissociation of P19-TH and proteins 14-3-3 may occur, or this complex may bind to α-synuclein, which leads to a decrease in TH activity [61]. Unfortunately, no studies have been found evaluating the level of 14-3-3 proteins during degeneration in acute PD modeling in mice. At the same time, data on the absence of changes or even a decrease in the expression of the gene and protein of α-synuclein in SN in the first hours and days after a similar administration of MPTP [41,62–64] do not support the above-mentioned hypothesis.

According to the data on the stoichiometry of phosphorylated TH forms in rats, the content of P40-TH in SN is the lowest compared to other forms [65,66], and the absence of changes in its content during the first day after MPTP administration indicates its secondary role in determination of TH activity during the period of neurodegeneration and neuroplasticity [25,27]. We also found no changes in the P31-TH level in the SN on the first day after MPTP administration. However, during the period of neuroplasticity (seven days and later after MPTP or 6-OHDA administration), an increase in the content of this form of TH was shown [25,27].

Thus, it can be concluded that during the degradation of the nigrostriatal system and the period of neuroplasticity, the level of DA in the SN is determined by the activity of the TH enzyme, but not by its content. P19-TH plays a role in the establishment of enzyme activity; however, further studies are required to establish the mechanism of this regulation.

4.2. Dopamine Synthesis in the Striatum

In this study, the concentration of DA was re-determined in the striatum. The change in DA concentration was the same as in previous studies with a similar scheme of MPTP administration [29,30]. However, the use of Shimadzu equipment with high resolution (Shimadzu Corporation, Japan) made it possible to identify differences not only between the DA level 2 h after two MPTP injections and all other conditions, but also between 3, 6, and 24 h after four MPTP injections (Figure 2A).

The concentration of DA in the striatum is the result of three coupled processes: synthesis, degradation, and reuptake from the synaptic cleft into vesicles involving DAT and VMAT2 to reuse DA. Notably, in the normal striatum, DA reuptake prevails over its degradation [7]. An indicator of DA synthesis is the content and activity of TH, which together make it possible to calculate the specific activity of the enzyme as the ratio of TH activity to its content [14,15].

According to the data, the total content of TH in the striatum remained at the control level during the period of degradation of the DA axon terminals (up to 6 h after four MPTP injections), and after 24 h it decreased by 40%. It is likely that the observed decrease in the TH level was associated with the retrograde spread of the neurodegenerative process from the axon terminals further along the axons towards the neuron bodies. Changes in the content of total TH also correlated with the level of P19-TH; however, the presence of a causal relationship between these indicators was not obvious.

Changes in TH activity in the striatum, as well as in the SN, did not correlate with the total TH content. At the same time, TH activity decreased by 10 times 2 h after two injections of MPTP and remained at this level up to 6 h after four injections. After 24 h, the TH content and its homospecific activity partially restored (up to 30% and up to 50%, respectively) [14]. Such a significant decrease in enzyme activity during the first 6 h probably occurs by the same mechanism as in the SN. This refers to the inhibition of TH by DA, the level of which increases in the cytosol after the entry of MPP+ into the synaptic terminals of DA axons. This is confirmed by a gradual decrease in the concentration of DA in the striatum, which "lags behind" the decrease in TH activity, as well as an increase in DA turnover after 3 h in such criteria as 3MT/DA and HVA/DA.

A more significant decrease in TH activity in the striatum than in the SN is associated with a difference in the level of DA in these two structures. The content of DA in the striatum is 650 times higher than in the SN (unpublished data), while the difference in TH content is not so high (three-to-five times higher in the striatum) [65,66].

A partial recovery of DA concentration and TH activity 24 h after four injections of MPTP is associated with the onset of the development of reparative processes, primarily the uptake of the neurotransmitter into synaptic vesicles that have already been released from MPP+. According to Fornai, 12 h after MPTP administration (4×20 mg/kg), MPP+ content in the striatum is 10% of the peak value, which occurs 1–2 h after MPTP administration [67]. This hypothesis is also confirmed by the gradual return of DA turnover in the striatum to the control value (HVA/DA and 3MT/DA).

Another mechanism that negatively affects TH activity is dephosphorylation. According to our data, 2 h after two MPTP injections, the content of P31-TH in the striatum decreased significantly. There is a subsequent decrease in contents of both this phosphorylated form of TH and P40-TH 3, 6, and 24 h after four MPTP injections. The results obtained are also consistent with data obtained 2 h after 15 mg/kg of MPTP and 9–16 days after 6-OHDA administration [26,27]. Considering that the amount of P31-TH is normally higher than that of P40-TH, and that the content of P31-TH after the induction of neurodegeneration decreases to a greater extent than that of P40-TH does, it can be concluded that the content of P31-TH to a greater extent determines the activity of TH in the striatum during the degradation of the nigrostriatal system. In addition to TH dephosphorylation during this critical period, an increase in enzyme nitration was shown with a similar scheme of MPTP administration (4×20 mg/kg), which also has an inhibitory effect on its activity [68].

Thus, a decrease in TH activity during the first 6 h after the last injection of MPTP is associated with its dephosphorylation at serine at position 31 (P31-TH) and partially at position 40 (P40-TH), as well as with cytosolic DA inhibition.

A partial recovery of TH activity and DA concentration was shown 24 h after four MPTP injections. Even though there were no significant differences between the content of P31-TH 6 and 24 h after four MPTP injections, the average values differed by a factor of two. Perhaps, if we observed a longer period after the administration of MPTP, when neuroplasticity continued to develop, significant differences would be shown. This would confirm the importance of the role of P31-TH for the activity of this enzyme not only during the period of degradation of the nigrostriatal system but also during neuroplasticity.

Thus, it was shown that TH activity in the striatum during the period of degradation of the nigrostriatal system and neuroplasticity was not determined by the total content of the enzyme, but largely depended on the level of P31-TH and, to a lesser extent, on P40-TH, as well as on the level of cytosolic DA. The role of P19-TH in determination of the total TH level in the retrograde degradation of DA fibers should be defined in future studies.

5. Conclusions

An analysis of DA synthesis in the nigrostriatal system during its degradation and in the first hours after its completion, i.e., at the beginning of neuroplasticity, in a model of the early clinical stage of PD, showed that:

1. DA content in the SN and the striatum did not depend on TH content but correlated with enzyme activity.
2. TH activity did not depend on the total protein content either in the SN or in the striatum.
3. TH activity in the SN was determined by the content of P19-TH; TH activity in the striatum was determined by P31-TH and P40-TH (to a lesser extent).

The data obtained indicated different regulation of DA synthesis in DA neuron bodies and their axon terminals. These data should be taken into account for the further development of symptomatic pharmacotherapy aimed at increasing TH activity.

Supplementary Materials: The following supporting information can be downloaded at: https://www.mdpi.com/article/10.3390/brainsci12060779/s1, Figure S1: Western blot representation of TH, P31-TH, P40-TH and P19-TH immunoreactivity and Ponceau staining in the striatum of the control, 2 h after 2×12 mg/kg of MPTP and 3, 6, 24 h after 4×12 mg/kg of MPTP; Figure S2: Western blot representation of TH, P31-TH, P40-TH and P19-TH immunoreactivity and Ponceau staining in the SN of the control and 3, 6, 24 h after 4×12 mg/kg of MPTP.

Author Contributions: A.K.: Methodology, investigation, formal analysis, writing—original draft, writing—review and editing. L.A.: investigation, validation. E.P.: investigation. A.B.: investigation. M.V.U.: editing. All authors have read and agreed to the published version of the manuscript.

Funding: This study was funded by RSF (project No. 20-75-00110).

Institutional Review Board Statement: Experimental procedures were carried out in accordance with the NIH Guide for the Care and Use of Laboratory Animals and were approved by the Animal Care and Use Committee of the Koltzov Institute of Developmental Biology RAS (No. 44 from 24 December 2020 and No. 55 from 9 December 2021).

Data Availability Statement: The data presented in this study are available on request from the corresponding author. The data are not publicly available due to legal issues.

Conflicts of Interest: The authors declare no conflict of interests.

References

1. Kaufman, S. The structure of the phenylalanine-hydroxylation cofactor. *Proc. Natl. Acad. Sci. USA* **1963**, *50*, 1085–1093. [CrossRef] [PubMed]
2. Nagatsu, T.; Levitt, M.; Udenfriend, S. Tyrosine hydroxylase. The initial step in norepinephrine biosynthesis. *J. Biol. Chem.* **1964**, *239*, 2910–2917. [CrossRef]
3. Matsuura, S.; Sugimoto, T.; Murata, S.; Sugawara, Y.; Iwasaki, H. Stereochemistry of biopterin cofactor and facile methods for the determination of the stereochemistry of a biologically active 5,6,7,8-tetrahydropterin. *J. Biochem.* **1985**, *98*, 1341–1348. [CrossRef] [PubMed]
4. Kaakkola, S.; Männistö, P.T.; Nissinen, E. Striatal membrane-bound and soluble catechol-O-methyl-transferase after selective neuronal lesions in the rat. *J. Neural Transm.* **1987**, *69*, 221–228. [CrossRef] [PubMed]
5. Nakamura, S.; Akiguchi, I.; Kimura, J. Topographic distributions of monoamine oxidase-B-containing neurons in the mouse striatum. *Neurosci. Lett.* **1995**, *184*, 29–31. [CrossRef]
6. Francis, A.; Pearce, L.B.; Roth, J.A. Cellular localization of MAO A and B in brain: Evidence from kainic acid lesions in striatum. *Brain Res.* **1985**, *334*, 59–64. [CrossRef]
7. Yavich, L.; Forsberg, M.M.; Karayiorgou, M.; Gogos, J.A.; Männistö, P.T. Site-specific role of catechol-O-methyltransferase in dopamine overflow within prefrontal cortex and dorsal striatum. *J. Neurosci.* **2007**, *27*, 10196–10209. [CrossRef]
8. Hirsch, L.; Jette, N.; Frolkis, A.; Steeves, T.; Pringsheim, T. The Incidence of Parkinson's Disease: A Systematic Review and Meta-Analysis. *Neuroepidemiology* **2016**, *46*, 292–300. [CrossRef]
9. Dauer, W.; Przedborski, S. Parkinson's disease: Mechanisms and models. *Neuron* **2003**, *39*, 889–909. [CrossRef]
10. Bernheimer, H.; Birkmayer, W.; Hornykiewicz, O.; Jellinger, K.; Seitelberger, F. Brain dopamine and the syndromes of Parkinson and Huntington. Clinical, morphological and neurochemical correlations. *J. Neurol. Sci.* **1973**, *20*, 415–455. [CrossRef]
11. Agid, Y. Parkinson's disease: Pathophysiology. *Lancet* **1991**, *337*, 1321–1324. [CrossRef]
12. Bezard, E.; Dovero, S.; Prunier, C.; Ravenscroft, P.; Chalon, S.; Guilloteau, D.; Crossman, A.R.; Bioulac, B.; Brotchie, J.M.; Gross, C.E. Relationship between the appearance of symptoms and the level of nigrostriatal degeneration in a progressive 1-methyl-4-phenyl-1,2,3,6-tetrahydropyridine-lesioned macaque model of Parkinson's disease. *J. Neurosci.* **2001**, *21*, 6853–6861. [CrossRef] [PubMed]
13. Kordower, J.H.; Olanow, C.W.; Dodiya, H.B.; Chu, Y.; Beach, T.G.; Adler, C.H.; Halliday, G.M.; Bartus, R.T. Disease duration and the integrity of the nigrostriatal system in Parkinson's disease. *Brain* **2013**, *136 Pt 8*, 2419–2431. [CrossRef] [PubMed]
14. Nagatsu, T.; Yamaguchi, T.; Rahman, M.K.; Trocewicz, J.; Oka, K.; Hirata, Y.; Nagatsu, I.; Narabayashi, H.; Kondo, T.; Iizuka, R. Catecholamine-related enzymes and the biopterin cofactor in Parkinson's disease and related extrapyramidal diseases. *Adv. Neurol.* **1984**, *40*, 467–473.
15. Mogi, M.; Harada, M.; Kiuchi, K.; Kojima, K.; Kondo, T.; Narabayashi, H.; Rausch, D.; Riederer, P.; Jellinger, K.; Nagatsu, T. Homospecific activity (activity per enzyme protein) of tyrosine hydroxylase increases in parkinsonian brain. *J. Neural Transm.* **1988**, *72*, 77–82. [CrossRef]
16. Dunkley, P.R.; Bobrovskaya, L.; Graham, M.E.; von Nagy-Felsobuki, E.I.; Dickson, P.W. Tyrosine hydroxylase phosphorylation: Regulation and consequences. *J. Neurochem.* **2004**, *91*, 1025–1043. [CrossRef]
17. Dunkley, P.R.; Dickson, P.W. Tyrosine hydroxylase phosphorylation in vivo. *J. Neurochem.* **2019**, *149*, 706–728. [CrossRef]
18. Daubner, S.C.; Lauriano, C.; Haycock, J.W.; Fitzpatrick, P.F. Site-directed mutagenesis of serine 40 of rat tyrosine hydroxylase. Effects of dopamine and cAMP-dependent phosphorylation on enzyme activity. *J. Biol. Chem.* **1992**, *267*, 12639–12646. [CrossRef] [PubMed]
19. Haycock, J.W.; Wakade, A.R. Activation and multiple-site phosphorylation of tyrosine hydroxylase in perfused rat adrenal glands. *J. Neurochem.* **1992**, *58*, 57–64. [CrossRef]
20. Haycock, J.W.; Lew, J.Y.; Garcia-Espana, A.; Lee, K.Y.; Harada, K.; Meller, E.; Goldstein, M. Role of serine-19 phosphorylation in regulating tyrosine hydroxylase studied with site- and phosphospecific antibodies and site-directed mutagenesis. *J. Neurochem.* **1998**, *71*, 1670–1675. [CrossRef]
21. Sutherland, C.; Alterio, J.; Campbell, D.G.; le Bourdellès, B.; Mallet, J.; Haavik, J.; Cohen, P. Phosphorylation and activation of human tyrosine hydroxylase in vitro by mitogen-activated protein (MAP) kinase and MAP-kinase-activated kinases 1 and 2. *Eur. J. Biochem.* **1993**, *217*, 715–722. [CrossRef] [PubMed]
22. Bobrovskaya, L.; Dunkley, P.R.; Dickson, P.W. Phosphorylation of Ser19 increases both Ser40 phosphorylation and enzyme activity of tyrosine hydroxylase in intact cells. *J. Neurochem.* **2004**, *90*, 857–864. [CrossRef] [PubMed]
23. Nakashima, A.; Ohnuma, S.; Kodani, Y.; Kaneko, Y.S.; Nagasaki, H.; Nagatsu, T.; Ota, A. Inhibition of deubiquitinating activity of USP14 decreases tyrosine hydroxylase phosphorylated at Ser19 in PC12D cells. *Biochem. Biophys. Res. Commun.* **2016**, *472*, 598–602. [CrossRef] [PubMed]
24. Naskar, A.; Prabhakar, V.; Singh, R.; Dutta, D.; Mohanakumar, K.P. Melatonin enhances L-DOPA therapeutic effects, helps to reduce its dose, and protects dopaminergic neurons in 1-methyl-4-phenyl-1,2,3,6-tetrahydropyridine-induced Parkinsonism in mice. *J. Pineal Res.* **2015**, *58*, 262–274. [CrossRef] [PubMed]
25. Liu, M.; Hunter, R.; Nguyen, X.V.; Kim, H.C.; Bing, G. Microsomal epoxide hydrolase deletion enhances tyrosine hydroxylase phosphorylation in mice after MPTP treatment. *J. Neurosci. Res.* **2008**, *86*, 2792–2801. [CrossRef]

26. Boukhzar, L.; Hamieh, A.; Cartier, D.; Tanguy, Y.; Alsharif, I.; Castex, M.; Arabo, A.; el Hajji, S.; Bonnet, J.J.; Errami, M.; et al. Selenoprotein T Exerts an Essential Oxidoreductase Activity That Protects Dopaminergic Neurons in Mouse Models of Parkinson's Disease. *Antioxid. Redox Signal.* **2016**, *24*, 557–574. [CrossRef]
27. Salvatore, M.F. ser31 Tyrosine hydroxylase phosphorylation parallels differences in dopamine recovery in nigrostriatal pathway following 6-OHDA lesion. *J. Neurochem.* **2014**, *129*, 548–558. [CrossRef]
28. Ugrumov, M.V.; Khaindrava, V.G.; Kozina, E.A.; Kucheryanu, V.G.; Bocharov, E.V.; Kryzhanovsky, G.N.; Kudrin, V.S.; Narkevich, V.B.; Klodt, P.M.; Rayevsky, K.S.; et al. Modeling of presymptomatic and symptomatic stages of parkinsonism in mice. *Neuroscience* **2011**, *181*, 175–188. [CrossRef]
29. Kolacheva, A.A.; Kozina, E.A.; Volina, E.V.; Ugryumov, M.V. Time course of degeneration of dopaminergic neurons and respective compensatory processes in the nigrostriatal system in mice. *Dokl. Biol. Sci.* **2014**, *456*, 160–164. [CrossRef]
30. Kolacheva, A.A.; Ugrumov, M.V. A Mouse Model of Nigrostriatal Dopaminergic Axonal Degeneration as a Tool for Testing Neuroprotectors. *Acta Naturae* **2021**, *13*, 110–113. [CrossRef]
31. Kozina, E.A.; Khakimova, G.R.; Khaindrava, V.G.; Kucheryanu, V.G.; Vorobyeva, N.E.; Krasnov, A.N.; Georgieva, S.G.; Goff, L.K.; Ugrumov, M.V. Tyrosine hydroxylase expression and activity in nigrostriatal dopaminergic neurons of MPTP-treated mice at the presymptomatic and symptomatic stages of parkinsonism. *J. Neurol. Sci.* **2014**, *340*, 198–207. [CrossRef] [PubMed]
32. Carlsson, A.; Lindqvist, M. In-vivo measurements of tryptophan and tyrosine hydroxylase activities in mouse brain. *J. Neural Transm.* **1973**, *34*, 79–91. [CrossRef] [PubMed]
33. Smith, P.K.; Krohn, R.I.; Hermanson, G.T.; Mallia, A.K.; Gartner, F.H.; Provenzano, M.D.; Fujimoto, E.K.; Goeke, N.M.; Olson, B.J.; Klenk, D.C. Measurement of protein using bicinchoninic acid. *Anal. Biochem.* **1985**, *150*, 76–85. [CrossRef]
34. Aldridge, G.M.; Podrebarac, D.M.; Greenough, W.T.; Weiler, I.J. The use of total protein stains as loading controls: An alternative to high-abundance single-protein controls in semi-quantitative immunoblotting. *J. Neurosci. Methods* **2008**, *172*, 250–254. [CrossRef]
35. Romero-Calvo, I.; Ocón, B.; Martínez-Moya, P.; Suárez, M.D.; Zarzuelo, A.; Martínez-Augustin, O.; de Medina, F.S. Reversible Ponceau staining as a loading control alternative to actin in Western blots. *Anal. Biochem.* **2010**, *401*, 318–320. [CrossRef]
36. Eaton, S.L.; Roche, S.L.; Hurtado, M.L.; Oldknow, K.J.; Farquharson, C.; Gillingwater, T.H.; Wishar, T.M. Total protein analysis as a reliable loading control for quantitative fluorescent Western blotting. *PLoS ONE* **2013**, *8*, e72457. [CrossRef]
37. de Iuliis, A.; Grigoletto, J.; Recchia, A.; Giusti, P.; Arslan, P. A proteomic approach in the study of an animal model of Parkinson's disease. *Clin. Chim. Acta* **2005**, *357*, 202–209. [CrossRef]
38. Jackson-Lewis, V.; Przedborski, S. Protocol for the MPTP mouse model of Parkinson's disease. *Nat. Protoc.* **2007**, *2*, 141–151. [CrossRef]
39. Kozina, E.A.; Kim, A.R.; Kurina, A.Y.; Ugrumov, M.V. Cooperative synthesis of dopamine by non-dopaminergic neurons as a compensatory mechanism in the striatum of mice with MPTP-induced Parkinsonism. *Neurobiol. Dis.* **2017**, *98*, 108–121. [CrossRef]
40. Mingazov, E.R.; Khakimova, G.R.; Kozina, E.A.; Medvedev, A.E.; Buneeva, O.A.; Bazyan, A.S.; Ugrumov, M.V. MPTP Mouse Model of Preclinical and Clinical Parkinson's Disease as an Instrument for Translational Medicine. *Mol. Neurobiol.* **2018**, *55*, 2991–3006. [CrossRef]
41. Kühn, K.; Wellen, J.; Link, N.; Maskri, L.; Lübbert, H.; Stichel, C.C. The mouse MPTP model: Gene expression changes in dopaminergic neurons. *Eur. J. Neurosci.* **2003**, *17*, 1–12. [CrossRef] [PubMed]
42. Fornai, F.; Schlüter, O.M.; Lenzi, P.; Gesi, M.; Ruffoli, R.; Ferrucci, M.; Lazzeri, G.; Busceti, C.L.; Pontarelli, F.; Battaglia, G.; et al. Parkinson-like syndrome induced by continuous MPTP infusion: Convergent roles of the ubiquitin-proteasome system and alpha-synuclein. *Proc. Natl. Acad. Sci. USA* **2005**, *102*, 3413–3418. [CrossRef] [PubMed]
43. Morfini, G.; Pigino, G.; Opalach, K.; Serulle, Y.; Moreira, J.E.; Sugimori, M.; Llinás, R.R.; Brady, S.T. 1-Methyl-4-phenylpyridinium affects fast axonal transport by activation of caspase and protein kinase C. *Proc. Natl. Acad. Sci. USA* **2007**, *104*, 2442–2447. [CrossRef] [PubMed]
44. Cartelli, D.; Casagrande, F.; Busceti, C.L.; Bucci, D.; Molinaro, G.; Traficante, A.; Passarella, D.; Giavini, E.; Pezzoli, G.; Battaglia, G.; et al. Microtubule alterations occur early in experimental parkinsonism and the microtubule stabilizer epothilone D is neuroprotective. *Sci. Rep.* **2013**, *3*, 1837. [CrossRef] [PubMed]
45. Kitayama, S.; Wang, J.B.; Uhl, G.R. Dopamine transporter mutants selectively enhance MPP+ transport. *Synapse* **1993**, *15*, 58–62. [CrossRef] [PubMed]
46. Gainetdinov, R.R.; Fumagalli, F.; Jones, S.R.; Caron, M.G. Dopamine transporter is required for in vivo MPTP neurotoxicity: Evidence from mice lacking the transporter. *J. Neurochem.* **1997**, *69*, 1322–1325. [CrossRef]
47. del Zompo, M.; Piccardi, M.P.; Ruiu, S.; Quartu, M.; Gessa, G.L.; Vaccari, A. Selective MPP+ uptake into synaptic dopamine vesicles: Possible involvement in MPTP neurotoxicity. *Br. J. Pharmacol.* **1993**, *109*, 411–414. [CrossRef]
48. Takahashi, N.; Miner, L.L.; Sora, I.; Ujike, H.; Revay, R.S.; Kostic, V.; Jackson-Lewis, V.; Przedborski, S.; Uhl, G.R. VMAT2 knockout mice: Heterozygotes display reduced amphetamine-conditioned reward, enhanced amphetamine locomotion, and enhanced MPTP toxicity. *Proc. Natl. Acad. Sci. USA* **1997**, *94*, 9938–9943. [CrossRef]
49. Staal, R.G.; Sonsalla, P.K. Inhibition of brain vesicular monoamine transporter (VMAT2) enhances 1-methyl-4-phenylpyridinium neurotoxicity in vivo in rat striata. *J. Pharmacol. Exp. Ther.* **2000**, *293*, 336–342.
50. Goldstein, D.S.; Sullivan, P.; Holmes, C.; Miller, G.W.; Alter, S.; Strong, R.; Mash, D.C.; Kopin, I.J.; Sharabi, Y. Determinants of buildup of the toxic dopamine metabolite DOPAL in Parkinson's disease. *J. Neurochem.* **2013**, *126*, 591–603. [CrossRef]

51. Choi, S.J.; Panhelainen, A.; Schmitz, Y.; Larsen, K.E.; Kanter, E.; Wu, M.; Sulzer, D.; Mosharov, E.V. Changes in neuronal dopamine homeostasis following 1-methyl-4-phenylpyridinium (MPP+) exposure. *J. Biol. Chem.* **2015**, *290*, 6799–6809. [CrossRef] [PubMed]
52. Gordon, S.L.; Quinsey, N.S.; Dunkley, P.R.; Dickson, P.W. Tyrosine hydroxylase activity is regulated by two distinct dopamine-binding sites. *J. Neurochem.* **2008**, *106*, 1614–1623. [CrossRef] [PubMed]
53. Briggs, G.D.; Gordon, S.L.; Dickson, P.W. Mutational analysis of catecholamine binding in tyrosine hydroxylase. *Biochemistry* **2011**, *50*, 1545–1555. [CrossRef] [PubMed]
54. Lou, H.; Montoya, S.E.; Alerte, T.N.; Wang, J.; Wu, J.; Peng, X.; Hong, C.S.; Friedrich, E.E.; Mader, S.A.; Pedersen, C.J.; et al. Serine 129 phosphorylation reduces the ability of alpha-synuclein to regulate tyrosine hydroxylase and protein phosphatase 2A in vitro and in vivo. *J. Biol. Chem.* **2010**, *285*, 17648–17661. [CrossRef] [PubMed]
55. Elgenaidi, I.S.; Spiers, J.P. Regulation of the phosphoprotein phosphatase 2A system and its modulation during oxidative stress: A potential therapeutic target? *Pharmacol. Ther.* **2019**, *198*, 68–89. [CrossRef]
56. Su, J.; Zhang, J.; Bao, R.; Xia, C.; Zhang, Y.; Zhu, Z.; Lv, Q.; Qi, Y.; Xue, J. Mitochondrial dysfunction and apoptosis are attenuated through activation of AMPK/GSK-3β/PP2A pathway in Parkinson's disease. *Eur. J. Pharmacol.* **2021**, *907*, 174202. [CrossRef]
57. Li, X.; Yang, W.; Li, X.; Chen, M.; Liu, C.; Li, J.; Yu, S. Alpha-synuclein oligomerization and dopaminergic degeneration occur synchronously in the brain and colon of MPTP-intoxicated parkinsonian monkeys. *Neurosci. Lett.* **2020**, *716*, 134640. [CrossRef]
58. Choi, H.K.; Won, L.A.; Kontur, P.J.; Hammond, D.N.; Fox, A.P.; Wainer, B.H.; Hoffmann, P.C.; Heller, A. Immortalization of embryonic mesencephalic dopaminergic neurons by somatic cell fusion. *Brain Res.* **1991**, *552*, 67–76. [CrossRef]
59. Choi, H.K.; Won, L.; Roback, J.D.; Wainer, B.H.; Heller, A. Specific modulation of dopamine expression in neuronal hybrid cells by primary cells from different brain regions. *Proc. Natl. Acad. Sci. USA* **1992**, *89*, 8943–8947. [CrossRef]
60. Wang, J.; Lou, H.; Pedersen, C.J.; Smith, A.D.; Perez, R.G. 14-3-3zeta contributes to tyrosine hydroxylase activity in MN9D cells: Localization of dopamine regulatory proteins to mitochondria. *J. Biol. Chem.* **2009**, *284*, 14011–14019. [CrossRef]
61. Perez, R.G.; Waymire, J.C.; Lin, E.; Liu, J.J.; Guo, F.; Zigmond, M.J. A role for alpha-synuclein in the regulation of dopamine biosynthesis. *J. Neurosci.* **2002**, *22*, 3090–3099. [CrossRef] [PubMed]
62. Rudenok, M.M.; Alieva, A.K.; Starovatykh, J.S.; Nesterov, M.S.; Stanishevskaya, V.A.; Kolacheva, A.A.; Ugryumov, M.V.; Slominsky, P.A.; Shadrina, M.I. Expression analysis of genes involved in mitochondrial biogenesis in mice with MPTP-induced model of Parkinson's disease. *Mol. Genet. Metab. Rep.* **2020**, *23*, 100584. [CrossRef] [PubMed]
63. Vila, M.; Vukosavic, S.; Jackson-Lewis, V.; Neystat, M.; Jakowec, M.; Przedborski, S. Alpha-synuclein up-regulation in substantia nigra dopaminergic neurons following administration of the parkinsonian toxin MPTP. *J. Neurochem.* **2000**, *74*, 721–729. [CrossRef] [PubMed]
64. Xu, Z.; Cawthon, D.; McCastlain, K.A.; Slikker, W., Jr.; Ali, S.F. Selective alterations of gene expression in mice induced by MPTP. *Synapse* **2005**, *55*, 45–51. [CrossRef] [PubMed]
65. Salvatore, M.F.; Garcia-Espana, A.; Goldstein, M.; Deutch, A.Y.; Haycock, J.W. Stoichiometry of tyrosine hydroxylase phosphorylation in the nigrostriatal and mesolimbic systems in vivo: Effects of acute haloperidol and related compounds. *J. Neurochem.* **2000**, *75*, 225–232. [CrossRef] [PubMed]
66. Salvatore, M.F.; Pruett, B.S. Dichotomy of tyrosine hydroxylase and dopamine regulation between somatodendritic and terminal field areas of nigrostriatal and mesoaccumbens pathways. *PLoS ONE* **2012**, *7*, e29867. [CrossRef]
67. Fornai, F.; Alessandrì, M.G.; Torracca, M.T.; Bassi, L.; Corsini, G.U. Effects of noradrenergic lesions on MPTP/MPP+ kinetics and MPTP-induced nigrostriatal dopamine depletions. *J. Pharmacol. Exp. Ther.* **1997**, *283*, 100–107.
68. Ara, J.; Przedborski, S.; Naini, A.B.; Jackson-Lewis, V.; Trifiletti, R.R.; Horwitz, J.; Ischiropoulos, H. Inactivation of tyrosine hydroxylase by nitration following exposure to peroxynitrite and 1-methyl-4-phenyl-1,2,3,6-tetrahydropyridine (MPTP). *Proc. Natl. Acad. Sci. USA* **1998**, *95*, 7659–7663. [CrossRef]

brain sciences

MDPI

Article

Not just a Snapshot: An Italian Longitudinal Evaluation of Stability of Gut Microbiota Findings in Parkinson's Disease

Rocco Cerroni [1,*], Daniele Pietrucci [2,3], Adelaide Teofani [4], Giovanni Chillemi [2], Claudio Liguori [1], Mariangela Pierantozzi [1], Valeria Unida [4], Sidorela Selmani [5], Nicola Biagio Mercuri [6,7] and Alessandro Stefani [1]

1 UOSD Parkinson's Center, Department of Systems Medicine, University of Rome Tor Vergata, 00133 Rome, Italy; dott.claudioliguori@yahoo.it (C.L.); pierantozzim@gmail.com (M.P.); stefani@uniroma2.it (A.S.)
2 Department for Innovation in Biological, Agro-Food and Forest Systems (DIBAF), University of Tuscia, 01100 Viterbo, Italy; daniele.pietrucci@unitus.it (D.P.); gchillemi@unitus.it (G.C.)
3 Institute of Biomembranes, Bioenergetics and Molecular Biotechnologies, IBIOM, Consiglio Nazionale della Ricerca (CNR), 70126 Bari, Italy
4 Department of Biology, University of Rome Tor Vergata, 00133 Rome, Italy; adelaide.teofani@uniroma2.it (A.T.); valeria.unida@gmail.com (V.U.)
5 Department of Neuroscience, Unizkm, 1001 Tirana, Albania; sidorela.selmani@live.com
6 Department of Systems Medicine, University of Rome Tor Vergata, 00133 Rome, Italy; mercurin@med.uniroma2.it
7 Istituto di Ricovero e Cura a Carattere Scientifico (IRCCS) "Fondazione Santa Lucia", 00179 Rome, Italy
* Correspondence: rocco.cerroni@gmail.com

Abstract: Most research analyzed gut-microbiota alterations in Parkinson's disease (PD) through cross-sectional studies, as single snapshots, without considering the time factor to either confirm methods and findings or observe longitudinal variations. In this study, we introduce the time factor by comparing gut-microbiota composition in 18 PD patients and 13 healthy controls (HC) at baseline and at least 1 year later, also considering PD clinical features. PD patients and HC underwent a fecal sampling at baseline and at a follow-up appointment. Fecal samples underwent sequencing and 16S rRNA amplicons analysis. Patients'clinical features were valued through Hoehn&Yahr (H&Y) staging-scale and Movement Disorder Society Unified PD Rating Scale (MDS-UPDRS) Part-III. Results demonstrated stability in microbiota findings in both PD patients and HC over a period of 14 months: both alfa and beta diversity were maintained in PD patients and HC over the observation period. In addition, differences in microbiota composition between PD patients and HC remained stable over the time period. Moreover, during the same period, patients did not experience any worsening of either staging or motor impairment. Our findings, highlighting the stability and reproducibility of the method, correlate clinical and microbiota stability over time and open the scenario to more extensive longitudinal evaluations.

Keywords: Parkinson's disease; gut microbiota; gut–brain axis; dysbiosis

Citation: Cerroni, R.; Pietrucci, D.; Teofani, A.; Chillemi, G.; Liguori, C.; Pierantozzi, M.; Unida, V.; Selmani, S.; Mercuri, N.B.; Stefani, A. Not just a Snapshot: An Italian Longitudinal Evaluation of Stability of Gut Microbiota Findings in Parkinson's Disease. *Brain Sci.* **2022**, *12*, 739. https://doi.org/10.3390/brainsci12060739

Academic Editors: Christina Piperi, Chiara Villa and Yam Nath Paudel

Received: 26 April 2022
Accepted: 1 June 2022
Published: 4 June 2022

Publisher's Note: MDPI stays neutral with regard to jurisdictional claims in published maps and institutional affiliations.

Copyright: © 2022 by the authors. Licensee MDPI, Basel, Switzerland. This article is an open access article distributed under the terms and conditions of the Creative Commons Attribution (CC BY) license (https://creativecommons.org/licenses/by/4.0/).

1. Introduction

The concept of "gut–brain axis", a bidirectional channel of influence and communication between the brain and the enteric nervous system (ENS), was first introduced in the late 2000s [1]. This axis, sustained by neurons of the sympathetic and parasympathetic nervous systems, as well as by circulating hormones, neuro-modulatory molecules, and stress-related gastrointestinal mediators, has been linked to inflammatory bowel diseases, neuropsychiatric syndromes, and neurodegenerative diseases [2]. Later evidence rightfully included the gut microbiota in this axis, turning it into the "microbiota–gut–brain axis"; indeed, resident bacteria may influence the gut and, consequently, the brain [3]. On the other hand, metabolites produced by the brain can influence the gut and, consequently, the gut microbiota [4]. The gut microbiota is an heterogenous and symbiotic community composed of about 100 trillion bacteria inhabiting the human intestine, which influences both intestinal physiology and dysfunctional processes through its metabolic activities and host interactions [5]. The whole genetic asset of gut microbiota, called a microbiome, contains more than 3 million unique genes, 150 fold the number of human genes [6]. A dysbiotic gut microbiota may influence the progression of central nervous system (CNS) diseases [7]. The role of the gut microbiota has been observed in multiple sclerosis [8,9], Alzheimer's disease [10], autism spectrum disorder [11], and depression [12]. The involvement of the gut microbiota was observed in Parkinson's disease (PD) using animal models [13] and through clinical observations; Braak and co-authors first postulated the presence of a *"noxa patogena"* ascending from the intestine or nasal cavity [14,15]. Today, dysbiosis of gut microbiota is commonly considered a well-established non-motor feature in PD, participating in both disease pathogenesis and, likely, clinical presentation [16].

In the last years, several studies investigated the specific role of altered gut microbiota in PD [17]. Different research groups, frequently utilizing fecal samples and sequencing, documented significant differences in gut microbial composition between PD patients and healthy subjects [18]. Scientific evidence from different parts of the world showed that PD patients have a higher relative abundance of bacteria from the genera *Akkermansia*, *Lactobacillus*, and *Bifidobacterium* and lower relative abundances of *Prevotella*, *Faecalibacterium*, *Bacteroidetes*, and *Blautia genera* [19–22]. Other groups found differences in microbiota composition between healthy subjects and PD patients, as well as between individuals with different PD phenotypes of symptomology [23,24]. Amongst the most consistent results, it is of note the reduction in "homeostatic" short chain fatty acids (SCFA)-producing families, like Lachnospiraceae and Prevotellaceae, as well as the increase in pro-inflammatory lipopolysaccharides (LPS)-producing families, like Enterobacteriaceae and Verrucomicrobiaceae [25–27]. In line with these studies, our group also investigated the presence of gut dysbiosis in PD patients [28]: our previous data showed that Enterobacteriaceae, Lactobacillaceae, and Enterococcaceae families were more abundant in PD fecal samples, whilst Lachnospiraceae families were decreased in PD patients compared to healthy subjects. Although most studies in literature linked intestinal dysbiosis to the pathogenesis of PD, a causal liaison between a dysbiotic gut microbiota and the development of PD is still far from being established [16]. More intriguingly, almost all PD microbiota studies were *"single snapshots"*, namely cross-sectional studies investigating intestinal microbial composition in PD patients and healthy subjects at a specific time, although considering possible confounders; indeed, only a few studies showed a longitudinal approach.

In this study, we introduce the time factor; we compare gut microbiota composition in a group of PD patients and healthy control subjects at baseline and at least 1 year later, also considering confounding factors and patients' clinical features.

Our aim is to investigate the stability and, thus, the reproducibility of gut microbial findings in PD patients in a relatively short time period, to give greater dignity to the differences highlighted, without neglecting clinical aspects.

2. Materials and Methods

2.1. Population and Study Design

We recruited, during scheduled visits, outpatients with PD afferent at the Movement Disorders Center of the Neurological Clinic of the University of Rome "Tor Vergata" between January 2017 and January 2020. All patients met the Movement Disorders Criteria for PD [29]; PD diagnosis was done by at least two board-certified neurologists.

Patients were also required to meet the following criteria: (1) no cognitive impairment, as defined by a Montreal Cognitive Assessment (MoCA) score above 25; (2) no chronic gastrointestinal (GI) disease, including malabsorption; (3) no clinical history of major intestinal surgery, gastric lesions, or gastric resection; and (4) complete agreement with the study design. Exclusion criteria were: (1) other concomitant neurologic and/or psychiatric diseases; (2) systemic and/or neurologic inflammatory, infectious, or autoimmune diseases;

(3) atypical parkinsonism syndromes and vascular parkinsonism; (4) acute GI inflammatory diseases or any other GI disease in the last 26 weeks; (5) use of domperidone, or any drug potentially affecting GI motility and integrity in the last 12 weeks; (6) use, in the last 12 weeks, of pre-probiotics or therapy based upon steroids, nonsteroidal anti-inflammatory drugs, or antibiotics; and (7) anamnesis suggestive of GI cancer pathology. We also recruited a healthy controls (HC) group, composed mainly of patients' cohabiting relatives or partners, to minimize lifestyle-confounding factors.

The exclusion criteria for HC were: (1) presence of any neurological disease; (2) presence of acute or chronic GI diseases both during the study and in medical history; (3) medical history of major intestinal surgery, gastric lesions, or gastric resection; and (4) use in the last 12 weeks of any drug potentially affecting GI function and integrity (steroids, NSAIDs, drugs with pro-kinetic anti-kinetic function on the intestinal motility, anti-acid drugs, pre/probiotics, antibiotics).

Patients and HC underwent a fecal sampling at baseline and at a follow-up appointment at least 12 months later. PD patients were clinically evaluated, both at the baseline and at the follow-up visit, through Hoehn & Yahr staging scale (H&Y) and Movement Disorders Society Unified Parkinson's Disease Rating Scale Part III (MDS-UPDRS Part III), performed during the best ON time, to estimate disease progression and degree of motor impairment, respectively. All study participants gave written informed consent after receiving an extensive disclosure of the study purposes and risks. Ethics Committee of Fondazione PTV Policlinico Tor Vergata approved the trial (RS 73/18).

2.2. Sequencing and 16S rRNA Amplicons Analysis

Fecal samples were collected and analyzed as described in our previous study [28]. Briefly, fecal samples were collected using the pre-analytical sample processing (PSP) stool collection tubes. These tubes ensure the storage and the conservation of the DNA at ambient temperature for at least three months. Samples were processed following the manufacturer's instructions during the timeframe that allows the extraction of a relevant DNA yield. First, stool samples were lysed at 95 °C, and PCR inhibitors and cell debris were removed. Samples were then treated with Proteinase K at 70 °C. Next, a DNA purification step was performed using a spin column system. Finally, the DNA was extracted and quantified using a NanoDrop spectrophotometer ND1000 (Termofisher, Waltham, Massachusetts, USA). The paired-end sequencing of the V3-V4 region of the 16S gene was performed using an Illumina MiSeq 2 × 300 bp (Illumina Inc., San Diego, CA, USA). Raw sequencing data are available from the Sequencing Read Archive (SRA) database [30] with BioProject ID: PRJNA510730. Bioinformatic data analysis was performed using the following software: Fatsqc vr. 0.11.9, Cutadapt vr. 2.9, and QIIME 2.0 [31]. Fastqc and Cutadapt were used to assess the read quality and the presence of Illumina sequencing adapter, respectively. The QIIME 2.0 pipeline removed chimeras from sequencing data, merged reads, and grouped 16S reads in amplicon sequence variants (ASVs). In detail, the QIIME pipeline was used to call the DADA2 [32] software to cluster the read in ASVs, and the taxonomy of representative sequences was assessed using the Q2-feature-classifier [31] and the SILVA database vr. 138 [33].

2.3. Statistical Analysis

Statistical analyses were performed in R vr. 3.5.3 (https://www.R-project.org/ accessed on 25 April 2022), using the following packages: vegan, phyloseq, and DESeq2. Phyloseq was used to import the file generated by the QIIME 2.0 pipeline in R, and DESeq2 was used to normalize samples according to McMurdie et al. [34]. The vegan package was used to compute richness (number of observed species, Chao1, and ACE indices) and alpha diversity indices (Shannon and Simpson indices). Phyloseq was used to evaluate four beta diversity indices (Bray-Curtis and Camberra distances, weighted and unweighted Unifrac). To search for differences in microbiota structure, a PERMANOVA test with 9999 permutations was performed. Statistical differences of taxa abundances between patients and controls over time were evaluated using a repeated-measure ANOVA. In detail, we used the following test: Taxa ~ Status + Time + Status:Time + Error (Individual_ID).

This model allows searching for taxa, which can change in abundance depending on the status (PD or HC), depending on the time of the measurement (the first measure or the follow-up), or a combination of both effects (status and time). The same model was also applied to alpha diversity metrics. All the p-values were corrected for multiple testing, and only tests with a false discovery rate less than 0.05 were considered statistically significant. In the case of significant bacterial families after the ANOVA test, we performed post-hoc comparisons to take into account differences among groups. Furthermore, to make a comparison between clinical data - Levodopa Equivalent Daily Dose (LEDD), H&Y and MDS-UPDRS Part III - at baseline and at follow-up, we used a paired t-test, after assessing the normal distribution of data using the Shapiro–Wilk test. The t-test was used to highlight differences in clinical data between the patients at the baseline and at the follow-up. The first group used in the test was represented by the clinical values of patients at the baseline. The second group was represented by the clinical values of the same patients at the follow-up.

3. Results

3.1. Recruitment and Demographics

We recruited 31 participants: 18 PD patients and 13 HC. PD patients and HC underwent a fecal sampling at baseline and at a follow-up appointment an average of 14 (SD ± 1.8) months later. PD patients also underwent a clinical examination performed by H&Y and MDS-UPDRS Part III scales at baseline and at the follow-up visit. The main demographic and clinical variables are reported in Table 1.

Table 1. Demographic and clinical data of our population.

		PD	HC
Tot		18	13
Sex (F/M) [1]		08/10	06/07
Age (years)		63.5 ± 8.1	62.8 ± 7.8
Dis. Duration (months)	**Baseline**	35.7 ± 10	
H&Y	**Baseline**	2.05 ± 0.6	
H&Y	**Follow-up**	2.15 ± 0.5	
MDS UPDRS-III	**Baseline**	23.6 ± 6.2	
MDS UPDRS-III	**Follow-up**	24.3 ± 5.6	
LEDD	**Baseline**	450.3 ± 40.2	
LEDD	**Follow-up**	479.2 ± 25.4	

[1] Abbreviations: M = male; F = female; H&Y = Hoehn & Yahr staging score; MDS UPDRS-III = Movement Disorders Society Unified Parkinson's Disease Rating Sale Part III score; LEDD = Levodopa Equivalent Daily Dose.

PD patients and HC did not differ in terms of gender and age. Levodopa equivalent daily dose (LEDD) did not change significantly at the follow-up from baseline (paired *t*-test *p*-value = 0.349).

3.2. Alfa and Beta Diversity

According to statistical analysis, microbial composition over time showed interesting results both in alfa diversity and in beta diversity. In our study, alpha diversity did not show any statistical differences at baseline and at follow-up in both PD patients and HC according to two different metrics (Shannon and Simpson indices). We performed a repeated-measure ANOVA in order to evaluate the differences between the PD patients and HC between the baseline and the follow-up. The ANOVA results did not show differences considering PD status (PD or HC), time (baseline and follow-up), or the combined effect of time and

PD status. These results indicate that the gut microbiota alpha diversity did not show any significant differences after the follow-up (Figure 1). Even not significant, we observed a reduction of both indices in both patients and controls at the follow-up visit. Considering the richness indices (number of observed species, Chao 1, and ACE indices), we identified a significant effect related to the time, indicating a reduction of richness in both PD patients and HC at the follow-up visit (Figure 1). This result suggests a decrease in the number of species in both patients and controls, independent of the pathology.

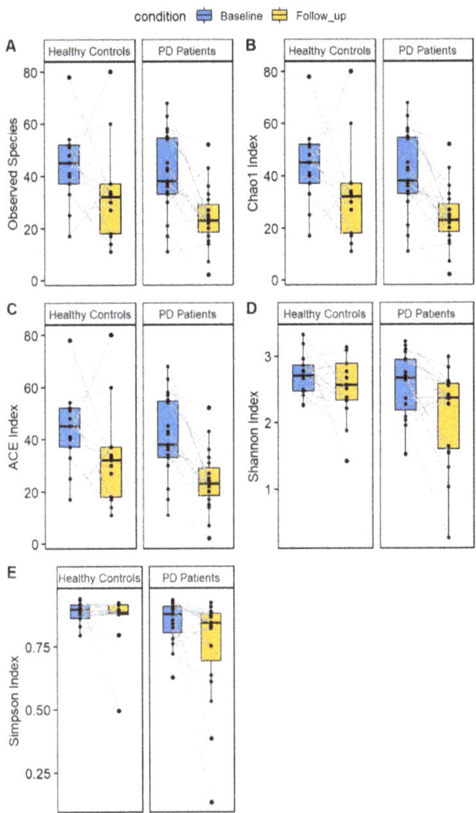

Figure 1. Alpha diversity indices evaluated for Parkinson's disease patients and healthy controls. The following indices are reported: number of observed species (**A**), Chao1 index (**B**), ACE index (**C**), Shannon Index (**D**), and Simpson Index (**E**). For both patients and controls, the values at the first visit ("baseline", colored in blue) and the follow-up (colored in yellow) are reported.

Our results also showed that the beta diversity was maintained in both PD patients and HC over time. More specifically, using the Bray–Curtis dissimilarity, no detectable differences were identified when analyzing the gut microbiota combining the effect of time (follow-up and baseline) and pathological status (PD or HC). In fact, the PERMANOVA test was not significant when both the effects were considered (p-value > 0.05 with 9999 permutations) (Figure 2A). Furthermore, observing the heatmap reported in Figure 2B, no clear pattern can be identified regarding the difference in time. Instead, some samples of the same patients cluster together (i.e., sample id 2, 3, 15, 18, 20, 22), independent of the analysis time. This scheme indicates that the microbiota has remained almost stable during the baseline and the follow-up analyses. Moreover, the baseline differences between patients and HC were unchanged at follow-up.

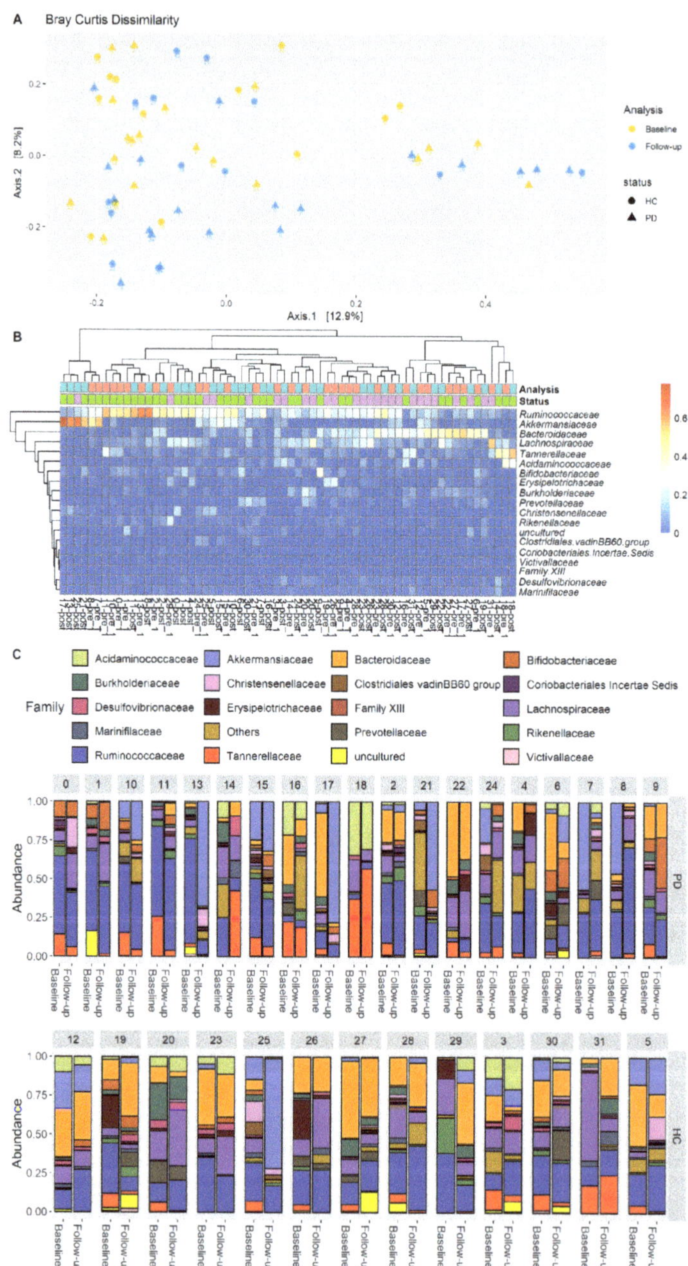

Figure 2. Beta diversity evaluated for Parkinson's disease patients and healthy controls. (A) Principal coordinate analysis. Each patient is represented by an anonymized ID; the shape indicates the microbiota evaluation (circle = baseline, triangle = follow-up). (B) Heatmap of Bray–Curtis distances between patients (columns) and the main bacterial families (rows). The cell reports the relative abundance of each family in all samples. Samples are reported in columns using the following color scheme: PD = green, HC = purple, baseline = cyan, follow-up = pink. (C) Barplot representing the relative abundance of microbial families across all samples. Each subject is represented with two barplots, one for the baseline and one for the follow-up test.

3.3. Families

After observing the alpha and beta diversity results, we decided to perform the most detailed analysis to identify specific taxa whose abundance may vary between baseline and follow-up tests. The beta diversity indicated that the structure of the microbiota did not change over time; however, the alpha diversity showed a reduction of richness in both patients and controls at the follow-up visit. Consequently, we performed the repeated ANOVA test on the bacterial families to identify the taxa which could be responsible for this trend. Through the analysis of differential abundance, no statistically significant changes in the composition of microbial families emerged due to the combined effect of pathology and time. Only without considering the effect of the disease in the repeated-measure ANOVA test did we identify families changing over time in the same way in PD patients and HC. In total, six families decreased at the follow-up visit compared to the baseline in both patients and HC. These families belong mainly to the Phylum Bacteroidota (A. Bacteroidaceae, B. Rikenellaceae, C. Barnesiellaceae, D. Marinifilaceae, E. Tannerellaceae) but also to Firmicutes (F. Lachnospiraceae). The relative abundance of these families is reported in Figure 3 and Table 2. The relative abundance of two families also varied among groups, independent of time, suggesting a difference between groups. In detail, the Lachnospiraceae and Bacteroidaceae families were less abundant in PD patients compared to controls. Although these results were not significant after adjusting for multiple testing, they indicated a specific trend.

Table 2. Bacterial families identified by the repeated-measure ANOVA test. The relative abundance is reported for HC and PD patients at baseline and follow-up tests. The p-values reported refer to the significance between HC and PD patients (status), follow-up and baseline (time), or considering the effect of status and time. All the p-values are corrected for multiple testing.

	Abundance at Baseline		Abundance at Follow-up		Repeated-Measure ANOVA Results		
	HC	PD	HC	PD	p-Value Status	p-Value Time	p-Value Status: Time
Bacteroidaceae	2.74×10^{-1}	2.15×10^{-1}	2.32×10^{-1}	1.28×10^{-1}	8.16×10^{-2}	5.99×10^{-3} *	9.80×10^{-1}
Tannerellaceae	2.58×10^{-2}	2.74×10^{-2}	1.71×10^{-2}	2.21×10^{-2}	1.00×10^{0}	2.12×10^{-2} *	9.80×10^{-1}
Rikenellaceae	4.95×10^{-3}	7.55×10^{-3}	5.02×10^{-3}	5.85×10^{-3}	8.16×10^{-2}	2.12×10^{-2} *	9.82×10^{-1}
Marinifilaceae	5.04×10^{-3}	4.98×10^{-3}	5.51×10^{-3}	2.81×10^{-3}	8.22×10^{-1}	2.65×10^{-2} *	9.89×10^{-1}
Lachnospiraceae	1.96×10^{-1}	1.58×10^{-1}	1.94×10^{-1}	1.41×10^{-1}	8.45×10^{-2}	1.23×10^{-3} *	9.80×10^{-1}
Barnesiellaceae	1.85×10^{-2}	1.60×10^{-2}	1.57×10^{-2}	1.29×10^{-2}	6.60×10^{-1}	1.11×10^{-2} *	9.81×10^{-1}

Finally, we processed data regarding the clinical aspects of PD patients over time. The analysis of patients' clinical variables showed no significant differences in H&Y (paired t-test p-value = 0.349) and MDS-UPDRS Part III (paired t-test p-value = 0.395) scales between baseline and follow-up visit scores."*" indicates a p-value lower than 0.05.

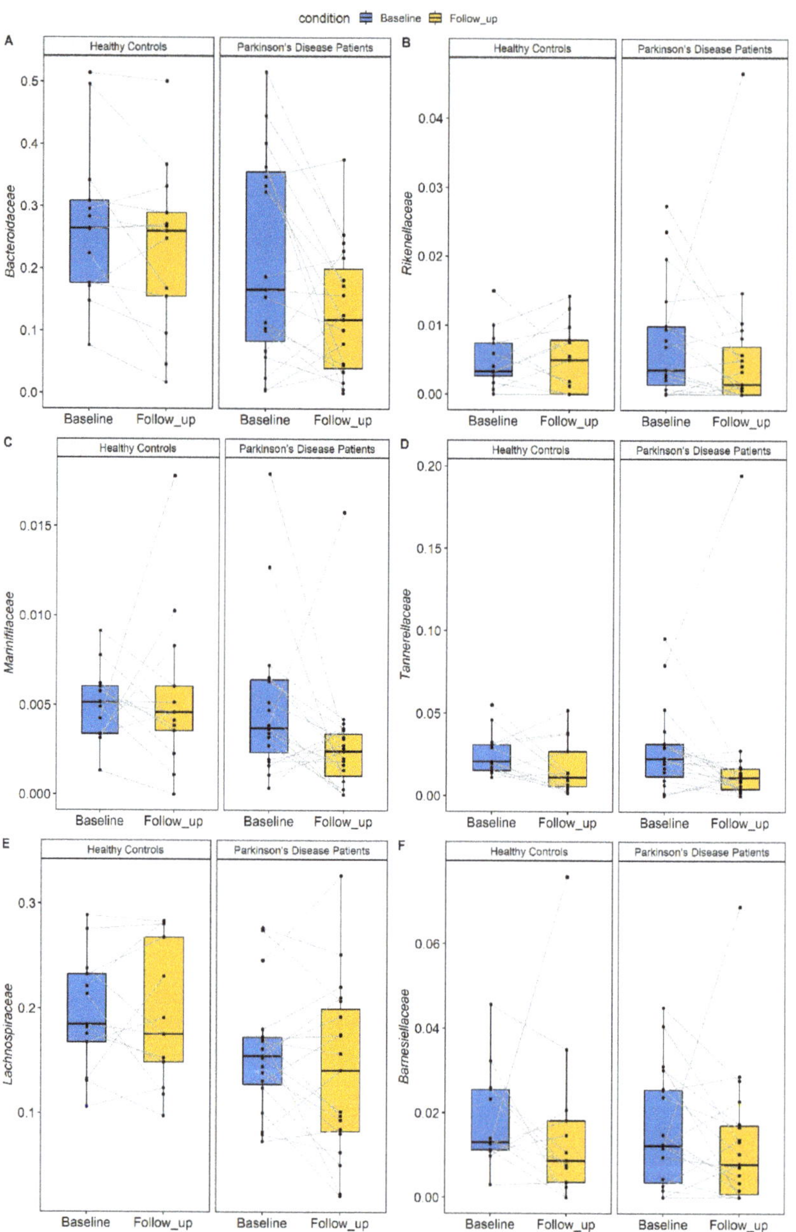

Figure 3. Taxa identified by the repeated-measure ANOVA test. All the families are reduced in both patients and controls at the follow-up (colored in yellow) compared to the baseline (colored in blue), independent of the health status (PD or HC). The effect of time is statistically significant after performing a multiple test correction.

4. Discussion

Our study demonstrated stability in microbiota findings both in PD patients and in HC over a period of approximately 14 months. Not only did microbiota composition remain stable in both PD patients and HC but also, and more intriguingly, differences in microbiota composition between PD patients and HC remained stable over time.

Our results cover both alpha and beta diversity, as well as microbial families.

Alpha diversity is commonly considered an expression of the number of species in each sample; its stability in our data highlights how our microbiota analysis is effective in showing biodiversity over time. We did not observe differences in two different alpha diversity metrics at the baseline and follow-up. However, we have observed, in a recent meta-analysis, that the alpha diversity alone does not represent a good microbiota marker for PD [35]. Furthermore, de la Cuesta-Zuluaga J and colleagues showed that the alpha diversity reaches a plateau in elderly individuals [36]. Although not statistically significant, we observed a reduction for both Shannon and Simpson indices in both patients and controls. We performed the analysis by using the richness indices (number of observed species, Chao1, and ACE) and identified a reduction of richness in both patients and controls.

Conversely, beta diversity is an expression of gut microbiota community structure. In our study, this community structure was maintained in both PD patients and HC. Furthermore, the baseline differences between PD patients and HC remained unchanged at follow-up; hence we can assume that the composition in community structure of microbiota between PD and HC is preserved over time.

Regarding microbiota families, the analysis of the differential abundance did not find statistically significant changes in the composition of the families due to the combined effect of pathology and time; this data further substantiates the stability of the microbiota over time, which is greatly influenced, among other factors, by fetal and perinatal life [37]. By excluding the effect of the disease, we identified families that vary over time in the same way in patients and controls. This trend is probably attributable to the influence of external factors (lifestyle and diet changes) and does not affect the microbiota stability. For example, a reduction of the microbiota diversity was observed due to seasonality [38].

Finally, the trend observed in Lachnospiraceae and Bacteroidaceae families, to be less abundant in PD patients than in controls, highlights, once again, the presence of differences in the microbiota structure between PD patients and HC, as previously reported [28,35].

Our study also considered PD clinical aspects in terms of progression and motor deficits. Interestingly, during the observation period, patients did not experience any worsening either in terms of staging or in terms of motor impairment. Indeed, our data showed no significant differences in H&Y and MDS-UPDRS Part III scores between the baseline and follow-up visit, although this does not exclude minimal changes in single items of MDS-UPDRS-III. Furthermore, it should be noted that the relatively short observation period did not allow for remarkable changes to the therapeutic regimen. In our opinion, these results allow us to correlate clinical and microbiota stability over time, reinforcing the pathogenic role of gut microbiota change in PD patients. Otherwise, the presence of clinical deterioration during the observation period with an unaffected microbiota would have weakened this link. Our results are in line with a previous study by Aho and colleagues, in which the stability of the fecal microbiota was established after two years [39]. Of note is also the stability of microbiota findings in the HC group. Dealing with a microbiota that shows stability and reproducibility is the basis for each disease-modifying therapy targeting microbiota.

The main objective of our research was to demonstrate the stability of the microbiota composition over a short time and the reproducibility of the method of analysis; therefore, data from other longitudinal studies investigating disease progression related to microbiota changes are not comparable [40]. However, the stability of both gut microbiota composition and clinical features in PD patients over a relatively short time period does not exclude that microbiota may vary over a longer period of time or with the worsening of the disease and/or with the intervention of a trigger factor.

The main limitations of this study are the relatively short observation period and the sample size, which was sufficient to make inferences but which would have had a better impact if larger. However, although small, our cohort is characterized by rather homogeneous characteristics. Looking ahead, our study lays the foundation for longitudinal evaluations with a wider observation time window and a larger cohort of patients, to correlate any worsening of the disease with further alterations of the microbiota over time.

5. Conclusions

In our study, we do not identify differences in the gut microbiota (beta diversity) structure at the follow-up in both PD patients and HC, which remained stable for both patients and controls. These results suggest that the gut microbiota may remain stable over a period of 14 months. We identified a reduction in some richness indices, in both patients and controls, indicating that some species may reduce their abundance in the gut microbiota. We confirmed this analysis with a differential abundance test on the bacterial families using a repeated-measure ANOVA. Seven families showed a decreasing trend in PD patients, which was also identified in HC. Consequently, these differences may be caused by other external factors (i.e., alimentations) unrelated to PD progression. Moreover, the PD cohort showed, at follow-up, the same degree of disease progression (H&Y) and motor impairment (MDS-UPDRS Part III) with respect to the baseline, supporting the idea that, in PD, microbial stability and disease stability could be correlated. Dysbiosis of gut microbiota can be considered a relevant and reliable feature of PD that can provide insights into the disease pathophysiology. For future studies, a crucial step will be the fecal sampling and analysis at multiple timepoints in disease progression, a wider observation window, and a larger patient population with different H&Y and MDS-UPDRS Part III scores. These longitudinal data could identify physio-pathological correlations between the variation in microbiota composition and the PD progression. Last but not least, these recent results on gut microbiota stability create an opportunity for new studies aimed at understanding whether and to what extent therapeutic interventions (levodopa, iCOMT, and/or the initiation of advanced therapies such as Levodopa Carbidopa Intestinal gel) can play an active role in modifying the gut microbiota.

Author Contributions: Conceptualization, R.C. and A.S.; methodology, R.C. and D.P.; software, D.P. and A.T.; validation, R.C. and D.P.; formal analysis, R.C., C.L. and D.P.; investigation, R.C., D.P., C.L., V.U. and A.T.; resources, R.C., V.U., C.L., M.P. and A.S.; data curation, R.C., D.P., S.S., V.U., S.S., M.P. and A.T.; writing—original draft preparation, R.C. and D.P.; writing—review and editing, R.C., D.P., M.P., N.B.M., G.C. and A.S.; visualization, D.P.; supervision, M.P. and A.S.; project administration, A.S.; funding acquisition, A.S. and N.B.M. All authors have read and agreed to the published version of the manuscript.

Funding: This research was funded by FONDAZIONE G.B. BARONI: MercuriNfondazionebaroni18–Studio dei parassiti intestinali prevalenti nelle feci dei pazienti afflitti da Malattia di Parkinson–n 1010302, 1030103, 1030106.

Institutional Review Board Statement: The study was conducted in accordance with the Declaration of Helsinki and approved by the Ethics Committee of Fondazione PTV Policlinico Tor Vergata (protocol code RS 73/18, date of approval 15/11/2018).

Informed Consent Statement: Informed consent was obtained from all subjects involved in the study.

Data Availability Statement: Raw sequencing data are available at Sequencing Read Archive (SRA) database with BioProject ID: PRJNA510730.

Conflicts of Interest: The authors declare no conflict of interest. The funders had no role in the design of the study; in the collection, analyses, or interpretation of data; in the writing of the manuscript; or in the decision to publish the results.

References

1. Rhee, S.H.; Pothoulakis, C.; Mayer, E.A. Principles and clinical implications of the brain-gut-enteric microbiota axis. *Nat. Rev. Gastroenterol. Hepatol.* **2009**, *6*, 306–314. [CrossRef]
2. Mayer, E.A. Gut feelings: The emerging biology of gut-brain communication. *Nat. Rev. Neurosci.* **2011**, *12*, 453–466. [CrossRef] [PubMed]

3. Yarandi, S.S.; Peterson, D.A.; Treisman, G.J.; Moran, T.H.; Pasricha, P.J. Modulatory effects of gut microbiota on the central nervous system: How the gut could play a role in neuropsychiatric health and disease. *J. Neurogastroenterol. Motil.* **2016**, *22*, 201–212. [CrossRef] [PubMed]
4. Molina-Torres, G.; Rodriguez-Arrastia, M.; Roman, P.; Sanchez-Labraca, N.; Cardona, D. Stress and the gut microbiota-brain axis. *Behav Pharmacol.* **2019**, *30*, 187–200. [CrossRef] [PubMed]
5. Lozupone, C.A.; Stombaugh, J.I.; Gordon, J.I.; Jansson, J.K.; Knight, R. Diversity, stability and resilience of the human gut microbiota. *Nature* **2012**, *7415*, 220–230. [CrossRef]
6. Proctor, L.M. The Human Microbiome Project in 2011 and beyond. *Cell Host Microbe.* **2011**, *10*, 287–291. [CrossRef]
7. Sorboni, S.G.; Moghaddam, H.S.; Jafarzadeh-Esfehani, R.; Soleimanpour, S. A Comprehensive Review on the Role of the Gut Microbiome in Human Neurological Disorders. *Clin. Microbiol. Rev.* **2022**, *35*, e0033820. [CrossRef]
8. Ochoa-Repáraz, J.; Kirby, T.O.; Kasper, L.H. The Gut Microbiome and Multiple Sclerosis. *Cold Spring Harb. Perspect. Med.* **2018**, *8*, a029017. [CrossRef]
9. Cox, L.M.; Maghzi, A.H.; Liu, S.; Tankou, S.K.; Dhang, F.H.; Willocq, V.; Song, A.; Wasén, C.; Tauhid, S.; Chu, R.; et al. Gut Microbiome in Progressive Multiple Sclerosis. *Ann. Neurol.* **2021**, *89*, 1195–1211. [CrossRef]
10. Kowalski, K.; Mulak, A. Brain-Gut-Microbiota Axis in Alzheimer's Disease. *J. Neurogastroenterol. Motil.* **2019**, *25*, 48–60. [CrossRef]
11. Garcia-Gutierrez, E.; Narbad, A.; Rodríguez, J.M. Autism Spectrum Disorder Associated With Gut Microbiota at Immune, Metabolomic, and Neuroactive Level. *Front. Neurosci.* **2020**, *14*, 578666. [CrossRef] [PubMed]
12. Liang, S.; Wu, X.; Hu, X.; Wang, T.; Jin, F. Recognizing Depression from the Microbiota-Gut-Brain Axis. *Int. J. Mol. Sci.* **2018**, *19*, 1592. [CrossRef] [PubMed]
13. Sampson, T.R.; Debelius, J.W.; Thron, T.; Janssen, S.; Shastri, G.G.; Ilhan, Z.E.; Challis, C.; Schretter, C.E.; Rocha, S.; Gradinaru, V.; et al. Gut Microbiota Regulate Motor Deficits and Neuroinflammation in a Model of Parkinson's Disease. *Cell* **2016**, *167*, 1469–1480. [CrossRef] [PubMed]
14. Braak, H.; Rüb, U.; Gai, W.P.; Del Tredici, K. Idiopathic Parkinson's disease: Possible routes by which vulnerable neuronal types may be subject to neuroinvasion by an unknown pathogen. *J. Neural Transm.* **2003**, *110*, 517–536. [CrossRef]
15. Braak, H.; de Vos, R.A.; Bohl, J.; Del Tredici, K. Gastric alpha-synuclein immunoreactive inclusions in Meissner's and Auerbach's plexuses in cases staged for Parkinson's disease-related brain pathology. *Neurosci. Lett.* **2006**, *396*, 67–72. [CrossRef] [PubMed]
16. Keshavarzian, A.; Engen, P.; Bonvegna, S.; Cilia, R. The gut microbiome in Parkinson's disease: A culprit or a bystander? *Prog. Brain Res.* **2020**, *252*, 357–450. [CrossRef] [PubMed]
17. Boertien, J.M.; Pereira, P.A.B.; Aho, V.T.E.; Scheperjans, F. Increasing Comparability and Utility of Gut Microbiome Studies in Parkinson's Disease: A Systematic Review. *J. Park Dis.* **2019**, *9*, S297–S312. [CrossRef]
18. Klann, E.M.; Dissanayake, U.; Gurrala, A.; Farrer, M.; Shukla, A.W.; Ramirez-Zamora, A.; Mai, V.; Vedam-Mai, V. The Gut-Brain Axis and Its Relation to Parkinson's Disease: A Review. *Front Aging Neurosci.* **2022**, *13*, 782082. [CrossRef]
19. Scheperjans, F.; Aho, V.; Pereira, P.A.; Koskinen, K.; Paulin, L.; Pekkonen, E.; Haapaniemi, E.; Kaakkola, S.; Eerola-Rautio, J.; Pohja, M.; et al. Gut microbiota are related to Parkinson's disease and clinical phenotype. *Mov. Disord.* **2015**, *30*, 350–358. [CrossRef]
20. Keshavarzian, A.; Green, S.J.; Engen, P.A.; Voigt, R.M.; Naqib, A.; Forsyth, C.B.; Mutlu, E.; Shannon, K.M. Colonic bacterial composition in Parkinson's disease. Colonic bacterial composition in Parkinson's disease. *Mov. Disord.* **2015**, *30*, 1351–1360. [CrossRef]
21. Li, W.; Wu, X.; Hu, X.; Wang, T.; Liang, S.; Duan, Y.; Jin, F.; Qin, B. Structural changes of gut microbiota in Parkinson's disease and its correlation with clinical features. *Sci. China Life Sci.* **2017**, *60*, 1223–1233. [CrossRef] [PubMed]
22. Barichella, M.; Severgnini, M.; Cilia, R.; Cassani, E.; Bolliri, C.; Caronni, S.; Ferri, V.; Cancello, R.; Ceccarani, C.; Faierman, S.; et al. Unraveling gut microbiota in Parkinson's disease and atypical parkinsonism. *Mov. Disord.* **2019**, *34*, 396–405. [CrossRef] [PubMed]
23. Heintz-Buschart, A.; Pandey, U.; Wicke, T.; Sixel-Döring, F.; Janzen, A.; Sittig-Wiegand, E.; Trenkwalder, C.; Oertel, W.H.; Mollenhauer, B.; Wilmes, P. The nasal and gut microbiome in Parkinson's disease and idiopathic rapid eye movement sleep behavior disorder. *Mov. Disord.* **2018**, *33*, 88–98. [CrossRef] [PubMed]
24. Hill-Burns, E.M.; Debelius, J.W.; Morton, J.T.; Wissemann, W.T.; Lewis, M.R.; Wallen, Z.D.; Peddada, S.D.; Factor, S.A.; Molho, E.; Zabetian, C.P.; et al. Parkinson's disease and Parkinson's disease medications have distinct signatures of the gut microbiome. *Mov. Disord.* **2017**, *32*, 739–749. [CrossRef]
25. Hasegawa, S.; Goto, S.; Tsuji, H.; Okuno, T.; Asahara, T.; Nomoto, K.; Shibata, A.; Fujisawa, Y.; Minato, T.; Okamoto, A.; et al. Intestinal dysbiosis and lowered serum lipopolysaccharide-binding protein in Parkinson's disease. *PLoS ONE* **2015**, *10*, e0142164. [CrossRef]
26. Unger, M.M.; Spiegel, J.; Dillmann, K.U.; Grundmann, D.; Philippeit, H.; Bürmann, J.; Faßbender, K.; Schwiertz, A.; Schäfer, K.H. Short chain fatty acids and gut microbiota differ between patients with Parkinson's disease and age-matched controls. *Parkinsonism Relat. Disord.* **2016**, *32*, 66–72. [CrossRef]
27. Bedarf, J.R.; Hildebrand, F.; Coelho, L.P.; Sunagawa, S.; Bahram, M.; Goeser, F.; Bork, P.; Wüllner, U. Functional implications of microbial and viral gut metagenome changes in early stage L-DOPA-naïve Parkinson's disease patients. *Genome Med.* **2017**, *9*, 39. [CrossRef]
28. Pietrucci, D.; Cerroni, R.; Unida, V.; Farcomeni, A.; Pierantozzi, M.; Mercuri, N.B.; Biocca, S.; Stefani, A.; Desideri, A. Dysbiosis of gut microbiota in a selected population of Parkinson's patients. *Parkinsonism Relat. Disord.* **2019**, *65*, 124–130. [CrossRef]

29. Postuma, R.B.; Berg, D.; Stern, M.; Poewe, W.; Olanow, C.W.; Oertel, W.; Obeso, J.; Marek, K.; Litvan, I.; Lang, A.E.; et al. MDS clinical diagnostic criteria for Parkinson's disease. *Mov. Disord.* **2015**, *30*, 1591–1601. [CrossRef]
30. Leinonen, R.; Sugawara, H.; Shumway, M. The sequence read archive. *Nucleic Acids Res.* **2011**, *39*, D19–D21. [CrossRef]
31. Bolyen, E.; Rideout, J.R.; Dillon, M.R.; Bokulich, N.A.; Abnet, C.C.; Al-Ghalith, G.A.; Alexander, H.; Alm, E.J.; Arumugam, M.; Asnicar, F.; et al. Reproducible, interactive, scalable and extensible microbiome data science using QIIME 2. *Nat. Biotechnol.* **2019**, *37*, 852–857. [CrossRef] [PubMed]
32. Callahan, B.J.; Mcmurdie, P.J.; Rosen, M.J.; Han, A.W.; Johnson, A.J.A.; Holmes, S.P. DADA2: High-resolution sample inference from Illumina amplicon data. *Nat. Methods* **2016**, *13*, 581–583. [CrossRef] [PubMed]
33. Quast, C.; Pruesse, E.; Yilmaz, P.; Gerken, J.; Schweer, T.; Yarza, P.; Peplies, J.; Glockner, F.O. The SILVA ribosomal RNA gene database project: Improved data processing and web-based tools. *Nucleic Acids Res.* **2013**, *41*, D590–D596. [CrossRef] [PubMed]
34. McMurdie, P.J.; Holmes, S. Waste not, want not: Why rarefying microbiome data is inadmissible. *PLoS Comput. Biol.* **2014**, *10*, e1003531. [CrossRef] [PubMed]
35. Plassais, J.; Gbikpi-Benissan, G.; Figarol, M.; Scheperjans, F.; Gorochov, G.; Derkinderen, P.; Cervino, A.C.L. Gut microbiome alpha-diversity is not a marker of Parkinson's disease and multiple sclerosis. *Brain Commun.* **2021**, *3*, fcab113. [CrossRef]
36. De la Cuesta-Zuluaga, J.; Kelley, S.T.; Chen, Y.; Escobar, J.S.; Mueller, N.T.; Ley, R.E.; McDonald, D.; Huang, S.; Swafford, A.D.; Knight, R.; et al. Age- and Sex-Dependent Patterns of Gut Microbial Diversity in Human Adults. *mSystems* **2019**, *4*, 4. [CrossRef]
37. Ghaisas, S.; Maher, J.; Kanthasamy, A. Gut microbiome in health and disease: Linking the microbiome-gut-brain axis and environmental factors in the pathogenesis of systemic and neurodegenerative diseases. *Pharmacol. Ther.* **2016**, *158*, 52–62. [CrossRef]
38. Davenport, E.R.; Mizrahi-Man, O.; Michelini, K.; Barreiro, L.B.; Ober, C.; Gilad, Y. Seasonal variation in human gut microbiome composition. *PLoS ONE* **2014**, *9*, e90731. [CrossRef]
39. Aho, V.T.E.; Pereira, P.A.B.; Voutilainen, S.; Paulin, L.; Pekkonen, E.; Auvinen, P.; Scheperjans, F. Gut microbiota in Parkinson's disease: Temporal stability and relations to disease progression. *EBioMedicine* **2019**, *44*, 691–707. [CrossRef]
40. Minato, T.; Maeda, T.; Fujisawa, Y.; Tsuji, H.; Nomoto, K.; Ohno, K.; Hirayama, M. Progression of Parkinson's disease is associated with gut dysbiosis: Two-year follow-up study. *PLoS ONE* **2017**, *12*, e0187307. [CrossRef]

Article

Shared Genetic Background between Parkinson's Disease and Schizophrenia: A Two-Sample Mendelian Randomization Study

Kiwon Kim [1,†], Soyeon Kim [2,†], Woojae Myung [3,*], Injeong Shim [2], Hyewon Lee [4,5], Beomsu Kim [2], Sung Kweon Cho [6], Joohyun Yoon [3], Doh Kwan Kim [7] and Hong-Hee Won [2,*]

1 Department of Psychiatry, Kangdong Sacred Heart Hospital, College of Medicine, Hallym University, Sungan-ro, Kangdong-gu, Seoul 05355, Korea; kkewni@gmail.com
2 Samsung Medical Center, Department of Digital Health, Samsung Advanced Institute for Health Sciences and Technology (SAIHST), Sungkyunkwan University, Seoul 06351, Korea; soyeon2019@skku.edu (S.K.); injeong.shim@gmail.com (I.S.); kyce@hanmail.net (B.K.)
3 Department of Neuropsychiatry, Seoul National University Bundang Hospital, Seongnam 13620, Korea; jy3020@tc.columbia.edu
4 Department of Health Administration and Management, College of Medical Sciences, Soonchunhyang University, Asan 31538, Korea; woniggo@gmail.com
5 Department of Software Convergence, Graduate School, Soonchunhyang University, Asan 31538, Korea
6 Department of Pharmacology, School of Medicine, Ajou University, Worldcup-ro, Yeongtong-gu, Suwon 16499, Korea; wontan2000@gmail.com
7 Samsung Medical Center, Department of Psychiatry, School of Medicine, Sungkyunkwan University, Seoul 06351, Korea; paulkim@skku.edu
* Correspondence: wjmyung@snubh.org (W.M.); wonhh@skku.edu (H.-H.W.); Tel.: +82-(31)-787-7430 (W.M.); +82-(2)-2148-7566 (H.-H.W.); Fax: +82-(31)-787-4058 (W.M.); +82-(2)-3410-0534 (H.-H.W.)
† These individuals contributed equally to this work.

Citation: Kim, K.; Kim, S.; Myung, W.; Shim, I.; Lee, H.; Kim, B.; Cho, S.K.; Yoon, J.; Kim, D.K.; Won, H.-H. Shared Genetic Background between Parkinson's Disease and Schizophrenia: A Two-Sample Mendelian Randomization Study. *Brain Sci.* **2021**, *11*, 1042. https://doi.org/10.3390/brainsci11081042

Academic Editors: Christina Piperi, Chiara Villa and Yam Nath Paudel

Received: 27 June 2021
Accepted: 30 July 2021
Published: 6 August 2021

Publisher's Note: MDPI stays neutral with regard to jurisdictional claims in published maps and institutional affiliations.

Copyright: © 2021 by the authors. Licensee MDPI, Basel, Switzerland. This article is an open access article distributed under the terms and conditions of the Creative Commons Attribution (CC BY) license (https://creativecommons.org/licenses/by/4.0/).

Abstract: *Background and objectives*: Parkinson's disease (PD) and schizophrenia often share symptomatology. Psychotic symptoms are prevalent in patients with PD, and similar motor symptoms with extrapyramidal signs are frequently observed in antipsychotic-naïve patients with schizophrenia as well as premorbid families. However, few studies have examined the relationship between PD and schizophrenia. We performed this study to evaluate whether genetic variants which increase PD risk influence the risk of developing schizophrenia, and vice versa. *Materials and Methods*: Two-sample Mendelian randomization (TSMR) with summary statistics from large-scale genome-wide association studies (GWAS) was applied. Summary statistics were extracted for these instruments from GWAS of PD and schizophrenia; *Results*: We found an increase in the risk of schizophrenia per one-standard deviation (SD) increase in the genetically-predicted PD risk (inverse-variance weighted method, odds ratio = 1.10; 95% confidence interval, 1.05–1.15; $p = 3.49 \times 10^{-5}$). The association was consistent in sensitivity analyses, including multiple TSMR methods, analysis after removing outlier variants with potential pleiotropic effects, and analysis after applying multiple GWAS subthresholds. No relationships were evident between PD and smoking or other psychiatric disorders, including attention deficit hyperactivity disorder, autism spectrum disorder, bipolar affective disorder, major depressive disorder, Alzheimer's disease, or alcohol dependence. However, we did not find a reverse relationship; genetic variants increasing schizophrenia risk did not alter the risk of PD; *Conclusions*: Overall, our findings suggest that increased genetic risk of PD can be associated with increased risk of schizophrenia. This association supports the intrinsic nature of the psychotic symptom in PD rather than medication or environmental effects. Future studies for possible comorbidities and shared genetic structure between the two diseases are warranted.

Keywords: Parkinson's disease; schizophrenia; Mendelian randomization; genetics

1. Introduction

Parkinson's disease (PD) is the second most common progressive neurodegenerative disease. PD is characterized by tremors, bradykinesia, rigidity, and posture instability [1]. Motor symptoms have been the primary focus in the diagnosis and the treatment target of patients with PD. However, non-motor symptoms have recently been described and include depression, anxiety, sleep problems, bowel and bladder habit changes, autonomic disturbances, and sensory complaints. Psychiatric symptoms that cause pronounced distress in patients with PD are also common and are closely related to low life satisfaction and quality of life [2]. Anxiety symptoms in PD have been reported by up to 30% of patients [3], and are accompanied by depression, somatic symptoms, and hostility-irritability. Suicide risk in PD patients is also higher than in the general population [4,5].

Psychotic symptoms are also frequently observed non-motor symptoms of PD, with a prevalence of 20% to 70%. Symptoms of the psychosis spectrum in early PD consist of minor experiences, such as passage and presence of hallucinations, illusions, and formed hallucinations, including recurring visual hallucinations with insight preserved [6]. Delusions and hallucinations can occur in the later stage of PD [7,8]. However, the biological mechanisms of psychotic symptoms in PD—whether based on PD itself, on cognitive decline, or on medication side effect—are not clearly understood. The primary treatment of choice for motor symptoms in PD is dopamine agonist. The onset of PD psychosis is suggested by the hyper-regulation of dopamine elicited by medication, which can be relieved by dose reduction. However, there has been no evidence of a direct causal relationship between pharmacological treatments or medication dose and psychotic symptoms in PD [9]. There is still a need for further investigation to clarify the causal relationship between PD psychosis and dopamine-related pathology.

In schizophrenia, although positive psychotic symptoms and negative symptoms as deficit are the usual described core features, motor symptoms have also been recently highlighted [10]. Motor symptoms, called parkinsonian symptoms, are frequently observed both in premorbid families of schizophrenia and in naïve patients for antipsychotics [11].

Despite the similarity between observed symptomatology of PD and schizophrenia, several components, including differences in the age of onset and incompatibility in dopamine hypothesis, and possible adverse effects of medications, have hampered research on an association between PD and schizophrenia. Only a few fragmented studies on the association between PD and schizophrenia have been reported [12,13].

It is difficult to establish the genetic association between PD and schizophrenia by observational epidemiological approaches. One promising approach to investigating the association is Mendelian randomization (MR) using genetic variants as the instrumental variables [14]. Association between genetic variants and disease outcome state can provide evidence with minimizing confounding factors, including age, medication, or environmental exposures.

We hypothesized that PD and schizophrenia have shared genetic background and the risk for PD has a causal effect on the risk for schizophrenia and *vice versa*. We tested the hypothesis using two-sample MR methods with summary statistics from large-scale genome-wide association studies (GWAS) of PD and schizophrenia.

2. Materials and Methods

2.1. Datasets

For two-sample MR, we obtained summary statistics from large-scale GWAS of PD, schizophrenia, and other psychiatric diseases or related traits, including bipolar disorder, major depressive disorder, Alzheimer's disease, alcohol dependence, and smoking. The GWASs of other psychiatric disorders or traits were used to test whether the genetic variants for PD act on schizophrenia through other biological pathways (horizontal pleiotropy) other than the direct effect of PD. They are available on the LD Hub website (http://ldsc.broadinstitute.org/gwashare/; accessed on 1 May 2020), and details are listed in Supplementary Table S1. We also used the information of recently identified single nucleotide polymorphisms (SNPs) from GWAS of PD [15]. For SNP selection, we conducted the following steps. First, we obtained SNPs as instrumental variants that were significantly associated with one disease or trait (exposure) at the P-thresholds for suggestive significance or at a stricter level ($p < 5 \times 10^{-8}$–5×10^{-6}). The relaxed thresholds have been considered in previous MR studies, given an exposure GWAS with a small number of genome-wide significant SNPs ($p < 5 \times 10^{-8}$) such as PD GWAS in this study [16,17]. Second, to ensure that the instruments for exposure were independent of each other, we performed linkage disequilibrium (LD) clumping with a 10 Mb window size and LD value ($r^2 < 0.001$) using data from European individuals from the 1000 Genomes Project Phase 3. Third, we extracted the same SNPs from GWAS of other diseases or traits (outcome), and by comparing the frequencies and effect sizes of the same SNPs on the exposure with those on the outcome, we removed SNPs with ambiguous alleles from the set of instruments.

2.2. Statistical Analyses

One of the underlying assumptions of the examination of the relationship using MR is that the instrumental variants have no pleiotropic effects. The effect estimate can be severely biased if the genetic variants extracted as instrumental variables violate this assumption. Therefore, we applied the MR-PRESSO method to detect and remove instruments with potential pleiotropic effects to eliminate the bias [18]. The MR-PRESSO consists of a global test, outlier test, and distortion test. Through a three-step procedure, we detected outlier variants having pleiotropic effects and excluded these in the subsequent analysis. We performed two-sample Mendelian randomization (TSMR) to infer the association of an exposure on an outcome using summary statistics from GWAS [19]. We performed TSMR analyses using the MR-Base software (http://www.mrbase.org/; accessed on 1 May 2020) that provides various functions for combining, harmonizing, and utilizing GWAS summary statistics. Multiple methods for TSMR have been developed and are different from each other in terms of sensitivity to heterogeneity, bias, and power. We selected the inverse-variance weighted (IVW) method as our main TSMR method because it provides reliable results in the presence of heterogeneity in an MR analysis and is appropriate when using a large number of SNPs [20–22]. The standard error of the IVW effect was estimated using a multiplicative random effects model. We performed leave-one-out analysis to test if the results were derived from any particular SNP. A forest plot was used to visualize heterogeneity between instruments due to horizontal pleiotropy and the contribution of each instrument to the overall estimate [19]. We also used the MR Egger regression (MR-Egger) and the weighted median (WM) for sensitivity analyses [23,24]. Since these two methods provide reliable causal estimates in the presence of a violation of MR assumptions, WM and MR-Egger have been used as sensitivity analyses in MR studies [23,25]. Since horizontal pleiotropy can be a confounding factor in MR, we tested pleiotropy by performing MR-Egger with the intercept unconstrained. The intercept of the MR-Egger shows a statistical estimate of the presence of directional pleiotropy.

3. Results

3.1. Causality between PD Risk and Schizophrenia Risk

To investigate the shared genetic background between PD risk and schizophrenia risk, we selected four genetic variants that were significantly associated with risk of PD from GWAS with 1713 cases with PD and 3978 controls as instruments (Table 1) [26]. We extracted summary statistics for these instruments from GWAS of schizophrenia in the Psychiatric Genomics Consortium (PGC) with 35,476 cases and 46,839 controls [27]. The same alleles of the four SNPs increased both the risk of PD and the risk of schizophrenia. We annotated the four instrumental variants using Variant Effect Predictor (VEP) [28]. Three of these SNPs were located at the *CDH8*, *SNCA*, and *WNT3* genes, with no known gene near the other SNP.

Table 1. List of the SNPs associated with risk for Parkinson's disease (PD) and their associations with risk for schizophrenia (SCZ).

SNP	Chromosome	Position (hg19)	Gene Region	Effective Allele	PD			SCZ		
					Beta	SE of Beta	*p* Value	Beta	SE of Beta	*p* Value
rs4889730	17	21717727	None	G	−0.32	0.05	2.83×10^{-11}	−0.02	0.01	0.04
rs2736990	4	90678541	SNCA	G	0.24	0.04	5.69×10^{-9}	0.03	0.01	1.60×10^{-3}
rs3784847	16	61977449	CDH8	G	0.46	0.08	1.66×10^{-9}	0.02	0.02	0.22
rs415430	17	44859144	WNT3	C	−0.29	0.05	4.50×10^{-8}	−0.04	0.01	1.65×10^{-3}

SNP—single nucleotide polymorphism; Beta—ln (odds ratio); SE—standard error.

Two-sample MR using IVW revealed a causality of PD risk on schizophrenia risk (odds ratio [OR] per log odds increase in PD risk 1.10, 95% confidence interval [CI] 1.05–1.15, $p = 3.49 \times 10^{-5}$) (Figure 1A and Table 2). A funnel plot was constructed. In a funnel plot, each dot shows the proportion of the precision (1/standard error) to Wald ratios per SNP, and the vertical line indicates the MR estimates jointed by the four instruments (Figure 1C). We observed overall symmetry in the funnel plot (asymmetry represents heterogeneity driven by directional horizontal pleiotropy that violates MR assumptions) [19]. Even after relaxing the *p*-value threshold for defining the instruments for exposure and including more SNPs with less significant association with PD risk, a causal effect of PD on schizophrenia risk remained significant (Table 2). OR, 95% CIs, and *p*-values of IVW method are shown in Supplementary Table S2. The Cochrane *Q* statistics, *Q* value, and heterogeneity (I^2[%]) were 2.54, 0.47, and 0, respectively, which indicated little heterogeneity between instrumental variants in the MR analysis for the risk of PD (Figure 2A). Leave-one-out analysis showed that all the SNPs contributed to the association of PD with schizophrenia (Figure 2B). In contrast, TSMR analysis with schizophrenia risk as exposure and PD risk as outcome showed no evidence for the causal effect of schizophrenia risk on PD risk (Supplementary Tables S3 and S4).

Table 2. Two-sample Mendelian randomization results of Parkinson's disease and other psychiatric disorders or related traits using the inverse-variance weighted regression.

Subthreshold for PD GWAS *p*-Value	5×10^{-8}		5×10^{-7}		5×10^{-6}	
	p-Value	OR (95% CI)	*p*-Value	OR (95% CI)	*p*-Value	OR (95% CI)
SCZ	3.49×10^{-5}	1.10 (1.05–1.15)	7.00×10^{-7}	1.10 (1.06–1.14)	6.64×10^{-5}	1.06 (1.03–1.09)
ADHD	0.78	0.99 (0.91–1.07)	1.00	1.00 (0.93–1.07)	0.31	1.02 (0.98–1.05)
ASD	0.07	0.92 (0.83–1.00)	0.10	0.93 (0.85–1.01)	0.10	0.93 (0.85–1.01)
BP	0.22	1.10 (0.93–1.26)	0.11	1.10 (0.97–1.23)	0.16	1.05 (0.98–1.12)
MDD	0.92	1.00 (0.89–1.10)	0.67	0.98 (0.90–1.07)	0.22	0.97 (0.91–1.02)
AD	0.67	1.03 (0.91–1.15)	0.86	0.99 (0.86–1.11)	0.39	0.98 (0.92–1.03)
Alcohol dependence	0.91	0.98 (0.68–1.28)	0.80	0.97 (0.73–1.21)	0.58	0.96 (0.81–1.11)
Smoking	0.99	1.00 (0.64–1.36)	0.46	1.21 (0.60–1.82)	0.25	1.15 (0.87–1.43)
Cannabis use	0.005	1.02 (1.01–1.03)	0.15	1.01 (0.99–1.03)	0.19	1.01 (0.99–1.02)

PD—Parkinson's disease; GWAS—genome-wide association study; OR—odds ratio; CI—confidence interval; SCZ—schizophrenia; ADHD—attention deficit hyperactivity disorder; ASD—autism spectrum disorders; BP—bipolar disorder; MDD—major depressive disorder; AD—Alzheimer's disease.

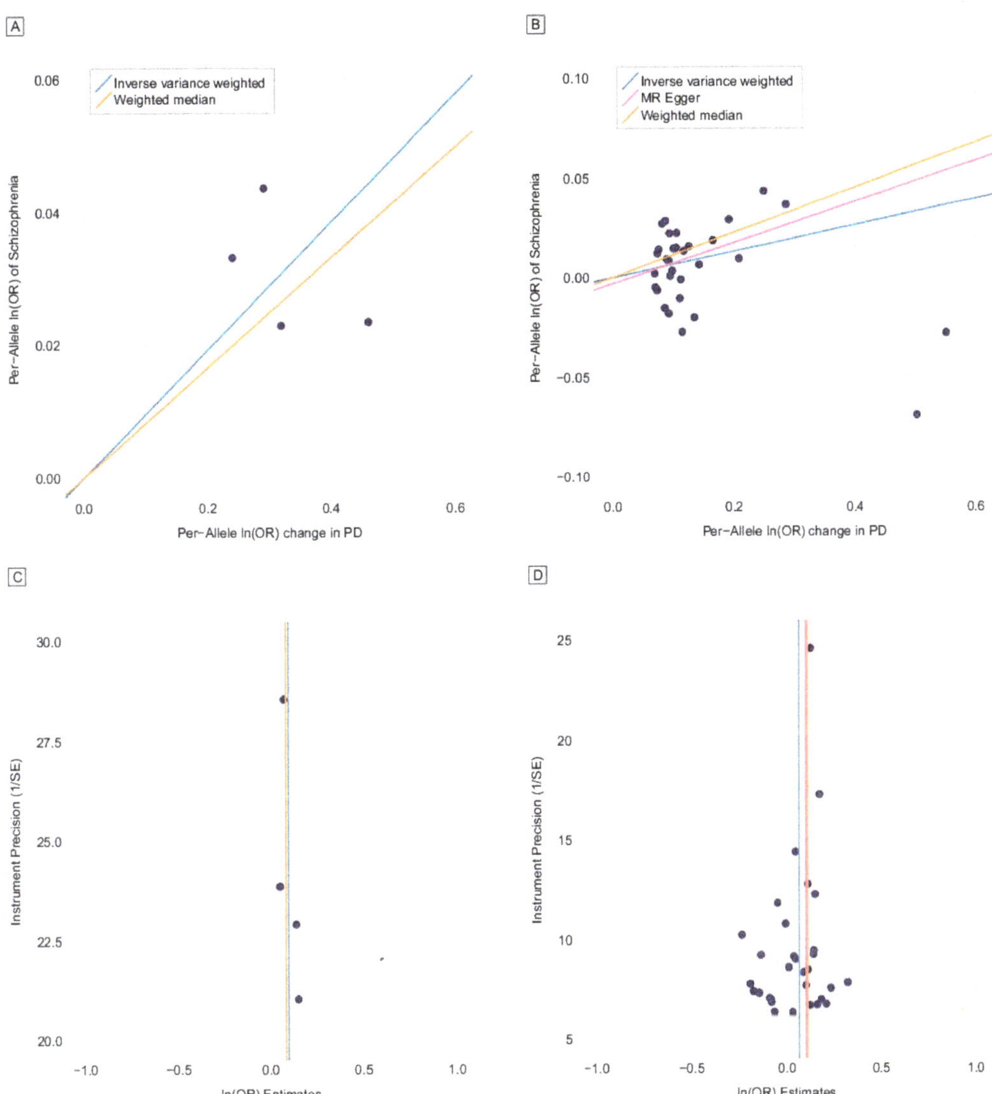

Figure 1. Mendelian randomization analysis for risk of Parkinson's disease (PD) on schizophrenia using genetic instruments. (**A**) Schizophrenia associations (scatter plot) using four genetic instruments (PD) [24]. (**B**) Schizophrenia associations (scatter plot) using 32 genetic instruments (PD) [15]. (**C**) Precision (funnel plot) using four genetic instruments (PD) [24]. (**D**) Precision (funnel plot) using 32 genetic instruments (PD) [15].

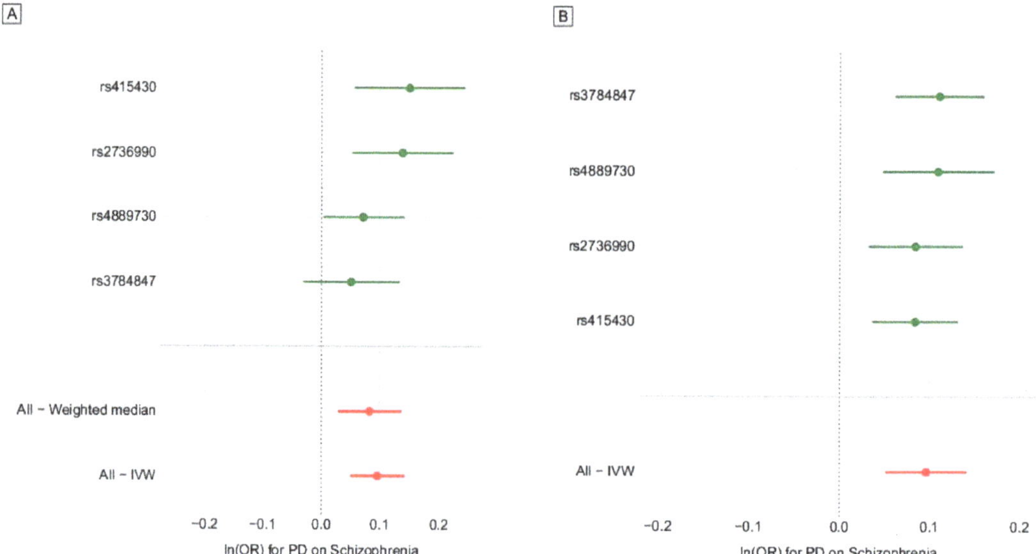

Figure 2. Forest and leave-one-out plots showing association with schizophrenia for each genetic instrument of Parkinson's disease (PD). (**A**) Forest plot using four genetic instruments (PD) [24]. (**B**) Leave-one-out plot using four genetic instruments (PD) [24].

In addition, we performed sensitivity analysis using WM and tested horizontal pleiotropy using MR-Egger (Supplementary Tables S2 and S5). The effect of PD on the risk of schizophrenia was significant in the WM analysis (OR 1.09, 95% CI 1.02–1.15, $p = 3.77 \times 10^{-3}$). MR-Egger showed no evidence of horizontal pleiotropy (intercept OR 1.05, 95% CI 0.99–1.11, $p = 0.25$).

We also tested newly discovered SNPs in a recent GWAS study with 6476 cases with PD and 302,042 controls [15]. The data included 34 SNPs associated with PD at the genome-wide significance level. Since two of the 34 variants showed pleiotropy when applying the outlier test of MR-PRESSO, we excluded them in the MR analysis. The IVW method using the 32 SNPs showed a positive causal effect (OR 1.07, 95% CI 1.02–1.12, $p = 1.81 \times 10^{-3}$) on schizophrenia risk per log odds increase in PD risk (Figure 1B,D). The effect of PD risk on the risk of developing schizophrenia was significant in IVW ($p = 1.81 \times 10^{-3}$) and WM ($p = 2.84 \times 10^{-5}$), but was marginally significant in MR-Egger ($p = 0.05$). However, both IVW and MR-Egger results were significant for SNPs passing GWAS sub-thresholds of $p < 5 \times 10^{-7}$ and $p < 5 \times 10^{-6}$ (Supplementary Tables S6 and S7). There was moderate heterogeneity between instrumental variants in the MR analysis for the risk of PD; the Cochran Q statistics, Q value, and heterogeneity (I^2[%]) were 45.43, 0.05, 32, respectively (Supplementary Figure S1A). However, the result of leave-one-out analysis for the risk of PD showed that single SNPs were not exclusively responsible for the associations of the risk of PD (Supplementary Figure S1B). MR-Egger suggested no evidence of pleiotropy (intercept OR 1.00, 95% CI 0.98–1.01, $p = 0.49$).

3.2. *Shared Genetic Background between PD and Other Psychiatric Disorders or Related Traits*

We performed TSMR for other psychiatric disorders to investigate if PD also has a shared genetic background with any other psychiatric disorders or related traits than schizophrenia. The other eight psychiatric disorders or related traits were not related to PD in the IVW MR results (Table 2). This implies that the SNPs selected as instruments showed no pleiotropy within the psychiatric traits. Even after mitigating the GWAS p-value threshold to define additional instrumental variants for PD, the association of PD risk on the seven psychiatric traits was not significant (Table 2). Our analysis showed no evidence for shared genetic background between PD and other psychiatric disorders.

4. Discussion

We used TSMR to evaluate the association between PD and schizophrenia. Our findings support a possible shared genetic background between PD and schizophrenia. The genetic variants increasing the risk of PD were likely to influence the increased risk of schizophrenia. However, we did not observe a reverse directional relationship; genetic variants increasing schizophrenia risk did not increase the risk of PD.

The relationship between PD and schizophrenia was significant in the IVW and WM analyses, but not in the MR-Egger regression approach. Of note, the causal inference by MR can be severely affected if a fundamental assumption of "no pleiotropy" is not satisfied. MR-Egger was developed to detect violations of this assumption and to provide a robust effect estimate when genetic instrumental variables showed pleiotropy. Our MR-Egger analyses suggested that there was no clear evidence for an influence of biological pleiotropy on the findings [23]. Because the power of MR-Egger was previously shown to be lower than other conventional methods, we used the IVW as the main MR method and the WM for sensitivity analysis. In addition, we also used the latest large-scale PD GWAS in 2017 [15] including more individuals than PD GWAS in 2009. By including more genetic variants and removing outlier variants with potential pleiotropic effects using MR-PRESSO, the effect of genetic risk of PD on the risk for schizophrenia appeared to be significant in the MR-Egger analysis. Moreover, our analysis using multiple GWAS subthresholds also confirmed evidence for an association between PD and risk of schizophrenia. Furthermore, we observed no significant relationship between PD and smoking, or with other psychiatric disorders, such as attention deficit hyperactivity disorder, autism spectrum disorder, bipolar affective disorder, major depressive disorder, Alzheimer's disease, or alcohol dependence. These results suggest that the effect of PD genetic risk on schizophrenia risk may not be via alternative risk factors, such as other psychiatric disorders or smoking [29]. Only the association between PD and cannabis use was significant at a threshold level (5×10^{-8}), which was not replicated when the weighted median method was used ($p > 0.05$). Although a recent study suggested the potential connection between cannabis use and schizophrenia [30], our results did not support the mediating effect of other mechanisms between PD and schizophrenia.

PD psychosis symptoms (A and B in Figure 3) range from mild psychotic symptoms, such as illusions or referential ideas (B in Figure 3), to prominent psychotic symptoms, including vivid hallucinations or systematized delusions that fulfilled characteristic symptoms of diagnostic criteria in schizophrenia (A in Figure 3) [31,32]. The unidirectionality of the causal effect of PD on schizophrenia implies that the psychotic symptoms in PD patients, which are similar to those in schizophrenia patients (A in Figure 3), were more likely to be due to the pathophysiology of the PD itself than dopamine agonist or environmental factors (Figure 3). This directionality could imply the possibility of genetic architecture related to PD bringing out psychotic symptoms, which could be interpreted as diagnosis of schizophrenia in clinical practice. However, the reverse relationship was not demonstrated: genetic architecture of schizophrenia was not observed to bring out PD symptoms, suggesting independent pathology of psychotic symptoms in PD from that of psychotic symptoms in schizophrenia. Even though psychotic symptoms observed in PD and schizophrenia share common clinical features, our finding suggests that genetic backgrounds in PD psychotic symptoms are independent of those in schizophrenia (A in Figure 3). This finding can also be supported by previously observed differences in dopamine related neuroimages (between PD and schizophrenia) and by novel, successful antipsychotic effect of targeting serotonin pathways on PD psychosis [33–35]. Clozapine is an effective antipsychotic for PD psychosis that displays low affinity for the dopamine receptor, but it has a selective effect on serotonin 5-HT2A and histamine H1 receptors [36]. In addition, pimavanserin, a recently approved drug for PD psychosis, is a selective serotonin 2A receptor inverse agonist [37]. Since schizophrenia is a subset of a psychotic disorder (Figure 3) and is not identical with the whole disease entity with a psychotic symptom, careful interpretation is needed. Further MR analyses between PD and broad psychosis phenotype [38] will be helpful to elucidate the underlying cause of psychotic symptoms in PD.

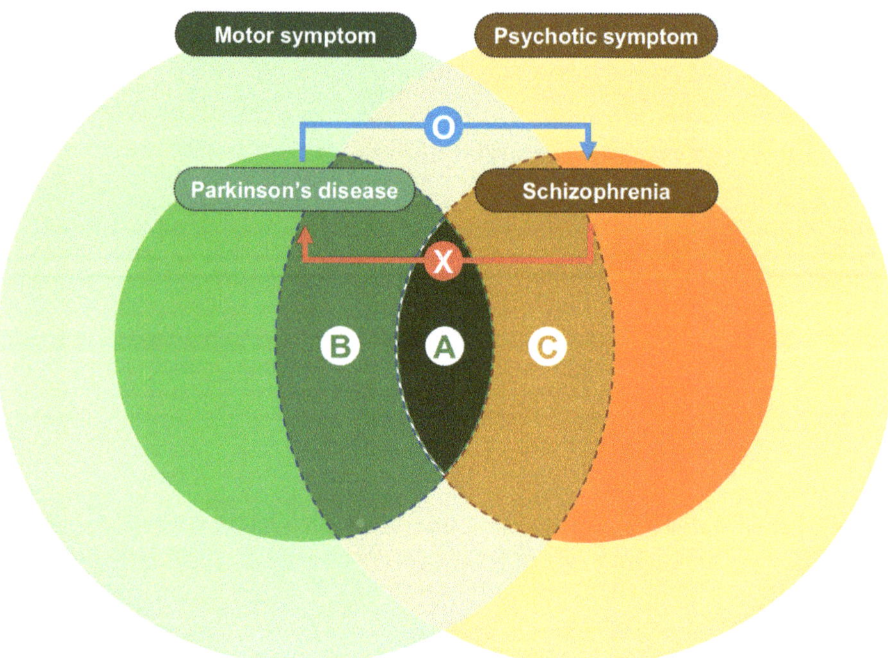

Figure 3. Schematic illustration of the association between Parkinson's disease and schizophrenia from two-sample Mendelian randomization. The blue arrow indicates the direction of causal relationship between Parkinson's disease and schizophrenia. The red arrow indicates an insignificant causal relationship between in reverse direction.

Motor symptoms in schizophrenia include extrapyramidal signs or saccadic eye movements, which are frequently observed in PD [39] (A in Figure 3) and hypokinesia, retarded catatonia, excited catatonia, echo-phenomena, or catalepsy, which are inter-related motor domains in schizophrenia [40]. Our results suggest that the motor symptom mimicking PD in schizophrenia might be related to the genetic risk of PD. Previous studies in drug-naïve schizophrenia patients with parkinsonism (A in Figure 3) have described decreased dopaminergic function in the striatum [41,42]. Considering that the main pathology of PD is dopamine depletion in the nigrostriatal pathway, these findings can be in line with our results.

PD is a neurodegenerative disorder that is frequent in middle-aged or elderly people, and schizophrenia is frequently diagnosed in individuals in their twenties and thirties [43]. However, the onset of the neurodegenerative process associated with PD remains uncertain [44]. Previous neuroimaging studies suggested that the period between the beginning of pathological changes and the onset of motor symptoms in PD would last from 3 to 6 years, although a prolonged phase may exist as long as 20 to 50 years [44–46]. In addition, non-motor symptoms, such as depression, rapid eye movement (REM) sleep behavior, or constipation, may precede the motor symptoms by up to 20 years [44]. Thus, it may be difficult to conclude that the onset of pathological changes of PD was later than the onset of schizophrenia. However, in consideration of the rapid progression of early-onset or young-onset PD (symptoms <50 years old), with short pre-symptomatic period and natural degenerative changes in individuals, these hypotheses need further research.

Epidemiological evidence of comorbidities of PD and schizophrenia is very rare due to obstacles in diagnosis [13]. For the diagnosis of schizophrenia, the psychotic symptoms are not attributable to the effect of medication or another medical condition, including PD or anti-Parkinson medications [32]. Moreover, there is no established diagnostic biomarker for schizophrenia [47]. As a result, only a small portion of patients with PD might be diagnosed with schizophrenia in clinical settings [48]. In addition, it is difficult to exclude drug-induced parkinsonism in patients with schizophrenia who have received anti-psychotics [49]. A case report of comorbid PD and schizophrenia presented two patients with significantly decreased dopamine transporter density in the striatum on F-18-N-(3-fluoropropyl)-2β-carbomethoxy-3β-(4-iodophenyl) nortropane positron emission tomography, which indicated genuine PD rather than drug-induced parkinsonism [13]. Further development of biomarkers for PD and schizophrenia and clinical attention for possible comorbidity of these diseases are needed.

Our study has several strengths. First, we utilized TSMR analysis using accurate estimates for the association of genetic variants with the studied traits from a very large number of samples. The increased statistical power is the key benefit of using summary statistics from the GWAS consortium in TSMR, particularly for the test of effects on dichotomous outcomes, such as PD or schizophrenia. In particular, MR analysis could be a promising method when exploring the relationship between two phenotypes with relatively rare incidence, obstacles in epidemiological studies for comorbidities, or lack of definite biomarkers. Our analyses in both directions were based on such vast samples, which might enable the identification of small effects that may exist in the context of these complex phenotypes. Second, our study revealed a robust shared genetic background between PD and schizophrenia across multiple complementary methods of TSMR analysis and multiple GWAS p-value cutoffs. In the presence of pleiotropic effects of instrumental genetic variants, MR-Egger provides a robust causal estimate, while the IWV and VM have better power for causal inference under the "no-pleiotropy" assumption. By applying MR-PRESSO, our analysis was less likely to violate the MR assumptions. We also showed that the causal estimates of polygenic variants passing sub-thresholds were consistent with those of GWAS variants.

Several limitations of our study must be acknowledged. We could not elaborate the specific symptomatology related to PD increasing risk of schizophrenia, because there is no currently available GWAS data for specific types of PD that are differentiated by weight of non-motor symptoms and motor symptoms. It is difficult to conduct subgroup analyses and effect moderation with publicly available GWAS summary statistics [50]. Further research can elaborate the specific symptomatology associated with increased risk of schizophrenia in PD, if there is independent GWAS data for each subphenotype of PD.

5. Conclusions

Our results support the view that an increased genetic risk of PD could be associated with an increased risk of schizophrenia. This finding suggests that the intrinsic pathophysiology of PD, rather than anti-Parkinson medication or environmental effects, has a more weighted effect in PD psychosis. In addition to the recent success of pimavanserine in PD psychosis, our finding provides more support for the independent pathology of PD psychosis and suggests a novel perspective on future treatment development. In addition, our results support the view that the motor symptoms in schizophrenia are causally related to the pathophysiology of PD. Thus, elaborate scales used to measure motor symptoms in PD could be applied with thorough assessment to measure motor symptoms in schizophrenia. The biological mechanisms of PD and schizophrenia need to be clarified, given their shared symptomatology and their significant impact on patients' quality of life [31]. Furthermore, additional clinical attention for possible comorbidity of these diseases is needed.

Supplementary Materials: The following are available online at https://www.mdpi.com/article/10.3390/brainsci11081042/s1, Figure S1: Forest and leave-one-out plots showing association with schizophrenia for each genetic instrument of Parkinson's disease (PD), Table S1: Genome-wide association studies used in this study, Table S2: Two-sample Mendelian randomization results of Parkinson's disease [2] (exposure trait) and schizophrenia [3] (outcome trait) using inverse variance weighting (IVW)and weighted median (WM), Table S3: Two-sample Mendelian randomization results of schizophrenia [3] (exposure trait) and Parkinson's disease [2] (outcome trait) using inverse variance weighting (IVW)and weighted median (WM), Table S4: Two-sample Mendelian randomization results of schizophrenia [3] (exposure trait) and Parkinson's disease [2] (outcome trait) using MR-Egger, Table S5: Two-sample Mendelian randomization results of Parkinson's disease [2] (exposure trait) and schizophrenia [3] (outcome trait) using MR-Egger, Table S6: Two-sample Mendelian randomization results of Parkinson's disease [1] (exposure trait) and schizophrenia [3] (outcome trait) using IVW and WM after removing outliers that exhibit pleiotropy detected by MR-PRESSO, Table S7: Two-sample Mendelian randomization results of Parkinson's disease [1] (exposure trait) and schizophrenia [3] (outcome trait) using MR-Egger after removing outliers that exhibit pleiotropy detected by MR-PRESSO.

Author Contributions: H.-H.W. and W.M. had full access to all of the data in the study and take responsibility for the integrity of the data and the accuracy of the data analysis. K.K. (Investigation, Visualization, Writing—original draft), S.K. (Data curation, Formal analysis, Investigation, Software, isualization, Writing—original draft), W.M. (Methodology, Project administration, Supervision, Writing—review & editing, Conceptualization), I.S. (Investigation), H.L. (Investigation), B.K. (Investigation), S.K.C. (Investigation), J.Y. (Writing—review & editing, Investigation), D.K.K. (Supervision, Writing—review & editing), H.-H.W. (Funding acquisition, Methodology, Project administration, Supervision, Writing—review & editing) All authors have read and agreed to the published version of the manuscript.

Funding: This paper was supported by Sungkyun Research Fund, Sungkyunkwan University, 2017. The funding source was not involved in the design and conduct of the study; collection, management, analysis, and interpretation of the data; and preparation, review, or approval of the manuscript.

Institutional Review Board Statement: This study was approved by the institutional review board of the Seoul National University Bundang Hospital (X-1902/524-903).

Informed Consent Statement: Informed consent was waived because of the retrospective nature of the study and the analysis used anonymous data.

Data Availability Statement: Data sharing not applicable. No new data were created or analyzed in this study.

Conflicts of Interest: The authors declare no conflict of interest.

References

1. Gelb, D.J.; Oliver, E.; Gilman, S. Diagnostic criteria for Parkinson disease. *Arch. Neurol.* **1999**, *56*, 33–39. [CrossRef] [PubMed]
2. Vescovelli, F.; Sarti, D.; Ruini, C. Well-being and distress of patients with Parkinson's disease: A comparative investigation. *Int. Psychogeriatr.* **2019**, *31*, 21–30. [CrossRef] [PubMed]
3. Mele, B.; Holroyd-Leduc, J.; Smith, E.E.; Pringsheim, T.; Ismail, Z.; Goodarzi, Z. Detecting anxiety in individuals with Parkinson disease: A systematic review. *Neurology* **2018**, *90*, e39–e47. [CrossRef]
4. Okbay, A.; Baselmans, B.M.; De Neve, J.E.; Turley, P.; Nivard, M.G.; Fontana, M.A.; Meddens, S.F.; Linner, R.K.; Rietveld, C.A.; Derringer, J.; et al. Genetic variants associated with subjective well-being, depressive symptoms, and neuroticism identified through genome-wide analyses. *Nat. Genet.* **2016**, *48*, 624–633. [CrossRef]
5. Turley, P.; Walters, R.K.; Maghzian, O.; Okbay, A.; Lee, J.J.; Fontana, M.A.; Nguyen-Viet, T.A.; Wedow, R.; Zacher, M.; Furlotte, N.A.; et al. Multi-trait analysis of genome-wide association summary statistics using MTAG. *Nat. Genet.* **2018**, *50*, 229–237. [CrossRef]
6. Ravina, B.; Marder, K.; Fernandez, H.H.; Friedman, J.H.; McDonald, W.; Murphy, D.; Aarsland, D.; Babcock, D.; Cummings, J.; Endicott, J.; et al. Diagnostic criteria for psychosis in Parkinson's disease: Report of an NINDS, NIMH work group. *Mov. Disord.* **2007**, *22*, 1061–1068. [CrossRef]
7. Inzelberg, R.; Kipervasser, S.; Korczyn, A.D. Auditory hallucinations in Parkinson's disease. *J. Neurol. Neurosurg. Psychiatry* **1998**, *64*, 533–535. [CrossRef] [PubMed]
8. Fenelon, G.; Mahieux, F.; Huon, R.; Ziegler, M. Hallucinations in Parkinson's disease: Prevalence, phenomenology and risk factors. *Brain* **2000**, *123*, 733–745. [CrossRef]

9. Goetz, C.G.; Pappert, E.J.; Blasucci, L.M.; Stebbins, G.T.; Ling, Z.D.; Nora, M.V.; Carvey, P.M. Intravenous levodopa in hallucinating Parkinson's disease patients: High-dose challenge does not precipitate hallucinations. *Neurology* **1998**, *50*, 515–517. [CrossRef] [PubMed]
10. Gabilondo, A.; Alonso-Moran, E.; Nuno-Solinis, R.; Orueta, J.F.; Iruin, A. Comorbidities with chronic physical conditions and gender profiles of illness in schizophrenia. Results from PREST, a new health dataset. *J. Psychosom. Res.* **2017**, *93*, 102–109. [CrossRef] [PubMed]
11. Walther, S.; Mittal, V.A. Motor System Pathology in Psychosis. *Curr. Psychiatry Rep.* **2017**, *19*, 97. [CrossRef] [PubMed]
12. Kim, T.J.; Lee, H.; Kim, Y.E.; Jeon, B.S. A case of Parkin disease (PARK2) with schizophrenia: Evidence of regional selectivity. *Clin. Neurol. Neurosurg.* **2014**, *126*, 35–37. [CrossRef]
13. Oh, J.; Shen, G.X.; Nan, G.X.; Kim, J.M.; Jung, K.Y.; Jeon, B. Comorbid schizophrenia and Parkinson's disease: A case series and brief review. *Neurol. Asia* **2017**, *22*, 139–142.
14. Smith, G.D.; Hemani, G. Mendelian randomization: Genetic anchors for causal inference in epidemiological studies. *Hum. Mol. Genet.* **2014**, *23*, R89–R98. [CrossRef]
15. Chang, D.; Nalls, M.A.; Hallgrimsdottir, I.B.; Hunkapiller, J.; van der Brug, M.; Cai, F.; International Parkinson's Disease Genomics Consortium; 23andMe Research Team; Kerchner, G.A.; Ayalon, G.; et al. A meta-analysis of genome-wide association studies identifies 17 new Parkinson's disease risk loci. *Nat. Genet.* **2017**, *49*, 1511–1516. [CrossRef] [PubMed]
16. Hartwig, F.P.; Borges, M.C.; Horta, B.L.; Bowden, J.; Smith, G.D. Inflammatory Biomarkers and Risk of Schizophrenia: A 2-Sample Mendelian Randomization Study. *JAMA Psychiatry* **2017**, *74*, 1226–1233. [CrossRef] [PubMed]
17. Sanna, S.; van Zuydam, N.R.; Mahajan, A.; Kurilshikov, A.; Vila, A.V.; Vosa, U.; Mujagic, Z.; Masclee, A.A.M.; Jonkers, D.; Oosting, M.; et al. Causal relationships among the gut microbiome, short-chain fatty acids and metabolic diseases. *Nat. Genet.* **2019**, *51*, 600–605. [CrossRef] [PubMed]
18. Verbanck, M.; Chen, C.Y.; Neale, B.; Do, R. Detection of widespread horizontal pleiotropy in causal relationships inferred from Mendelian randomization between complex traits and diseases. *Nat. Genet.* **2018**, *50*, 693–698. [CrossRef] [PubMed]
19. Hemani, G.; Zheng, J.; Elsworth, B.; Wade, K.H.; Haberland, V.; Baird, D.; Laurin, C.; Burgess, S.; Bowden, J.; Langdon, R.; et al. The MR-Base platform supports systematic causal inference across the human phenome. *Elife* **2018**, *7*, e34408. [CrossRef]
20. Dastani, Z.; Hivert, M.F.; Timpson, N.; Perry, J.R.; Yuan, X.; Scott, R.A.; Henneman, P.; Heid, I.M.; Kizer, J.R.; Lyytikainen, L.P.; et al. Novel loci for adiponectin levels and their influence on type 2 diabetes and metabolic traits: A multi-ethnic meta-analysis of 45,891 individuals. *PLoS Genet.* **2012**, *8*, e1002607. [CrossRef]
21. International Consortium for Blood Pressure Genome-Wide Association Studies. Genetic variants in novel pathways influence blood pressure and cardiovascular disease risk. *Nature* **2011**, *478*, 103–109. [CrossRef]
22. Burgess, S.; Butterworth, A.; Thompson, S.G. Mendelian randomization analysis with multiple genetic variants using summarized data. *Genet. Epidemiol.* **2013**, *37*, 658–665. [CrossRef]
23. Bowden, J.; Smith, G.D.; Burgess, S. Mendelian randomization with invalid instruments: Effect estimation and bias detection through Egger regression. *Int. J. Epidemiol.* **2015**, *44*, 512–525. [CrossRef] [PubMed]
24. Thomas, D.C.; Lawlor, D.A.; Thompson, J.R. Re: Estimation of bias in nongenetic observational studies using "Mendelian triangulation" by Bautista et al. *Ann. Epidemiol.* **2007**, *17*, 511–513. [CrossRef] [PubMed]
25. Bowden, J.; Smith, G.D.; Haycock, P.C.; Burgess, S. Consistent Estimation in Mendelian Randomization with Some Invalid Instruments Using a Weighted Median Estimator. *Genet. Epidemiol.* **2016**, *40*, 304–314. [CrossRef]
26. Simon-Sanchez, J.; Schulte, C.; Bras, J.M.; Sharma, M.; Gibbs, J.R.; Berg, D.; Paisan-Ruiz, C.; Lichtner, P.; Scholz, S.W.; Hernandez, D.G.; et al. Genome-wide association study reveals genetic risk underlying Parkinson's disease. *Nat. Genet.* **2009**, *41*, 1308–1312. [CrossRef]
27. Schizophrenia Working Group of the Psychiatric Genomics Consortium. Biological insights from 108 schizophrenia-associated genetic loci. *Nature* **2014**, *511*, 421–427. [CrossRef] [PubMed]
28. McLaren, W.; Gil, L.; Hunt, S.E.; Riat, H.S.; Ritchie, G.R.; Thormann, A.; Flicek, P.; Cunningham, F. The Ensembl Variant Effect Predictor. *Genome Biol.* **2016**, *17*, 122. [CrossRef]
29. Gage, S.H.; Jones, H.J.; Taylor, A.E.; Burgess, S.; Zammit, S.; Munafo, M.R. Investigating causality in associations between smoking initiation and schizophrenia using Mendelian randomization. *Sci. Rep.* **2017**, *7*, 40653. [CrossRef] [PubMed]
30. Gage, S.H.; Jones, H.J.; Burgess, S.; Bowden, J.; Smith, G.D.; Zammit, S.; Munafo, M.R. Assessing causality in associations between cannabis use and schizophrenia risk: A two-sample Mendelian randomization study. *Psychol. Med.* **2017**, *47*, 971–980. [CrossRef]
31. Ffytche, D.H.; Creese, B.; Politis, M.; Chaudhuri, K.R.; Weintraub, D.; Ballard, C.; Aarsland, D. The psychosis spectrum in Parkinson disease. *Nat. Rev. Neurol.* **2017**, *13*, 81–95. [CrossRef] [PubMed]
32. American Psychiatric Association. Diagnostic and Statistical Manual of Mental Disorders (DSM-5®): American Psychiatric Pub. 2013. Available online: https://dsm.psychiatryonline.org/doi/book/10.1176/appi.books.9780890425596 (accessed on 3 August 2021).
33. Hacksell, U.; Burstein, E.S.; McFarland, K.; Mills, R.G.; Williams, H. On the discovery and development of pimavanserin: A novel drug candidate for Parkinson's psychosis. *Neurochem. Res.* **2014**, *39*, 2008–2017. [CrossRef] [PubMed]
34. Ballanger, B.; Strafella, A.P.; van Eimeren, T.; Zurowski, M.; Rusjan, P.M.; Houle, S.; Fox, S.H. Serotonin 2A receptors and visual hallucinations in Parkinson disease. *Arch. Neurol.* **2010**, *67*, 416–421. [CrossRef] [PubMed]

35. Huot, P.; Johnston, T.H.; Darr, T.; Hazrati, L.N.; Visanji, N.P.; Pires, D.; Brotchie, J.M.; Fox, S.H. Increased 5-HT2A receptors in the temporal cortex of parkinsonian patients with visual hallucinations. *Mov. Disord.* **2010**, *25*, 1399–1408. [CrossRef] [PubMed]
36. Purkayastha, S.; Ford, J.; Kanjilal, B.; Diallo, S.; Del Rosario Inigo, J.; Neuwirth, L.; El Idrissi, A.; Ahmed, Z.; Wieraszko, A.; Azmitia, E.C.; et al. Clozapine functions through the prefrontal cortex serotonin 1A receptor to heighten neuronal activity via calmodulin kinase II-NMDA receptor interactions. *J. Neurochem.* **2012**, *120*, 396–407. [CrossRef]
37. Cummings, J.; Isaacson, S.; Mills, R.; Williams, H.; Chi-Burris, K.; Corbett, A.; Dhall, R.; Ballard, C. Pimavanserin for patients with Parkinson's disease psychosis: A randomised, placebo-controlled phase 3 trial. *Lancet* **2014**, *383*, 533–540. [CrossRef]
38. Psychosis Endophenotypes International Consortium; Wellcome Trust Case-Control Consortium; Bramon, E.; Pirinen, M.; Strange, A.; Lin, K.; Freeman, C.; Bellenguez, C.; Su, Z.; Band, G.; et al. A genome-wide association analysis of a broad psychosis phenotype identifies three loci for further investigation. *Biol. Psychiatry* **2014**, *75*, 386–397. [CrossRef]
39. Caldani, S.; Amado, I.; Bendjemaa, N.; Vialatte, F.; Mam-Lam-Fook, C.; Gaillard, R.; Krebs, M.O.; Bucci, M.P. Oculomotricity and Neurological Soft Signs: Can we refine the endophenotype? A study in subjects belonging to the spectrum of schizophrenia. *Psychiatry Res.* **2017**, *256*, 490–497. [CrossRef]
40. Morrens, M.; Docx, L.; Walther, S. Beyond boundaries: In search of an integrative view on motor symptoms in schizophrenia. *Front. Psychiatry* **2014**, *5*, 145. [CrossRef] [PubMed]
41. Mateos, J.J.; Lomena, F.; Parellada, E.; Mireia, F.; Fernandez-Egea, E.; Pavia, J.; Prats, A.; Pons, F.; Bernardo, M. Lower striatal dopamine transporter binding in neuroleptic-naive schizophrenic patients is not related to antipsychotic treatment but it suggests an illness trait. *Psychopharmacology* **2007**, *191*, 805–811. [CrossRef]
42. Schmitt, G.J.; Meisenzahl, E.M.; Frodl, T.; La Fougere, C.; Hahn, K.; Moller, H.J.; Dresel, S. Increase of striatal dopamine transmission in first episode drug-naive schizophrenic patients as demonstrated by [(123)I]IBZM SPECT. *Psychiatry Res.* **2009**, *173*, 183–189. [CrossRef]
43. Plaschke, R.N.; Cieslik, E.C.; Muller, V.I.; Hoffstaedter, F.; Plachti, A.; Varikuti, D.P.; Goosses, M.; Latz, A.; Caspers, S.; Jockwitz, C.; et al. On the integrity of functional brain networks in schizophrenia, Parkinson's disease, and advanced age: Evidence from connectivity-based single-subject classification. *Hum. Brain Mapp.* **2017**, *38*, 5845–5858. [CrossRef]
44. Gaig, C.; Tolosa, E. When does Parkinson's disease begin? *Mov. Disord.* **2009**, *24*, S656–S664. [CrossRef]
45. Vingerhoets, F.J.; Snow, B.J.; Lee, C.S.; Schulzer, M.; Mak, E.; Calne, D.B. Longitudinal fluorodopa positron emission tomographic studies of the evolution of idiopathic parkinsonism. *Ann. Neurol.* **1994**, *36*, 759–764. [CrossRef] [PubMed]
46. Scherman, D.; Desnos, C.; Darchen, F.; Pollak, P.; Javoy-Agid, F.; Agid, Y. Striatal dopamine deficiency in Parkinson's disease: Role of aging. *Ann. Neurol.* **1989**, *26*, 551–557. [CrossRef]
47. Cohen, B.M. Embracing Complexity in Psychiatric Diagnosis, Treatment, and Research. *JAMA Psychiatry* **2016**, *73*, 1211–1212. [CrossRef] [PubMed]
48. Gadit, A. Schizophrenia and Parkinson's disease: Challenges in management. *BMJ Case Rep.* **2011**, *2011*. [CrossRef]
49. Postuma, R.B.; Berg, D.; Stern, M.; Poewe, W.; Olanow, C.W.; Oertel, W.; Obeso, J.; Marek, K.; Litvan, I.; Lang, A.E.; et al. MDS clinical diagnostic criteria for Parkinson's disease. *Mov. Disord.* **2015**, *30*, 1591–1601. [CrossRef]
50. Lawlor, D.A. Commentary: Two-sample Mendelian randomization: Opportunities and challenges. *Int. J. Epidemiol.* **2016**, *45*, 908–915. [CrossRef] [PubMed]

Article

Homology Modelling, Molecular Docking and Molecular Dynamics Simulation Studies of CALMH1 against Secondary Metabolites of *Bauhinia variegata* to Treat Alzheimer's Disease

Noopur Khare [1,2], Sanjiv Kumar Maheshwari [1], Syed Mohd Danish Rizvi [3], Hind Muteb Albadrani [4], Suliman A. Alsagaby [4], Wael Alturaiki [4], Danish Iqbal [4,5,*], Qamar Zia [4,5], Chiara Villa [6], Saurabh Kumar Jha [7,8,9], Niraj Kumar Jha [7,8,9] and Abhimanyu Kumar Jha [7,*]

1 Institute of Biosciences and Technology, Shri Ramswaroop Memorial University, Barabanki 225003, Uttar Pradesh, India; noopur.khare2009@gmail.com (N.K.); sanjiv08@gmail.com (S.K.M.)
2 Department of Biotechnology, Dr. A.P.J. Abdul Kalam Technical University, Lucknow 226021, Uttar Pradesh, India
3 Department of Pharmaceutics, College of Pharmacy, University of Hail, Hail 2240, Saudi Arabia; sm.danish@uoh.edu.sa
4 Department of Medical Laboratory Sciences, College of Applied Medical Sciences, Majmaah University, Majmaah 11952, Saudi Arabia; h.albadrani@mu.edu.sa (H.M.A.); s.alsaqaby@mu.edu.sa (S.A.A.); w.alturaiki@mu.edu.sa (W.A.); qamarzia@mu.edu.sa (Q.Z.)
5 Health and Basic Sciences Research Center, Majmaah University, Al Majmaah 15341, Saudi Arabia
6 School of Medicine and Surgery, University of Milano-Bicocca, 20900 Monza, Italy; chiara.villa@unimib.it
7 Department of Biotechnology, School of Engineering and Technology, Sharda University, Greater Noida 201310, Uttar Pradesh, India; Saurabh.jha@sharda.ac.in (S.K.J.); nirajkumarjha2011@gmail.com (N.K.J.)
8 Department of Biotechnology, School of Applied & Life Sciences (SALS), Uttaranchal University, Dehradun 248007, Uttarakhand, India
9 Department of Biotechnology Engineering and Food Technology, Chandigarh University, Mohali 140413, Punjab, India
* Correspondence: da.mohammed@mu.edu.sa (D.I.); abhimanyu2006@gmail.com (A.K.J.)

Citation: Khare, N.; Maheshwari, S.K.; Rizvi, S.M.D.; Albadrani, H.M.; Alsagaby, S.A.; Alturaiki, W.; Iqbal, D.; Zia, Q.; Villa, C.; Jha, S.K.; et al. Homology Modelling, Molecular Docking and Molecular Dynamics Simulation Studies of CALMH1 against Secondary Metabolites of *Bauhinia variegata* to Treat Alzheimer's Disease. *Brain Sci.* **2022**, *12*, 770. https://doi.org/10.3390/brainsci12060770

Academic Editor: Claudio Torres

Received: 15 February 2022
Accepted: 31 May 2022
Published: 12 June 2022

Publisher's Note: MDPI stays neutral with regard to jurisdictional claims in published maps and institutional affiliations.

Copyright: © 2022 by the authors. Licensee MDPI, Basel, Switzerland. This article is an open access article distributed under the terms and conditions of the Creative Commons Attribution (CC BY) license (https://creativecommons.org/licenses/by/4.0/).

Abstract: Calcium homeostasis modulator 1 (CALHM1) is a protein responsible for causing Alzheimer's disease. In the absence of an experimentally designed protein molecule, homology modelling was performed. Through homology modelling, different CALHM1 models were generated and validated through Rampage. To carry out further in silico studies, through molecular docking and molecular dynamics simulation experiments, various flavonoids and alkaloids from *Bauhinia variegata* were utilised as inhibitors to target the protein (CALHM1). The sequence of CALHM1 was retrieved from UniProt and the secondary structure prediction of CALHM1 was done through CFSSP, GOR4, and SOPMA methods. The structure was identified through LOMETS, MUSTER, and MODELLER and finally, the structures were validated through Rampage. *Bauhinia variegata* plant was used to check the interaction of alkaloids and flavonoids against CALHM1. The protein and protein–ligand complex were also validated through molecular dynamics simulations studies. The model generated through MODELLER software with 6VAM A was used because this model predicted the best results in the Ramachandran plot. Further molecular docking was performed, quercetin was found to be the most appropriate candidate for the protein molecule with the minimum binding energy of −12.45 kcal/mol and their ADME properties were analysed through Molsoft and Molinspiration. Molecular dynamics simulations showed that CALHM1 and CALHM1–quercetin complex became stable at 2500 ps. It may be seen through the study that quercetin may act as a good inhibitor for treatment. With the help of an in silico study, it was easier to analyse the 3D structure of the protein, which may be scrutinized for the best-predicted model. Quercetin may work as a good inhibitor for treating Alzheimer's disease, according to in silico research using molecular docking and molecular dynamics simulations, and future in vitro and in vivo analysis may confirm its effectiveness.

Keywords: homology modelling; LOMETS; MUSTER; iGEMDOCK; AutoDock vina

1. Introduction

Alzheimer's disease is a long-term illness that causes brain cell loss and degeneration. The most common kind of dementia is Alzheimer's disease, which is characterised as a gradual loss of mental, communicative, and social abilities that makes it difficult for a person to operate independently [1]. Current Alzheimer's disease medicines may temporarily alleviate symptoms or delay the progression of the illness. Medications can often assist patients with Alzheimer's disease enhance their neuron function. For those with Alzheimer's disease, a variety of programs and services can be quite beneficial [2].

Intracellular calcium (Ca^{2+}) dynamics govern key neuronal functions such as neurotransmission, synaptic plasticity, learning, and memory, and signalling cascades, cytoskeleton modifications, synaptic function, and neuronal survival are all affected by changes in Ca^{2+} dynamics [3]. Several investigations have indicated the essential role of Ca^{2+} dysregulation in central Alzheimer's disease-related pathogenic processes since the first systematic hypothesis was proposed twenty years ago (AD). Disturbances in Ca^{2+} signals were discovered in the early stages of Alzheimer's disease, even before the build-up of amyloid ß-peptide (Aß), a clinical marker of the disease [4]. A growing body of evidence shows that mutations in AD-related genes such as presenilins, amyloid precursor protein, or apolipoprotein-E affect Ca^{2+} signalling, leading to apoptosis, synaptic plasticity failure, and neurodegeneration [5].

Ca^{2+}o (extracellular calcium) plays an important part in physiological processes. In a number of physiological and pathological circumstances, changes in Ca^{2+}o concentration ([Ca^{2+}o]) have been discovered to modify neuronal excitability, although the mechanisms by which neurons detect [Ca^{2+}]o remain unknown [6]. Calcium homeostasis modulator 1 (CALHM1) expression has been shown to generate cation currents in cells and enhance the concentration of cytoplasmic Ca^{2+} ([Ca^{2+}]) in response to Ca^{2+}o removal and subsequent addition. It is unclear if CALHM1 is a pore-forming ion channel or a modulator of endogenous ion channels. CALHM1 is also expressed in mouse cortical neurons, which respond to reduced [Ca^{2+}]o with increased conductivity and potential firing action, as well as a significantly higher [Ca^{2+}i] when Ca^{2+}o is withdrawn [7]. Those reactions, on the other hand, are significantly reduced in mouse neurons that have had CALHM1 genetically eliminated. These findings demonstrate that CALHM1 is an evolutionarily conserved family of ion channels that senses membrane voltage and external Ca^{2+} levels and plays a role in cortical neuronal excitability and Ca^{2+} homeostasis, notably in response to decreasing and restoring [Ca^{2+}o] [8].

The absence of an experimentally characterized structure has hampered progress in determining the function of CALHM1 in Alzheimer's disease. The two most common experimental approaches for determining the structure of proteins are X-ray crystallography and nuclear magnetic resonance (NMR) spectroscopy. These techniques, however, have requirements such as a high time and personnel needs [9]. Obtaining protein sequences, on the other hand, is much easier than obtaining protein structure, thanks to current sequencing methods. As a result, databases such as UniProt (https://www.uniprot.org/ (accessed on 7 January 2021)) and TrEMBL (Translated EMBL) (https://www.uniprot.org/statistics/TrEMBL (accessed on 8 January 2021)) include many protein sequences. In the late twentieth century, computational approaches for predicting the structure of proteins gave a sequence of amino acids. The information essential for a protein's correct folding is encoded in its amino acid sequence, according to research (Anfinsen's dogma). Homology modelling (based on sequence comparison) and threading are presently the most used computational approaches for predicting protein structure (based on sequence comparison) [10].

The goal of this research was to create useful models of the CALHM1 computational protein structure. As a result, additional research and analysis of CALHM1 function in Alzheimer's disease will be aided [11]. Comparative modelling was carried out in the absence of its experimentally deduced structure using the software programs MODELLER (https:/salilab.org/modeller/ (accessed on 15 January 2021)), LOMETS (Local MetaThreading Server) (https:/zhanglab.ccmb.med.umich.edu/LOMETS/ (accessed on 19 January 2021)) and MUSTER (MUlti-Sources ThreadER) (https:/zhanglab.cccmb.med.umich.edu/MUSTER/ (accessed on 30 January 2021)) [12]. RAMPAGE (Ramachandran Plot Assessment) (http:/mordred.bioc.cam.ac.uk/~{}rapper (accessed on 3 February 2021)) was then used to test the model structure. Using SPDBV (Swiss PDB Viewer) (https:/spdbv.vital-it.ch/ (accessed on 5 February 2021)) software, the energy minimisation of the four modelled structures was carried out. Using GOR4 (Garnier–Osguthorpe–Robson) (https:/npsa-prabi.ibcp.fr/NPSA/npsagor4.html (accessed on 5 February 2021)), CFSSP (Chou and Fasman Secondary Structure Prediction Server) (http:/www.biogem.org/tool/chou-fasman/ (accessed on 5 February 2021)), and SOPMA (Self-Optimized Prediction System with Alignment) algorithms (https:/npsa-prabi.ibcp.fr/cgi-bin/npsaautomat.pl?page=/NPSA/npsasopma.html (accessed on 5 February 2021)) also generated secondary protein structure [13].

2. Material and Methods

The protein structure prediction modelling for comparative modelling consisted of the following steps. Target identification came first, followed by alignments of the target and prototype sequences. The model was built when the alignment template procedure was completed. Finally, the model's strength, stearic collisions, and stability were evaluated.

2.1. Protein Sequence Retrieval

The CALHM1 protein sequence (accession number: Q8IU99 (CAHM1_HUMAN) was saved from the UniProt database (https://www.uniprot.org/ (accessed on 7 January 2021)) [14].

2.2. Protein Secondary Structure Prediction

The CALHM1 protein sequence (accession number: O43315 (AQP9 HUMAN)) was further subjected to secondary structure prediction on the Expasy server using GOR4, SOPMA and CFSSP [15].

2.3. Protein Tertiary Structure Prediction through Template Identification

A thorough search of the PDB (Protein Data Bank) (http://www.rcsb.org/ (accessed on 16 February 2021)) was conducted to search the most similar sequences already known for experimentally designed structures. The template protein structures (6VAM A and 6LMT A) were analysed as the most accurate template for identifying the three-dimensional protein structure based on various factors such as E-value, percentage identity, alignment score, and query coverage [16].

2.4. Modelling

The protein CALHM1 three-dimensional structure was calculated using MODELLER version 9.15, LOMETS, and the MUSTER server. MODELLER carries out a comparative modelling of the proteins according to the identified template. LOMETS is based on a metathreading technique for identifying the protein structure based on a template. MUSTER is based on a protein threading algorithm, which identifies PDB library template structures [17]. It generates sequence–template alignments with multiple structural data by combining different sequences. Several models were created through two templates, and a comparison of their DOPE score was selected for the best model [18]

2.5. Validation of the Structure

The RAMPAGE server was used to create Ramachandran plots in order to validate the predicted protein structures by looking at criteria such as preferred, allowed, and outside amino acid residue areas. The pdb files of the best target gene models predicted by MODELLER, LOMETS, and MUSTER were sent to the RAMPAGE service to create Ramachandran plots. Plots were identified for the anticipated structures, and the plots were matched to determine the best structure among the projected structures and further studies. Validation of the protein structure's quality was also carried out using the ProSA server [12].

2.6. Energy Minimisation of the Predicted Molecule

For SPDBV to achieve the lowest energy conformation, the identified and analysed models were subjected to energy minimisation.

2.7. Preparation of Ligand Molecule

The flavonoids and alkaloids structure of the plant *Bauhinia variegata* were retrieved in sdf format from the PubChem online database. The stem bark of the plant contains beta-carotene, quercetin, stigmasterol, hentriacontane, flavanone, isoquericetroside, kaempferol-3-glucoside, lupeol, myricetol, phenanthriquinone, quercitroside, rutoside, xanthophyll, dihydroquercetin, octacosanol, and beta-sitosterol. All the structures were retrieved in 3D structure in SDF format and were further converted into pdb format through online converting tool [19].

2.8. Initial Docking through iGEMDOCK Software

Initial docking was performed to screen the ligands on the basis of the binding energy. The docking process was carried out through iGEMDOCK version 2.1. The result was in the form of an electrostatic force, hydrogen bonds, and Van Der Waals forces [20]. The docking was performed between protein and ligand with a population size of 200 and the number of generations was 70 with 2 solutions [21].

2.9. Final Molecular Docking through AutoDock Vina and Drug Likeliness Property Analysis

The ligands were screened through iGEMDOCK and these screened ligands were tested against CALHM1 protein through AutoDock vina software. This software tool is freely available online. The protein CALHM1 was assigned with Kollman charges and polar hydrogens. The screened ligands were added with nonpolar hydrogen atoms and with Gasteiger partial charges. The torsion angles were allowed to rotate freely. A grid box of 80 × 80 × 80 Å was adjusted in such a manner that it was covering the target molecule to give the best docked result. The docking algorithm was adjusted to 100 runs. The default parameters were the Lamarckian genetic algorithm (LGA) and the empirical free energy function. The best-targeted molecule was screened further based on its minimal binding energy (Kcal/mol) [22].

Drug likeliness analysis was done through Molsoft (http://www.molsoft.com/ (accessed on 2 March 2021)) and Molinspiration (http://www.molinspiration.com/ (accessed on 2 March 2021)). Different properties of the screened ligand were analysed through pkCSM [23].

2.10. Molecular Dynamics Simulations

According to the molecular docking results, a molecular dynamics simulation was performed. The molecule which showed the minimum binding energy with the protein molecule was compared with the protein molecule for the dynamics study. The Groningen Machine for Chemical Simulations (GROMACS) 4.5.6 package was used to run molecular dynamics simulations. For the simulations, Gromacs was utilised to build the protein target and ligand file. With the help of an online server PRODRG2.5, the topology parameters of the ligand were generated [24]. The protein and ligand complex was placed inside the shell. The volume of the box was 284.14 nm^3 and the distance between the protein molecule and the box was kept at 1.0 nm. After adding 8 sodium ions to the shell, simple point charges and water molecules were neutralized. Energy minimisation was achieved using the steepest approach of 8 ps. The machine was balanced at 40 ps when the temperature was raised to 300 K. The simulations at 10 ns were performed at 1 bar and at the temperature of 300 K. Finally, an all-bond restriction was employed to keep the ligand from migrating during molecular dynamics [25].

3. Results and Discussion

3.1. Protein Sequence

The protein sequence of CALHM1 was retrieved in FASTA format from the UniProt database as shown in Figure 1.

```
        10         20         30         40         50
MMDKFRMIFQ FLQSNQESFM NGICGIMALA SAQMYSAFDF NCPCLPGYNA
        60         70         80         90        100
AYSAGILLAP PLVLFLLGLV MNNNVSMLAE EWKRPLGRRA KDPAVLRYMF
       110        120        130        140        150
CSMAQRALIA PVVWVAVTLL DGKCFLCAFC TAVPVSALGN GSLAPGLPAP
       160        170        180        190        200
ELARLLARVP CPEIYDGDWL LAREVAVRYL RCISQALGWS FVLLTTLLAF
       210        220        230        240        250
VVRSVRPCFT QAAFLKSKYW SHYIDIERKL FDETCTEHAK AFAKVCIQQF
       260        270        280        290        300
FEAMNHDLEL GHTHGTLATA PASAAAPTTP DGAEEEREKL RGITDQGTMN
       310        320        330        340
RLLTSWHKCK PPLRLGQEEP PLMGNGWAGG GPRPPRKEVA TYFSKV
```

Figure 1. Protein sequence of CALHM1.

3.2. Protein Secondary Structure Prediction

The secondary structure prediction of CALHM1 was carried out with the aid of methods such as Chou and Fasman Secondary Structure Prediction Server (CFSSP), Garnier–Osguthorpe–Robson (GOR4), and the Self-Optimised Prediction Method with Alignment (SOPMA). Information from GOR4, CFSSP, and SOPMA Expasy tools were obtained and the secondary structures such as alpha helix, beta strand, and random coil for the target CALHM1 were extracted.

Chou and Fasman Secondary Structure Prediction Server (CFSSP) is an empirical predictive tool of secondary protein structures. The method depends on analyses of the relative frequencies in alpha helices, beta sheets, and turns of each amino acid based on known protein structures solved with X-ray crystallography. The analysis of the CFSSP showed that CALHM1 consisted of 271 alpha helix, 248 extended strands, and 34 turns, as shown in Figure 2.

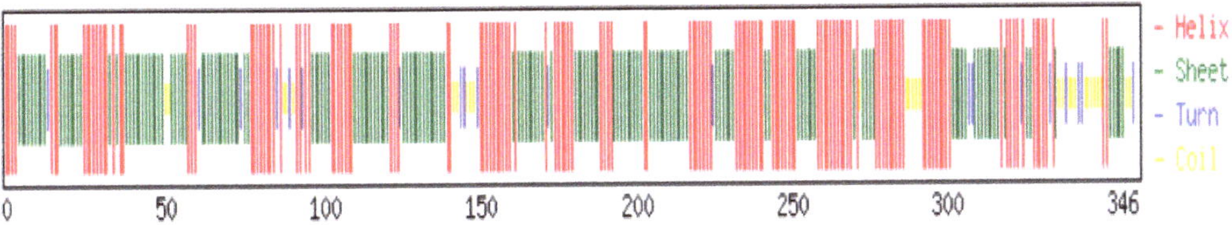

Figure 2. CFSSP result.

The Garnier–Osguthorpe–Robson (GOR4) method is a technique based on information theory to predict the secondary structures in proteins. It uses probability factors derived from empirical research of known tertiary protein structures solved using X-ray crystallography. The analysis of GOR4 showed that CALHM1 consisted of 133 alpha helix, 45 extended strands, and 168 random coils, as shown in Figure 3.

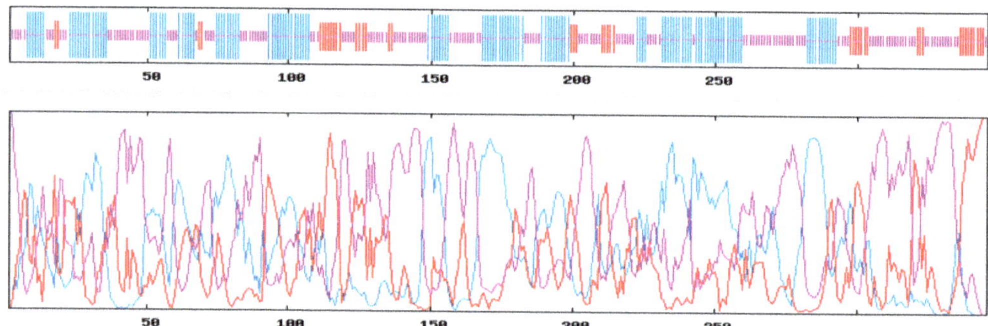

Figure 3. GOR4 result (dark blue is denoting alpha helix, light blue is denoting pi helix, dark pink is denoting beta bridge, red is denoting extended strands).

The Self-Optimised Prediction Method with Alignment (SOPMA) is an Expasy server protein-aided secondary structure prediction tool. Using consensus estimation from several alignments, the algorithm contributes to significant advances in protein secondary structure. Analysing SOPMA, CALHM1 consisted of 182 alpha helix, 34 extended strands, and 119 random coils as shown in Figure 4.

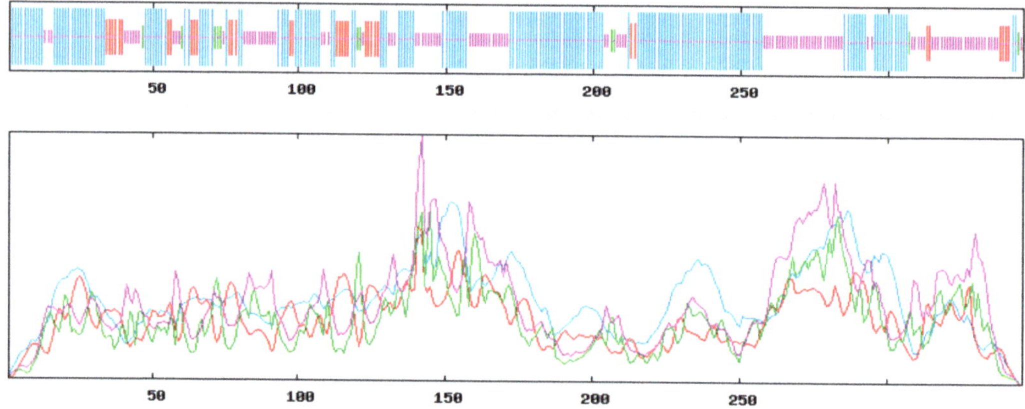

Figure 4. SOPMA result (dark blue is denoting alpha helix, green is denoting pi helix, dark pink is denoting beta bridge, red is denoting extended strands).

3.3. Template Identification

PDB Blast was performed to define CALHM1 modelling prototype structures for comparative homology modelling. We compared the templates and selected two of them (6VAM A and 6LMT A) based on their query cover, E-value and identity as shown in Table 1. Using MODELLER software, the two structures were downloaded from PDB for modelling the protein.

Table 1. CALHM1 BLAST parameters.

Query Cover	E-Value	Identity	Accession
99%	6×10^{-169}	68.36%	6VAM A
88%	4×10^{-139}	58.82%	6LMT A

3.4. Modelling through MODELLER

Through 6VAM A and 6LMT A prototype files, structures were modelled using MODELLER version 9.15 software for the protein CALHM1. Fifty models were produced using 6VAM A and 6LMT A modellers. With the help of the DOPE score as a criterion, we selected one best model for 6VAM A (Model 1) as shown in Figures 5 and 6LMT A (Model 2) as shown in Figure 6.

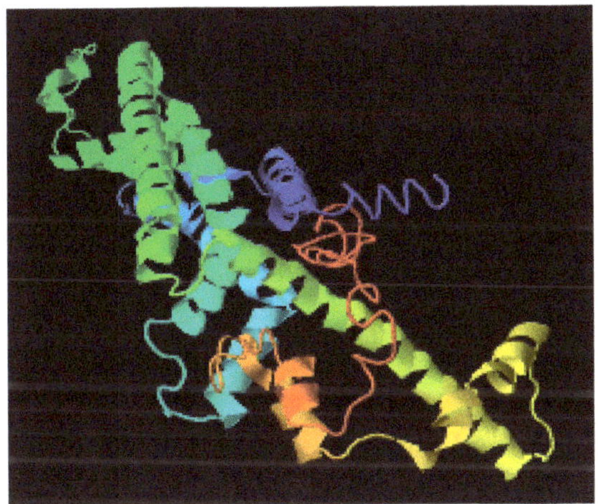

Figure 5. Best predicted model by MODELLER (6VAM A).

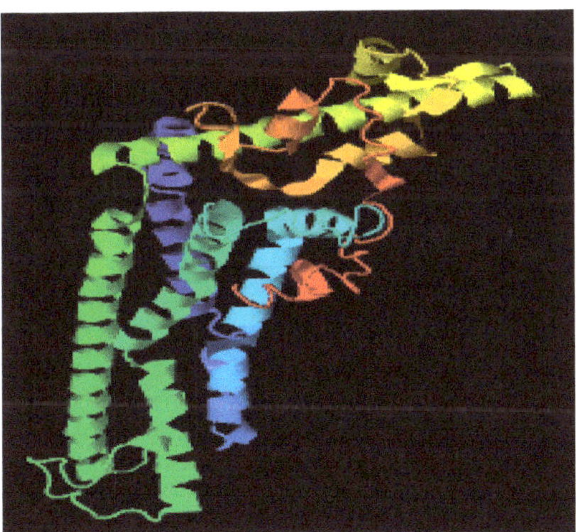

Figure 6. Best predicted model by MODELLER (6LMT A).

3.5. Structure Prediction through LOMETS Server

LOMETS server was used as a meta threading approach to identify the 3D structure of the given sequence. Ten protein structures were generated, and the best structures were further evaluated. Comparing the Z-score and maximum coverage, the best model among them was selected as shown in Figure 7.

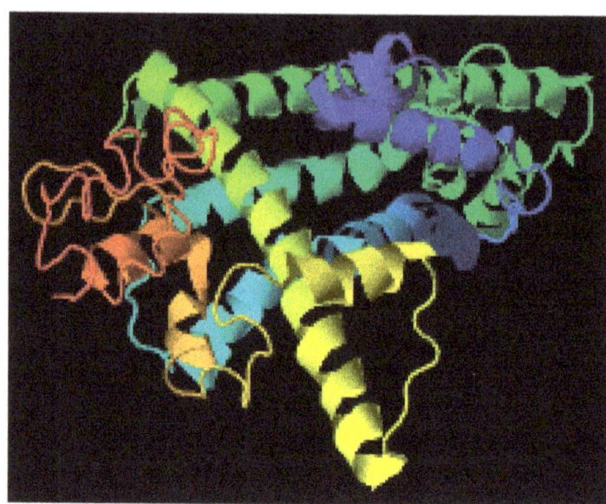

Figure 7. Best predicted model by LOMETS.

3.6. Protein Structure Prediction Using MUSTER Server

In addition, the MUSTER online server was used for protein treading. This server created ten different protein sequence models, among which the structure with the lowest Z-score and the maximum coverage was chosen as the fittest structure as shown in Figure 8.

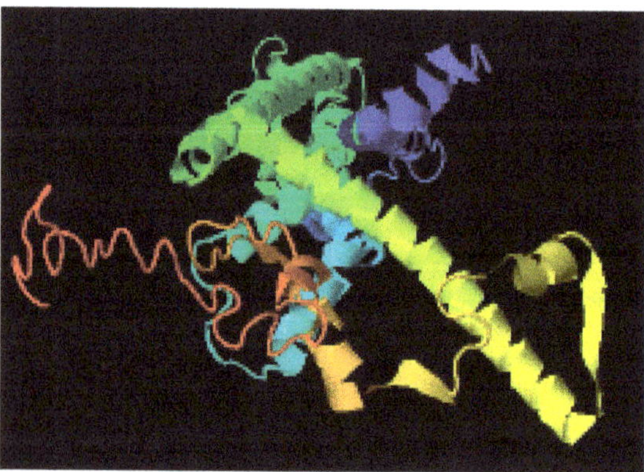

Figure 8. Best predicted model by MUSTER.

3.7. Structure Validation Using Ramachandran Plot

The selected four protein structures were uploaded to RAMPAGE to analyse the predicted structures, which produced the Ramachandran plots for the predicted protein structures as shown in Figure 9. In this plot, the amino acids (residues) have been identified in three distinct regions. The three distinct regions were favoured region, allowed region, and outlier region as shown in Table 2. Further structure validation was also performed through the ProSA server (https://prosa.services.came.sbg.ac.at/prosa.php (accessed on 3 February 2021)) to analyse the protein structure as shown in Figure 10.

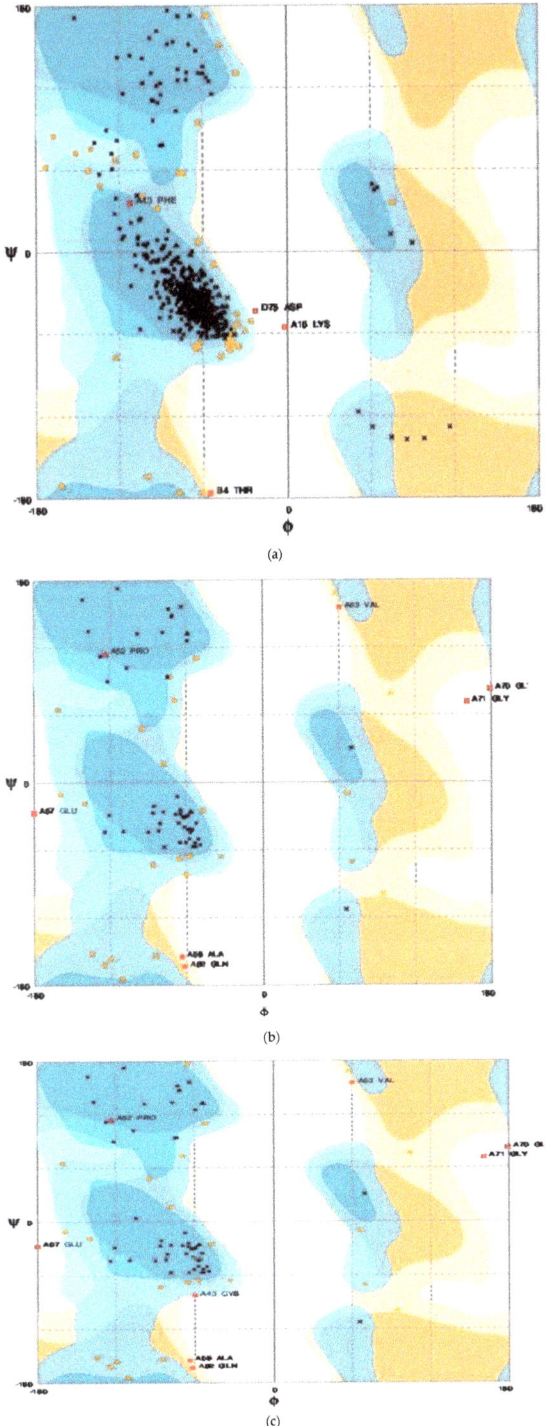

Figure 9. *Cont.*

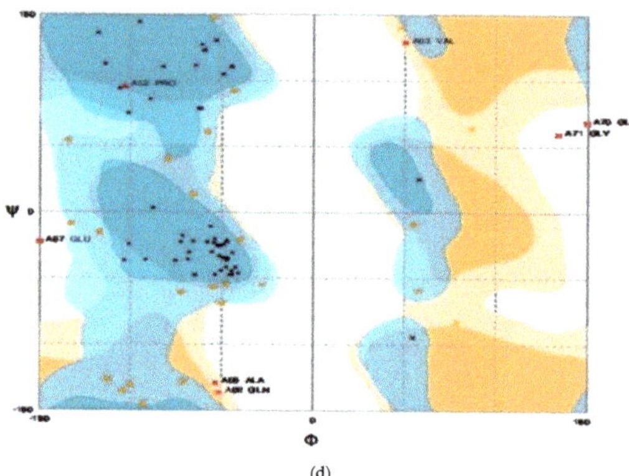

(d)

Figure 9. Ramachandran plots for (**a**) MODELLER (6VAM A); (**b**) MODELLER (6LMT A); (**c**) LO-ETS server; (**d**) MUSTER server.

Table 2. Ramachandran plot for each model.

	MODELLER		LOMETS	MUSTER
	6VAM A	**6LMT A**		
Favoured region	187	172	177	196
Allowed region	5	6	14	10
Outlier region	2	3	15	6

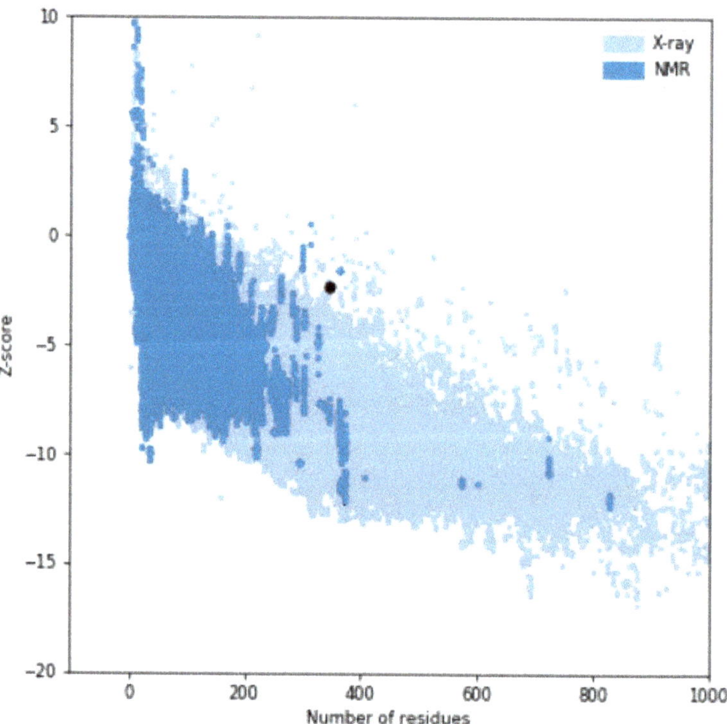

Figure 10. Quality of the protein structure (ProSA Server).

3.8. Energy Minimisation

The energy minimisation was carried through SPDBV (Swiss PDB Viewer) (https://spdbv.vital-it.ch/ (accessed on 5 February 2021)) software for the predicted models. The values retrieved from the energy minimisation were analysed to identify the best protein structure which was predicted as shown in Table 3. MODELLER 6VAM A had the least energy content (E = 2468.876 KJ/mol).

Table 3. Energy minimisation values through SPDBV.

	MODELLER		LOMETS	MUSTER
	6VAM A	6LMT A		
Energy (KJ/mol)	2468.876	5688.255	10,265.889	8714.236

3.9. Ligands from Bauhinia variegata

The three-dimensional structure in sdf format were downloaded from PubChem. The secondary metabolites were xanthophyll, beta-carotene, beta-sitosterol, dihydroquercetin, quercetin, stigma sterol, hentriacontane, octacosanol, flavanone, isoquericetroside, kaempeferol-3-glucoside, lupeol, myricetol, phenanthriquinone, quercitroside, and rutoside, as shown in Table 4.

Table 4. Compounds and their respective PubChem IDs.

Compound	PubChem ID
Hentriacontane	CID: 12410
Octacosanol	CID:68406
Stigmasterol	CID:5280794
Betasitosterol	CID:222284
Flavanone	CID:10251
Isoquericetroside	CID:5484006
Kaempeferol-3-glucoside	CID:6325460
Lupeol	CID:259846
Myricetol	CID:5281672
Phenanthriquinone	CID:6763
Quercitroside	CID:5280459
Rutoside	CID:5280805
Xanthophyll	CID:5281243
Beta- carotene	CID:5280489
Dihydroquercetin	CID:439533
Quercetin	CID:5280343

3.10. Screening of Ligands through iGEMDOCK

All the downloaded ligands were interacted with the protein molecule (CALHM1). Table 5 shows how the Van der Waal forces, binding energy, and hydrogen bond were used to filter the best docked molecules. The structure of the screened docked protein– ligands are shown in Figure 11.

Table 5. Initial docking by iGEMDOCK.

Ligands	Binding Energy	VDW	HBond
Quercetin (CID: 5280343)	−12.66	−22.13	−2.34
Dihydroquercetin (CID: 439533)	−10.30	−21.11	−2.18
Beta-carotene (CID: 5280489)	−10.26	−20.11	−3.42
Xanthophylls (CID: 5281243)	−8.20	−11.33	−4.57
Stigma sterol (CID: 5280794)	−7.80	−29.20	−7.6
Beta-sitosterol (CID: 222284)	−6.70	−30.29	−3.41

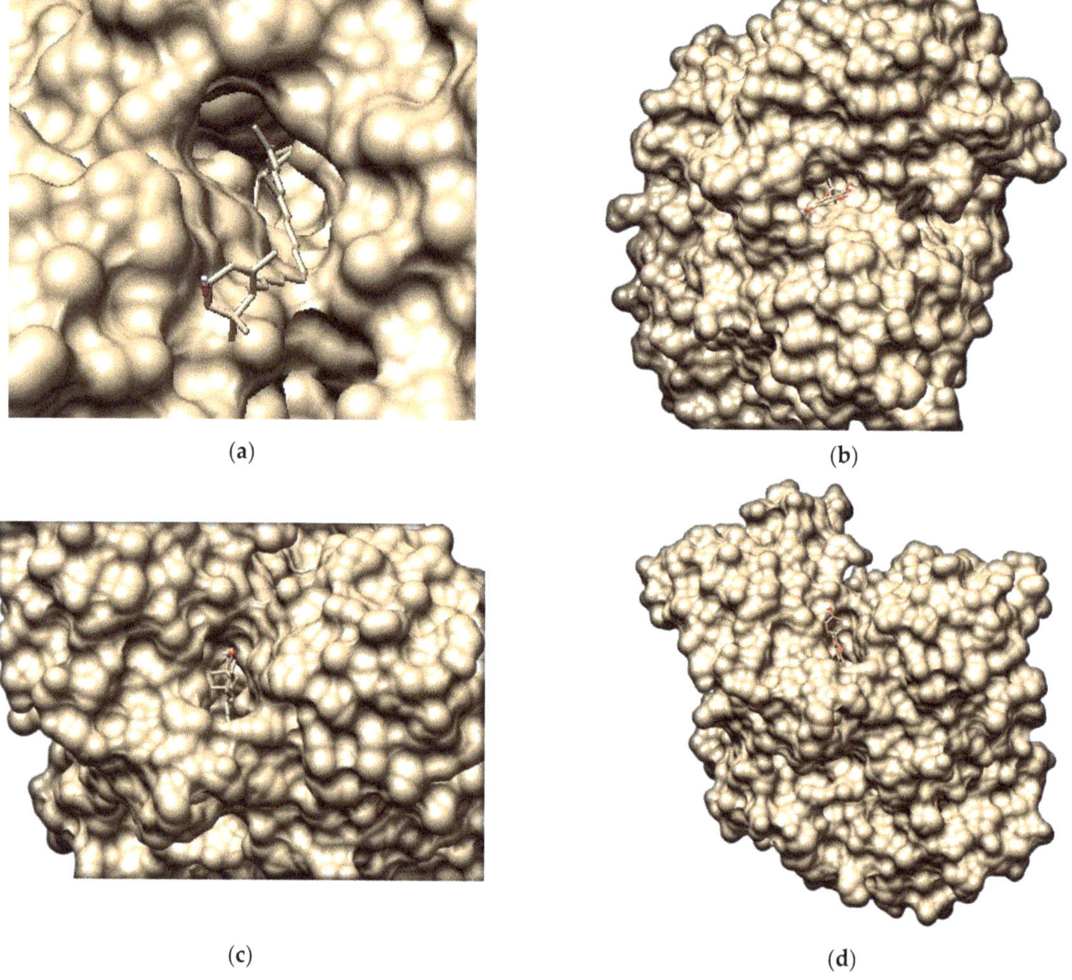

Figure 11. *Cont.*

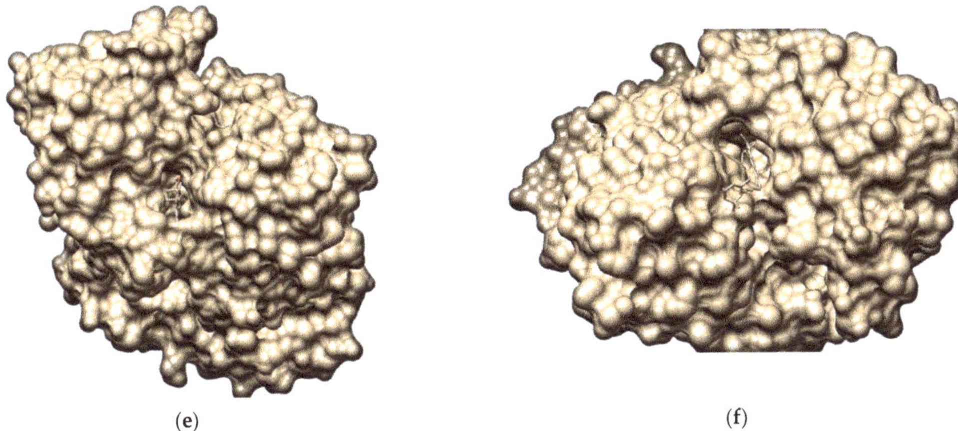

Figure 11. Molecular docking analysis: (**a**) pose view of CALHM1 with betacarotene; (**b**) pose view of CALHM1 with betasitosterol; (**c**) pose view of CALHM1 with quercetin; (**d**) pose view of CALHM1 with stigmasterol; (**e**) pose view of CALHM1 with xanthophyll; (**f**) pose view of CALHM1 with dihydroquercetin.

3.11. Molecular Docking Analysis through AutoDock Vina

Using the AutoDock vina program, the screened ligands were docked with CALHM1. The ligands were then sorted according to their minimal binding affinity. The optimum posture was determined based on the lowest binding affinity, as illustrated in Figure 12. Quercetin (CID: 5280343) exhibited the lowest binding affinity with CALHM1 according to molecular docking findings. The optimum energy value after comparing the poses of quercetin with CALHM1 was −12.45 Kcal/mol, as indicated in Table 6. The AutoDock vina energies were evaluated.

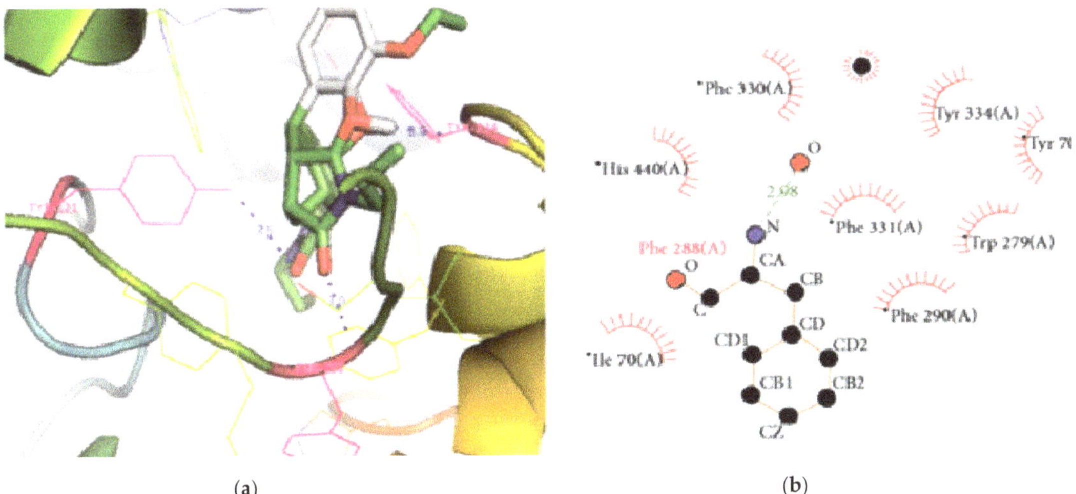

Figure 12. CALHM1 docking and Ligplot interaction with quercetin. (**a**) The hydrogen bond distance between the docked ligand and the active site is shown; (**b**) a two-dimensional depiction of a ligand and a protein residue.

Table 6. Docking result of quercetin against CALHM1.

S.No	Mode	Affinity (kcal/mol)	Distance From Best Mode RMSD l.b	Distance From Best Mode RMSD u.b
1.	1	−12.45	0.000	0.000
2.	2	−12.34	21.115	20.357
3.	3	−12.12	12.235	12.514
4.	4	−11.65	9.656	6.524
5.	5	−11.25	8.459	2.722
6.	6	−10.81	8.287	10.650
7.	7	−9.45	7.775	1.089
8.	8	−9.32	6.002	11.924
9.	9	−8.23	6.028	14.615

3.12. *Cheminformatics Properties and Lipinski's Rule of Five Validation of Quercetin*

The cheminformatics properties were studied for quercetin. The properties were molecular formula, molecular weight (g/mol), molar refractivity (cm^3), density (cm^3), drug likeness, etc., as shown in Table 7. The molecular weight of quercetin was 302.24 g/mol, the value of log P was 0.56, molecular refractivity was 122.60 cm^3. The comparative result of quercetin predicts it to be a good candidate. Quercetin was also validated for its Lipinski's rule of five properties, the value of quercetin predicted <10 hydrogen bond acceptors, <5 hydrogen bond donors, <500 g/mol molecular weight, <5 log P value [26].

Table 7. Cheminformatics properties of quercetin.

Molecular Formula	$C_{15}H_{10}O_7$
Molecular weight (g/mol)	302.24
Hydrogen bond acceptor	7
Hydrogen bond donor	5
Rotatable bonds	1
Log p	0.56
No of atoms	22
Polar surface area (A^2)	103.49 A^2
Molar refractivity (cm^3)	122.60
Density (cm^3)	1.23
Molar volume (cm^3)	268.73 cm^3
Drug likeness	1
Lipinski validation	yes
GPCR ligand	−0.06
Ion channel modulator	−0.19
Kinase inhibitor	0.28
Nuclear receptor ligand	0.36
Protease inhibitor	−0.25
Enzyme inhibitor	0.28

3.13. *Quercetin's Pharmacokinetic Properties*

Properties such as absorption, distribution, metabolism, excretion and toxicity properties (ADMET) were analysed for quercetin under pharmacokinetics.

The absorption property was analysed by intestinal water solubility at −2.925 log mol/L, and the intestinal solubility was found to be 96.902 percent, and the skin permeability value was −2.735 log Kp, which showed a strong quercetin structure that validated its good behaviour in terms of drug likeness. The blood brain barrier (BBB) and central nervous system (CNS) permeability values of quercetin were analysed for distribution properties with a weak BBB value of −1.098 log BB. However, the permeability value of the central nervous system (CNS) was −3.065 log PS. The CYP3A4 substrate, which is the isoform of cytochrome P450, confirmed the metabolism property. The property of excretion showed that the total clearance value was 0.407, which showed that quercetin had nontoxic actions, and nontoxicity was inferred [27]. Table 8 shows all of the results.

Table 8. Pharmacokinetic properties of quercetin.

S.No.	Property	Model Name	Predicted Value	Unit
1.	Absorption	Water solubility	−2.925	Numeric (log mol/L)
2.	Absorption	$Caco_2$ permeability	−0.229	Numeric (log Papp in 10–6 cm/s)
3.	Absorption	Intestinal absorption (human)	96.902	Numeric (% absorbed)
4.	Absorption	Skin permeability	−2.735	Numeric (log Kp)
5.	Absorption	P-glycoprotein substrate	Yes	Categorical (Yes/No)
6.	Absorption	P-glycoprotein I inhibitor	No	Categorical (Yes/No)
7.	Absorption	P-glycoprotein II inhibitor	No	Categorical (Yes/No)
8.	Distribution	VDss (human)	1.559	Numeric (log L/kg)
9.	Distribution	Fraction unbound (human)	0.206	Numeric (Fu)
10.	Distribution	BBB permeability	−1.098	Numeric (log BB)
11.	Distribution	CNS permeability	−3.065	Numeric (log PS)
12.	Metabolism	CYP2D6 substrate	No	Categorical (yes/no)
13.	Metabolism	CYP3A4 substrate	No	Categorical (yes/no)
14.	Metabolism	CYP1A2 inhibitor	Yes	Categorical (yes/no)
15.	Metabolism	CYP2C19 inhibitor	No	Categorical (yes/no)
16.	Metabolism	CYP2C9 inhibitor	No	Categorical (yes/no)
17.	Metabolism	CYP2D6 inhibitor	No	Categorical (yes/no)
18.	Metabolism	CYP3A4 inhibitor	No	Categorical (yes/no)
19.	Excretion	Total clearance	0.407	Numeric (log ml/min/kg)
20.	Excretion	Renal OCT2 substrate	No	Categorical (yes/no)
21.	Toxicity	AMES toxicity	No	Categorical (yes/no)
22.	Toxicity	Max. tolerated dose (human)	0.499	Numeric (log mg/kg/day)
23.	Toxicity	hERG I inhibitor	No	Categorical (yes/no)
24.	Toxicity	hERG II inhibitor	No	Categorical (yes/no)

Table 8. *Cont.*

S.No.	Property	Model Name	Predicted Value	Unit
25.	Toxicity	Oral rat acute toxicity (LD50)	2.471	Numeric (mol/kg)
26.	Toxicity	Oral rat chronic toxicity (LOAEL)	2.612	Numeric (log mg/kg_bw/day)
27.	Toxicity	Hepatotoxicity	No	Categorical (yes/no)
28.	Toxicity	Skin sensitisation	No	Categorical (yes/no)
29.	Toxicity	*T. pyriformis* toxicity	0.288	Numeric (log µg/L)
30.	Toxicity	Minnow toxicity	3.721	Numeric (log mM)

3.14. Molecular Dynamic Simulations Analysis

The protein target (CALHM1) along with quercetin and CALHM1 were selected for the molecular dynamics simulations to check the conformations. Using the GROMACS, the trajectories were analysed in terms of RMSD (root mean square deviation), Rg (radius of gyration), SASA (solvent accessible surface area), and RMSF (root mean square fluctuation).

3.15. Root Mean Square Deviation (RMSD)

The RMSD was used to identify the stability of unliganded CALHM1 and CALHM1 with quercetin. The system was in balance, with RMSD fluctuating about 2500 ps. As seen in Figure 13, the backbone atoms grew up to around 0.23 nm before stabilising until the end of the simulation, showing that the molecular system was then properly set.

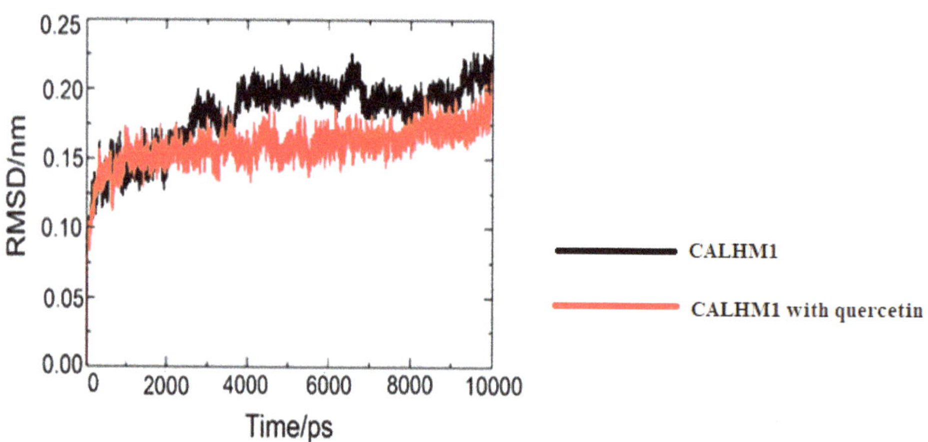

Figure 13. Time dependence of root mean square deviation. RMSD values for unliganded CALHM1 and CALHM1–quercetin complex.

3.16. Radius of Gyration

After around 2500 ps, the systems stabilised, indicating that the molecular dynamics simulation equilibrated. The CALHM1–quercetin binding was anticipated to be excellent based on the radius of gyration. As illustrated in Figure 14, the environment does not alter during contact.

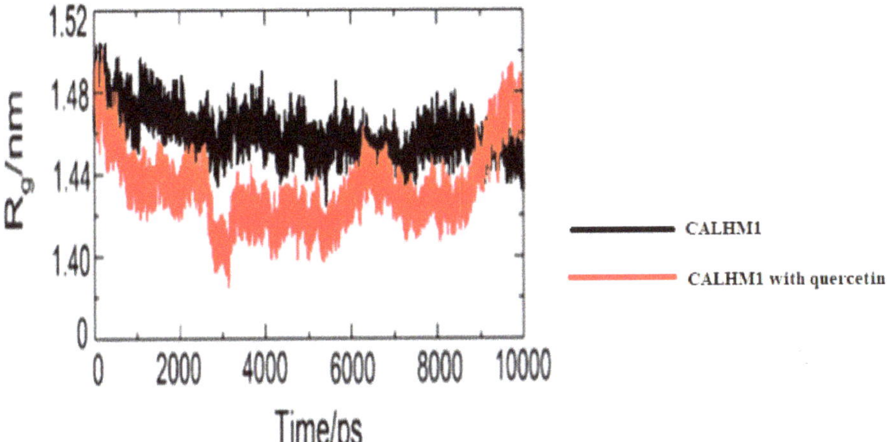

Figure 14. Radius of gyration (Rg) during 10,000 ps of MD simulation of unliganded CALHM1 and CALHM1–quercetin complex.

3.17. Solvent Accessible Surface Area (SASA)

The CALHM1 total solvent exposed surface area was displayed at 10,000 ps. The differences seen in the SASA graph were a bit similar to those of the radius of gyration. As demonstrated in Figure 15, it can be seen that there is a similarity between the SASA and the radius of gyration, which shows the accuracy in the simulation results.

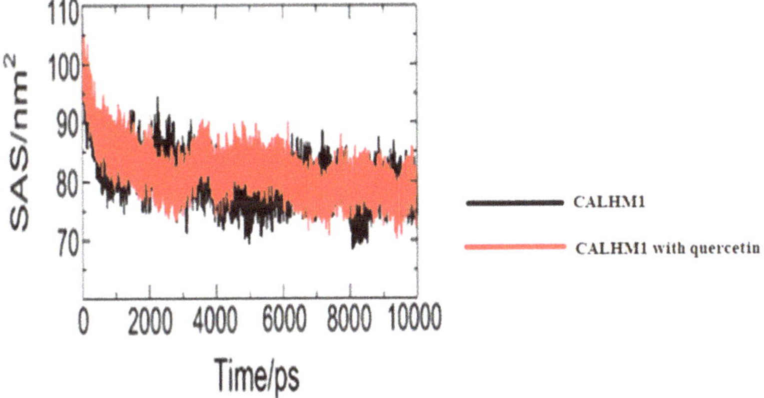

Figure 15. Solvent accessible surface area (SASA) during 10,000 ps of MD simulation of unliganded CALHM1 and CALHM1–quercetin complex.

3.18. Root Mean Square Fluctuation (RMSF)

The RMSF was used to investigate the mobility of CALHM1 residues in the presence and absence of ligands. The findings show that variations greater than 0.25 nm indicate residues located far from each ligand's binding sites. Furthermore, as seen in Figure 16, the residues in contact with the quercetin were the most stable and had the lowest RMSF values.

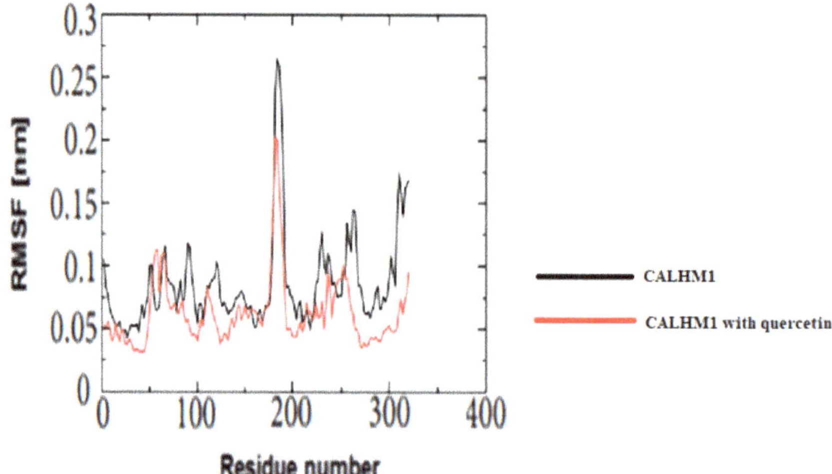

Figure 16. The RMSF values of unliganded CALHM1 and CALHM1–quercetin complex.

4. Conclusions

In recent years, there has been a huge growth in the use of various software and algorithms to predict protein structure using in silico methods. However, the precision of structure prediction, the severity of fold assignment errors, and the modelling of side chains and loops all need a significant body shift. CALHM1 (calcium homeostasis modulator 1) is involved in the pathogenesis of Alzheimer's disease. CALHM1's various models were developed and verified using homology modelling and Rampage. The interaction of alkaloids and flavonoids with CALHM1 was tested using the *Bauhinia variegata* plant. The model created with MODELLER software and 6VAM A was chosen because it performed best on the Ramachandran plot. Quercetin was shown to be the best choice for the protein molecule, with a minimum binding energy of −12.45 kcal/mol, and its ADME qualities were assessed using Molsoft and Molinspiration. At 2500 ps, CALHM1 and the CALHM1–quercetin combination became stable, according to molecular dynamics simulations. Finally, the in silico analysis suggested that quercetin might be a suitable therapeutic inhibitor. Quercetin may operate as a good inhibitor for treating Alzheimer's disease, according to in silico research using molecular docking and molecular dynamics simulations, and future in vitro and in vivo investigations may establish its therapeutic potential.

Author Contributions: N.K.—Performed the entire experiment. S.K.M., S.M.D.R., H.M.A., S.A.A., W.A., Q.Z. and C.V.—They all helped in performing the experiment. S.K.J. and N.K.J.—They both helped in designing and preparing the manuscript. D.I. and A.K.J.—They conceived the idea and prepared the manuscript. All authors have read and agreed to the published version of the manuscript.

Funding: This research was funded by a Deanship of Scientific Research at Majmaah University under project no. R-2022-177.

Institutional Review Board Statement: Not applicable.

Informed Consent Statement: Not applicable.

Data Availability Statement: Not applicable.

Acknowledgments: The authors are very thankful to the Institute of Technology and Management, Meerut, Uttar Pradesh, India, Shri Ramswaroop Memorial University, India and Institute of Applied Medicines and Research, Ghaziabad, India for providing a good platform to carry out the research work. The authors would also like to thank Deanship of Scientific Research at Majmaah University for supporting this work under project no. R-2022-177.

Conflicts of Interest: The authors declare no conflict of interest.

References

1. Chandra, P.M.; Venkateshwar, J. Biological evaluation of Schiff bases of new isatin derivatives for anti Alzheimer's activity. *Asian J. Pharm. Clin. Res.* **2014**, *7*, 114–117. Available online: https://innovareacademics.in/journals/index.php/ajpcr/article/view/966 (accessed on 31 December 2020).
2. Khare, N.; Maheshwari, S.K.; Jha, A.K. Screening and identification of secondary metabolites in the bark of *Bauhinia variegata* to treat Alzheimer's disease by using molecular docking and molecular dynamics simulations. *J. Biomol. Struct. Dyn.* **2020**, *39*, 5988–5998. [CrossRef]
3. Ma, Z.; Siebert, A.P.; Cheung, K.H.; Lee, R.J.; Johnson, B.; Cohen, A.S.; Foskett, J.K. Calcium homeostasis modulator 1 (CALHM1) is the pore-forming subunit of an ion channel that mediates extracellular Ca^{2+} regulation of neuronal excitability. *Proc. Natl. Acad. Sci. USA* **2012**, *109*, E1963–E1971. [CrossRef]
4. Syrjanen, J.L.; Michalski, K.; Chou, T.H.; Grant, T.; Rao, S.; Simorowski, N.; Furukawa, H. Structure and assembly of calcium homeostasis modulator proteins. *Nat. Struct. Mol. Biol.* **2020**, *27*, 150–159. [CrossRef] [PubMed]
5. Rubio, M.F.; Seto, S.N.; Pera, M.; Bosch, M.M.; Plata, C.; Belbin, O.; Soininen, H. Rare variants in calcium homeostasis modulator 1 (CALHM1) found in early onset Alzheimer's disease patients alter calcium homeostasis. *PLoS ONE* **2013**, *8*, e74203.
6. Ma, Z.; Tanis, J.E.; Taruno, A.; Foskett, J.K. Calcium homeostasis modulator (CALHM) ion channels. *Pflüg. Arch.-Eur. J. Physiol.* **2016**, *468*, 395–403. [CrossRef] [PubMed]
7. Taruno, A.; Vingtdeux, V.; Ohmoto, M.; Ma, Z.; Dvoryanchikov, G.; Li, A.; Koppel, J. CALHM1 ion channel mediates purinergic neurotransmission of sweet, bitter and umami tastes. *Nature* **2013**, *495*, 223–226. [CrossRef] [PubMed]
8. Nacmias, B.; Tedde, A.; Bagnoli, S.; Lucenteforte Cellini, E.; Piaceri, I.; Sorbi, S. Lack of implication for CALHM1 P86L common variation in Italian patients with early and late onset Alzheimer's disease. *J. Alzheimer's Dis.* **2010**, *20*, 37–41. [CrossRef]
9. Bigiani, A. Calcium homeostasis modulator 1-like currents in rat fungiform taste cells expressing amiloride-sensitive sodium currents. *Chem. Senses* **2017**, *42*, 343–359. [CrossRef] [PubMed]
10. Dreses, W.U.; Lambert, J.C.; Vingtdeux, V.; Zhao, H.; Vais, H.; Siebert, A.; Pasquier, F. A polymorphism in CALHM1 influences Ca^{2+} homeostasis, Aβ levels, and Alzheimer's disease risk. *Cell* **2008**, *133*, 1149–1161. [CrossRef]
11. Dreses, W.U.; Vingtdeux, V.; Zhao, H.; Chandakkar, P.; Davies, P.; Marambaud, P. CALHM1 controls the Ca^{2+}-dependent MEK, ERK, RSK and MSK signaling cascade in neurons. *J. Cell Sci.* **2013**, *126*, 1199–1206. [CrossRef] [PubMed]
12. Wu, S.; Zhang, Y. MUSTER: Improving protein sequence profile–profile alignments by using multiple sources of structure information. *Proteins* **2008**, *72*, 547–556. [CrossRef] [PubMed]
13. Orry, A.J.; Abagyan, R. *Homology Modeling: Methods and Protocols*; Humana Press: New York, NY, USA, 2012.
14. Karim, R.; Aziz, M.; Al, M.; Shatabda, S.; Rahman, M.S.; Mia, M. CoMOGrad and PHOG: From computer vision to fast and accurate protein tertiary structure retrieval. *Sci. Rep.* **2015**, *5*, 13275. [CrossRef]
15. Drozdetskiy, A.; Cole, C.; Procter, J.; Barton, G.J. JPred4: A protein secondary structure prediction server. *Nucleic Acids Res.* **2015**, *43*, W389–W394. [CrossRef]
16. Liu, J.; Wu, T.; Guo, Z.; Hou, J.; Cheng, J. Improving protein tertiary structure prediction by deep learning and distance prediction in CASP14. *Proteins* **2022**, *90*, 58–72. [CrossRef]
17. Waterhouse, A.; Bertoni, M.; Bienert, S.; Studer, G.; Tauriello, G.; Gumienny, R.; Schwede, T. SWISS-MODEL: Homology modelling of protein structures and complexes. *Nucleic Acids Res.* **2018**, *46*, W296–W303. [CrossRef] [PubMed]
18. Marhefka, C.A.; Moore, B.M.; Bishop, T.C.; Kirkovsky, L.; Mukherjee, A.; Dalton, J.T.; Miller, D.D. Homology modeling using multiple molecular dynamics simulations and docking studies of the human androgen receptor ligand binding domain bound to testosterone and nonsteroidal ligands. *J. Med. Chem.* **2001**, *44*, 1729–1740. [CrossRef] [PubMed]
19. Khare, P.; Kishore, K.; Sharma, D. A study on the standardization parameters of Bauhinia variegate. *Asian J. Pharm. Clin. Res.* **2017**, *10*, 133–136. [CrossRef]
20. Dallakyan, S.; Olson, A.J. Small-molecule library screening by docking with PyRx. In *Chemical Biology*; Humana Press: New York, NY, USA, 2015; pp. 243–250.
21. Nousheen, L.; Akkiraju, P.C.; Enaganti, S. Molecular docking mutational studies on human surfactant protein-D. *World J. Pharm. Res.* **2014**, *3*, 1140–1148.
22. Fuhrmann, J.; Rurainski, A.; Lenhof, H.P.; Neumann, D. A new Lamarckian genetic algorithm for flexible ligand-receptor docking. *J. Comput. Chem.* **2010**, *31*, 1911–1918. [CrossRef]
23. Pires, D.E.; Blundell, T.L.; Ascher, D.B. pkCSM. Predicting small molecule pharmacokinetic and toxicity properties using graphbased signatures. *J. Med. Chem.* **2015**, *58*, 4066–4072. [CrossRef] [PubMed]
24. Van, S.P.D.; Lindahl, E.; Hess, B.; Groenhof, G.; Mark, A.E.; Berendsen, H.J. GROMACS: Fast, flexible, and free. *J. Comput. Chem.* **2005**, *26*, 1701–1718.
25. Iqbal, D.; Khan, M.S.; Waiz, M.; Rehman, M.T.; Alaidarous, M.; Jamal, A.; Alothaim, A.S.; AlAjmi, M.F.; Alshehri, B.M.; Banawas, S.; et al. Exploring the Binding Pattern of Geraniol with Acetylcholinesterase through In Silico Docking, Molecular Dynamics Simulation, and In Vitro Enzyme Inhibition Kinetics Studies. *Cells* **2021**, *10*, 3533. [CrossRef]

26. Iqbal, D.; Rehman, M.T.; Bin Dukhyil, A.; Rizvi, S.M.D.; Al Ajmi, M.F.; Alshehri, B.M.; Alsaweed, M. High-Throughput Screening and Molecular Dynamics Simulation of Natural Product-like Compounds against Alzheimer's Disease through Multitarget Approach. *Pharmaceuticals* **2021**, *14*, 937. [CrossRef] [PubMed]
27. Ongey, E.L.; Yassi, H.; Pflugmacher, S.; Neubauer, P. Pharmacological and pharmacokinetic properties of lanthipeptides undergoing clinical studies. *Biotechnol. Lett.* **2017**, *39*, 473–482. [CrossRef] [PubMed]

Article

Transcriptome Sequencing Reveal That Rno-Rsf1_0012 Participates in Levodopa-Induced Dyskinesia in Parkinson's Disease Rats via Binding to Rno-mir-298-5p

Chun-Lei Han [1,2,†], Qiao Wang [2,3,†], Chong Liu [2,3], Zhi-Bao Li [2,3], Ting-Ting Du [2,3], Yun-Peng Sui [2,3], Xin Zhang [2,3], Jian-Guo Zhang [1,2,3], Yi-Lei Xiao [4,*], Guo-En Cai [5,6,*] and Fan-Gang Meng [1,2,3,7,*]

1 Department of Neurosurgery, Beijing Tiantan Hospital, Capital Medical University, Beijing 100070, China
2 Beijing Key Laboratory of Neurostimulation, Beijing 100070, China
3 Department of Functional Neurosurgery, Beijing Neurosurgical Institute, Capital Medical University, Beijing 100070, China
4 Department of Neurosurgery, Liaocheng People's Hospital, Liaocheng 252004, China
5 Department of Neurology, Fujian Medical University Union Hospital, Fuzhou 350001, China
6 Fujian Key Laboratory of Molecular Neurology, Institute of Clinical Neurology, Institute of Neuroscience, Fujian Medical University, Fuzhou 350122, China
7 Chinese Institute for Brain Research, Beijing 102206, China
* Correspondence: yileixiao@163.com (Y.-L.X.); cgessmu@fjmu.edu.cn (G.-E.C.); mengfg@ccmu.edu.cn (F.-G.M.)
† These authors contributed equally to this work.

Citation: Han, C.-L.; Wang, Q.; Liu, C.; Li, Z.-B.; Du, T.-T.; Sui, Y.-P.; Zhang, X.; Zhang, J.-G.; Xiao, Y.-L.; Cai, G.-E.; et al. Transcriptome Sequencing Reveal That Rno-Rsf1_0012 Participates in Levodopa-Induced Dyskinesia in Parkinson's Disease Rats via Binding to Rno-mir-298-5p. *Brain Sci.* **2022**, *12*, 1206. https://doi.org/10.3390/brainsci12091206

Academic Editors: Christina Piperi and Ashu Johri

Received: 10 August 2022
Accepted: 2 September 2022
Published: 7 September 2022

Publisher's Note: MDPI stays neutral with regard to jurisdictional claims in published maps and institutional affiliations.

Copyright: © 2022 by the authors. Licensee MDPI, Basel, Switzerland. This article is an open access article distributed under the terms and conditions of the Creative Commons Attribution (CC BY) license (https://creativecommons.org/licenses/by/4.0/).

Abstract: Levodopa-induced dyskinesia (LID) is a common complication of chronic dopamine replacement therapy in the treatment of Parkinson's disease (PD), and a noble cause of disability in advanced PD patients. Circular RNA (circRNA) is a novel type of non-coding RNA with a covalently closed-loop structure, which can regulate gene expression and participate in many biological processes. However, the biological roles of circRNAs in LID are not completely known. In the present study, we established typical LID rat models by unilateral lesions of the medial forebrain bundle and repeated levodopa therapy. High-throughput next-generation sequencing was used to screen circRNAs differentially expressed in the brain of LID and non-LID (NLID) rats, and key circRNAs were selected according to bioinformatics analyses. Regarding fold change ≥2 and $p < 0.05$ as the cutoff value, there were a total of 99 differential circRNAs, including 39 up-regulated and 60 down-regulated circRNAs between the NLID and LID groups. The expression of rno-Rsf1_0012 was significantly increased in the striatum of LID rats and competitively bound rno-mir-298-5p. The high expression of target genes PCP and TBP in LID rats also supports the conclusion that rno-Rsf1_0012 may be related to the occurrence of LID.

Keywords: Parkinson's disease; levodopa-induced dyskinesia; circular RNA; next-generation sequencing

1. Introduction

Parkinson's disease (PD) is a chronic, progressive disease mainly affecting middle-aged and elderly people. It is characterized by tremors, rigidity, decreased movement, abnormal postural reflex, and autonomic nervous dysfunctions. It is the second most common neurodegenerative disease in the world after Alzheimer's disease and affects approximately 1% of adults over age 60 [1]. The main pathological changes of PD include the loss of dopaminergic neurons and the formation of Lewy bodies in the substantia nigra of the midbrain. The pathogenesis of PD is unclear, but mitochondrial dysfunction, oxidative stress, altered protein handling, and inflammation may contribute to nigral dopaminergic cell death [2].

At present, levodopa is still the first choice for PD treatment. However, with progression of the disease, long-term levodopa treatment may show decreased efficacy and cause symptom fluctuations and motor complications. Studies have suggested that 8–45% of PD patients experienced dyskinesia after 4–6 years of treatment [3–5] with a disability as high as 43% [6]. It is believed that the occurrence and development of LID are related to the inherent lesions of the nigrostriatal regions and the pulsatile delivery of levodopa, involving a variety of neural signaling pathways and changes in brain network electrophysiological activities [7]. At present, the exact mechanism of LID is not clear, and effective treatment strategies are also lacking in clinical practice.

Circular RNA is a newly discovered non-coding RNA with a closed circular structure, which is resistant to exonuclease and abundant in the whole transcriptome. The functions of most circRNAs remain unexplored, but many circRNAs exert important biological functions by acting as microRNAs or protein inhibitors, to regulate protein functions or to be translated [8]. CircRNAs are evolutionarily conserved and are more stable than linear RNAs, so circRNAs have enormous potential to be diagnostic and prognostic biomarkers [9]. Studies have shown that circRNAs are closely related to brain development [10], nervous system tumors [11], Alzheimer's disease [12], and PD [13]. For example, Hanan et al. found that circSLC8A1, which carries seven binding sites for miR-128 and is strongly bound to the microRNA effector protein, Ago2, may affect oxidative stress in PD by regulating miR-128 [13].

There is growing evidence suggesting the potential role of circRNAs in nervous system disease. However, there are no reports on the role of circRNAs in LID. Here we studied the circRNA expression profile of the striatum of LID rats by high-throughput sequencing to identify circRNAs related to LID for further study.

2. Materials and Methods

2.1. Animals

Male specific pathogen-free (SPF) Sprague–Dawley (SD) rats (250−300 g) were obtained from Vital-River Experimental Animal Technology (Beijing, China). Animals were maintained in a temperature-controlled room on a 12/12 h light/dark cycle with ad libitum access to standard food and water. Animal experiments were conducted according to the Chinese Animal Welfare Act and Guidance for Animal Experimentation of Capital Medical University. The study protocol was approved by the Ethics Committee of Beijing Neurosurgical Institute, Capital Medical University (Protocol No.: AEEI-2018-200).

2.2. 6-OHDA Lesion and L-DOPA Administration

Rats were anesthetized with 2−3% isoflurane through an animal anesthesia ventilator system (RWD Life Science, Shenzhen, China) and placed in a stereotaxic frame (David Kopf Instruments, Tujunga, CA, USA). Based on previous PD rat model studies, rats were unilaterally lesioned by injection of 6-OHDA (12 μg/2.4 μL in 0.02% ascorbate in saline (162957, Sigma-Aldrich, St. Louis, MO, USA)) into the medial forebrain bundle (MFB) (from bregma: anterior posterior (AP): −3.6 mm, medial lateral (ML): −1.8 mm; dorsal ventral (DV): −8.2 mm from the skull)) using a Hamilton syringe (88000, Hamilton, Reno, NV, USA). The 6-OHDA was injected at a rate of 1 μL/min, and the needle was left in place for an additional 5 min to allow diffusion of 6-OHDA before being slowly retracted. To determine lesion efficacy, turning behavior was recorded 3 weeks later over a 90 min period after injection of apomorphine (0.5 mg/kg by subcutaneous injection (16094, Cayman, Ann Arbor, MI, USA)) [14]. The control rats received a sham lesion using saline.

Similar to human LID, repeated levodopa therapy induced abnormal involuntary movements (AIMs) (including dystonia, hyperkinesia, and/or stereotypies) in the PD model rats [15]. Starting 3 days after the turning behavior test, PD rats received methyl L-DOPA (6 mg/kg, 5 mg/mL; D9628-5G, Sigma-Aldrich) and peripheral decarboxylase inhibitor, benserazide (B7283-1G, Sigma-Aldrich, 12 mg/kg, 10 mg/mL) using single daily intraperitoneal injections for 21 days. The 6-OHDA-lesioned control rats received saline using the same protocol (Figure 1A). AIMs were rated every 3 days after L-DOPA therapy six times in 180 min [16]. For each AIM category (exhibition of axial, limb, oral-lingual, and locomotor movements), a severity score of 0–4 was assigned and summed for each time point. Rats with average AIMs >12 were assigned to the LID group, whereas those with no apparent dyskinesia and average AIM scores ≤12 constituted the NLID group. After the final AIM rating, the rats were euthanized for analyses. The control rats were fed for the same time without treatment and were euthanized.

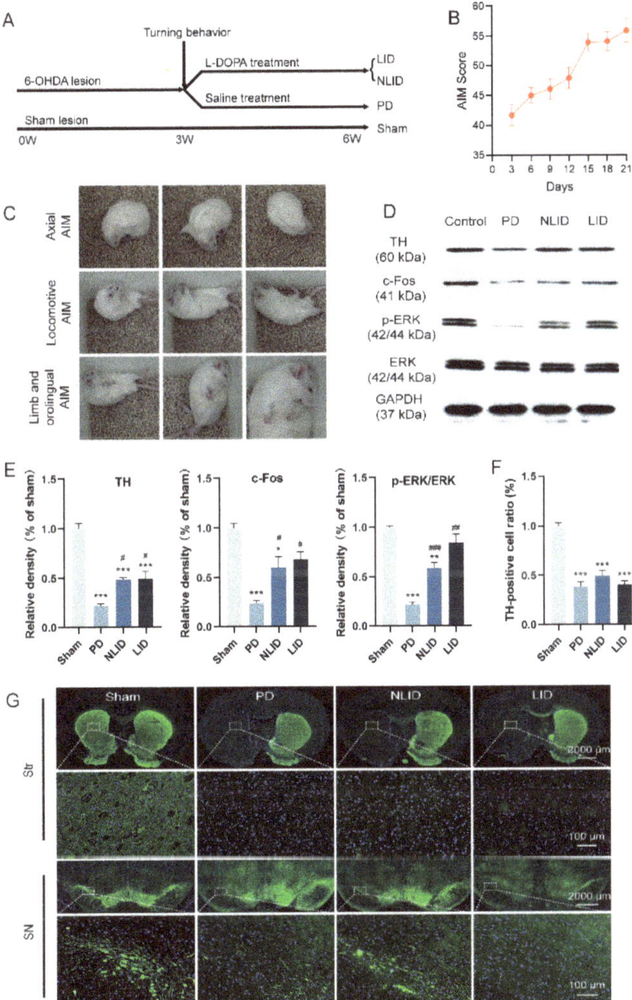

Figure 1. Validation of the levodopa-induced dyskinesia (LID) rat model. (**A**) Experimental timeline. A unilateral Parkinson's disease (PD) rat model was established by 6-OHDA stereotactic intracerebral injection of the right medial forebrain bundle. Contralateral rotation was induced by apomorphine intraperitoneal injection 3 weeks later. The validated PD rats were administered with L-DOPA and benserazide for 3 weeks to induce a rat model of dyskinesia. Involuntary movements (AIMs) were rated every 3 days during the treatment. (**B**) AIMs of LID rats at various times. At 30 min intervals, AIMs were rated for 60 s for each rat for a total of 180 min. For each AIM category, the scores were summed for each time point. Data are shown as mean ± SEM (n = 15). (**C**) The typical AIM of LID

rats. The limb AIM was manifested by tremor of the distal extremity and grasp, and contraction of the shoulder muscles, with hemiballismus. The axial AIM is manifested by torsion of the head, neck, and upper torso, or even loss of balance. The locomotive aim was shown as turning in circles to the opposite side of the injury. (**D**,**E**) Western blot of TH in the striatum and substantia nigra (SN) and c-FOS, p-ERK, and ERK in the striatum of control, PD, LID, and (non-LID) NLID rats (n = 3–5). (**F**) Ratio of TH-positive cells in the brain of control, PD, LID, and NLID rats. TH-positive cells were significantly decreased in the striatum and substantia nigra of PD, LID, and NLID rats. (**G**) Immunostaining of TH in the brain of control, PD, LID, and NLID rats. * $p < 0.05$, ** $p < 0.01$, *** $p < 0.001$ vs. sham group; # $p < 0.05$, ## $p < 0.01$, ### $p < 0.001$ vs. PD group.

2.3. Immunofluorescence Analysis

Rat brains were fixed with 4% paraformaldehyde and embedded with paraffin. Tissue sections (thickness) containing substantia nigra and striatum were incubated overnight at 4 °C with anti-tyrosine hydroxylase (TH) antibody (ab112, 1:700; Abcam, Cambridge, MA, USA). Immunolabeled sections were washed and incubated with goat secondary antibodies conjugated with Alexa Fluor 488 (ab150129, Merck Biosciences, Nottingham, UK). Sections were mounted with medium containing diamidino-2-phenylindole (DAPI) (H1200-10, Vector Laboratories, Burlingame, CA, USA). The images were analyzed using Pannoramic Viewer software (3D HISTECH, Budapest, Hungary).

2.4. Western Blotting

Striatum proteins of the Sham, PD, NLID, and LID rats were extracted using a protein extraction kit (GPP1814; GenePool, Beijing, China). Rabbit polyclonal anti-TH (ab112; 1:200), rabbit polyclonal anti-c-FOS (ab7963; 1:500) (both from Abcam); rabbit monoclonal anti-ERK1/2 (#4695; 1:1000), and rabbit monoclonal anti-p-ERK1/2 (#4377; 1:500) (both from Cell Signaling Technology, Danvers, MA, USA) were used as the primary antibodies. Rabbit monoclonal anti-GAPDH antibody (ab181602; 1:3000; Abcam) was used for the loading control. Protein band density was quantified using the Quantity One software (version 4.6.2; Bio-Rad, Hercules, CA, USA).

2.5. CircRNAs Extraction and Sequencing

Total RNA was extracted from the right striatum of LID and NLID rats using an RNeasy mini kit (Qiagen, Hilden, Germany) according to the manufacturer's instructions. Strand-specific libraries were prepared using the TruSeq Stranded Total RNA Sample Preparation kit (Illumina, San Diego, CA, USA). Qubit 2.0 fluorometry (Life Technologies, Carlsbad, CA, USA) was used to quantify the purified libraries. An Agilent 2100 bioanalyzer (Agilent Technologies, Santa Clara, CA, USA) was used to confirm the insert size and calculate the molar concentration. The library was diluted to 10 pM and then sequenced on the Illumina HiSeq X-ten system. Library construction and sequencing were performed by Shanghai Biotechnology Corp. (Shanghai, China).

2.6. Differential Expression Analysis of circRNAs

Clean reads were obtained by filtering-out rRNA reads, adapters, short fragments, and other low-quality reads from raw reads using Seqtk. Fragments per kilobase of transcript per million fragments mapped (FPKM) was used as an index to measure the expression levels of transcripts. Q30 was calculated to evaluate sequencing accuracy. Clean reads were compared to the reference genome Rnor 6.0 using BWA-MEM [17]. Circular RNA candidates were predicted by CIRI computational pipelines [18]. Perl scripts were used to classify the predicted circRNAs. Counts of reads mapping across an identified backsplice were normalized by read length and number of reads mapping.

Differentially expressed genes between the LID and NLID groups were identified using edgeR [19]. The significance threshold (p-value) was determined using false discovery rate (FDR). The fold change was calculated according to the spliced reads per billion mapping value. Differentially expressed genes were filtered by the criteria of $p \leq 0.05$ and fold change ≥2. The parental gene was obtained according to the position information of circRNAs.

2.7. GO and KEGG Analyses

Gene Ontology (GO) annotation and Kyoto Encyclopedia of Genes and Genomes (KEGG) pathway analysis were performed on the parental genes of differentially expressed circRNAs. Analysis of GO terms enrichment was performed using clusterProfiler [20]. The p-value and FDR of each function were calculated by Fisher's exact test and multiple comparison test to screen out the significant function represented by different genes. The selection criteria for significant GO were $p < 0.05$. The KEGG pathways were assessed using KOBAS software [21].

2.8. qRT-PCR

Total RNA was extracted using the Ultrapure RNA Kit (CWbio, Beijing, China) and the remaining genomic DNA was digested using the DNase I Kit (CWbio). The RNA samples were reverse-transcribed into cDNA by using the HiFi-MMLV cDNA First Strand Synthesis Kit (CWbio) according to the manufacturer's instructions. The qRT-PCR was performed with UltraSYBR Mixture (CWbio). The sequences of primers were listed in Table 1. Each sample was run in triplicate. GAPDH was used as a reference and the relative expression levels were calculated with the $2^{-\Delta\Delta Ct}$ method.

Table 1. Primer sequences used for qRT-PCR.

ID		Sequence of Primers
rno-Rsf1_0012	Forward	5′-GCCTTCCGAATCACCCAGAA-3′
	Reverse	5′-GAATCCATTGACCGCTCATCAG-3′
rno-Rims2_0060	Forward	5′-GGCTCACAAGACAGGATTCTATT-3′
	Reverse	5′-GCTTTCTGTCTGAAGGCATGT-3′
rno-N4bp1_0001	Forward	5′-GCCATTACGAGTACATCAAAGGG-3′
	Reverse	5′-AACACAGAGGTCAGCACAAGTA-3′
rno-Ick_0003	Forward	5′-AAGGACTGGCGTTCATTCACA-3′
	Reverse	5′-GATGGCAGCACCAGCACAA-3′
rno-Stk39_0001	Forward	5′-GCTCTTCTCTGCTGGCTTGG-3′
	Reverse	5′-GGCTTACCTTGGCTTTCTGGAA-3′
rno-Ell2_0005	Forward	5′-GGTGGGTGCTTGTTAAGTATATTAC-3′
	Reverse	5′-GCTGCTTGATCTTCTGATATTCTTG-3′
PCP4	Forward	5′-CTCACTGCCAGAGGAGGAATG-3′
	Reverse	5′-AATTCTTCTTGGACCTTCTTCTGC-3′
TBP	Forward	5′-CTTCAGTCCAATGATGCCTTACG-3′
	Reverse	5′-CTGCTGCTGCTGCTGTCTT-3′

2.9. Competing Endogenous RNAs Network

The CeRNA (competing endogenous RNA) network was constructed based on the relationships between circRNAs, miRNAs, and mRNAs. The miRNA binding sites on the circRNAs and target genes of miRNAs were analyzed using miRanda and TargetScan systems [22]. The potential target mRNAs were predicted by TargetScan and miRDB (http://www.mirdb.org/ accessed on 1 May 2022) [23]. Cytoscape (version 3.8.2; www.cytoscape.org, accessed on 1 May 2022) was used to build the network.

2.10. Fluorescence In Situ Hybridization (FISH)

FISH was performed to detect the subcellular location of rno-Rsf1_0012. The brain tissue was incubated in the fixative for 12 h, then dehydrated by gradient alcohol, followed by paraffin treatment and embedding. The paraffin sections (thickness) were sliced and incubated in a 62 °C oven for 2 h. Xylene and ethanol were used for dewaxing and dehydration, respectively. The slices were boiled in the retrieval solution for 10–15 min and cooled to room temperature. The tissue was digested with proteinase K (20 μg/mL) at 37 °C for 20 min. Endogenous peroxidase was blocked with 3% methanol-H_2O_2. After prehybridization, a rno-Rsf1_0012 probe hybridize solution with a concentration of 1 μM was added to each section and the sections were incubated in a humidity chamber and hybridized overnight at 42 °C. Blocking solution (rabbit serum) was added to the section after removing the hybridization solution, then anti-DIG-HRP was added and incubated at 37 °C for 40 min. TSA chromogenic reagent was added to the labeled tissue and reacted in the dark for 5 min at room temperature. Cell nuclei were stained with DAPI for 8 min in the dark. The slides were observed with an Eclipse Ci (Nikon, Tokyo, Japan). The sequence of the rno-Rsf1_0012 probe was 5′-DIG-GCC TTT GGG TTT TAC TAG TTC TGG GTG ATT CG-DIG-3′.

2.11. Dual-Luciferase Reporter Assay

The interaction among circRNAs of interest and the predicted miRNA was confirmed using the dual-luciferase reporter assay. The mutant sequence fragments were assembled. To confirm the target binding, the wild-type sequence fragments of rno-Rsf1_0012 (wt) and the mutant sequence fragments of rno-Rsf1_0012 (mut) containing the estimated binding position were inserted into the pGL4.74 vector. The vectors and rno-miR-298-5p or mimics-NC were co-transfected into HEK293 cells. A dual-luciferase reporter gene assay kit (Beyotime Biotechnology, Shanghai, China) was used to detect the luciferase activity.

2.12. Statistical Analysis

Statistical analyses were performed using Prism 9 software (GraphPad, La Jolla, CA, USA). Data were compared by Student's *t*-test (two groups) or by one-way ANOVA analysis of variance, followed by appropriate multiple comparisons tests (more than two groups). Data are expressed as the mean ± SEM.

3. Results

3.1. Validation of the LID Rat Models

SD rats were treated with apomorphine 3 weeks after surgery. Rats showing more than seven contralateral rotations/min were regarded as successful PD models. In this study, 43 SD rats were surgically treated, and 34 of them eventually became the PD model rats. The success of PD modeling was 79.1%. After chronic L-DOPA administration, 20 PD rats developed dyskinesia behavior and were assigned to the LID group (Figure 1C). Within 3 weeks, the AIM score of LID rats increased gradually (Figure 1B), while the control group treated with saline did not develop dyskinesia.

TH is a rate-limiting enzyme of catecholamine synthesis and a marker of dopaminergic neurons. TH activity, TH synthesis, and TH mRNA are decreased in the striatum of PD patients and animal models. TH immunofluorescence analysis of striatum and substantia nigra indicated that TH was significantly decreased in the ipsilateral side striatum and substantia nigra of PD, LID, and NLID rats, suggesting that dopaminergic neurons in the substantia nigra were lost and dopamine in the striatum was decreased after striatal 6-OHDA injection, while there was no significant change in the contralateral side (Figure 1F,G).

Western blot analysis of TH levels confirmed the results of TH immunohistochemistry; compared with the control group, the protein levels of TH of PD, LID, and NLID rats decreased significantly (Figure 1D,E). Immediate early genes (IEGs) can be activated transiently and rapidly in response to stimuli. IEGs coded proteins, including ΔFosB, FosB, and c-Fos are known as downstream signaling proteins of extracellular signal-regulated kinase (ERK) phosphorylation, which is hyperactivated in LID models and patients [24–27]. Therefore, we further assessed the expressions of c-Fos, phosphorylated (p-) ERK, and total ERK in the striatum of LID rats. As expected, c-Fos and p-ERK protein levels were reduced in PD rats compared with sham controls, while the expression levels of these two proteins in LID and NLID rats were significantly increased compared with PD rats, consistent with our previous report [28] (Figure 1D,E). Taken together, these results indicated that LID rat models were successfully established.

3.2. circRNAs Expression Profiles of LID and NLID Rats

Following 6-OHDA lesioning and L-DOPA administration, total RNA was isolated from striatum samples of three LID rats and three NLID rats for sequencing. The circRNAs and mRNAs that were differently expressed in LID and NLID rats were screened by high-throughput RNA sequencing to figure out the global circRNAs and mRNA landscape. In terms of the percentages of bases, they were evenly distributed in all samples. The percentages of clean reads in all samples were >93%, and the ratio of mapped reads of all samples was >92%. The reads derived from linear RNA and circRNA were unevenly located in all chromosomes, especially on ch1, ch2, and ch3 (Figure S1). The classification of predicted circRNAs indicated that most of them were ecircRNAs (Figure S2). According to the criterion of fold change $\geq$2 and $p < 0.05$, a total of 99 differentially expressed circRNAs (DEcircRNAs) were obtained between the LID and NLID groups, of which 39 were up-regulated and 60 were down-regulated (Table 2; Figure 2A,B).

Table 2. Top 10 up- and down-regulated DEcircRNAs obtained by sequencing.

ID	Type	Fold Change	Regulation
rno-Ick_0003	exon	13.13123	UP
rno-N4bp1_0001	exon	11.10113	UP
rno-Ell2_0005	exon	10.88539	UP
rno-Rims2_0060	exon	9.183134	UP
rno-Stk39_0001	exon	9.045527	UP
rno-Rsf1_0012	exon	8.201328	UP
rno-Chd2_0001	exon	4.98735	UP
rno-Trip12_0024	exon	4.913605	UP
rno-Arl8b_0001	exon	4.636859	UP
rno-Dmd_0004	exon	4.296785	UP
rno-Ralgps2_0004	exon	6.565114	DOWN
rno-Susd1_0002	exon	7.722112	DOWN
circRNA.15164	intergenic region	8.034623	DOWN
rno-Sergef_0005	exon	8.351556	DOWN
circRNA.4818	exon	8.367079	DOWN
rno-Kdm4c_0013	exon	8.40449	DOWN
rno-Pcsk5_0002	exon	8.931029	DOWN
rno-Prex2_0027	exon	9.574243	DOWN
rno-Rps6ka5 0004	exon	10.93379	DOWN
rno-Slc16a10_0001	exon	12.47348	DOWN

3.3. Functional Annotation of the Host Genes of DEcircRNAs

GO enrichment analysis revealed that the host genes of DEcircRNAs mainly participated in protein ubiquitination (GO terms: protein ubiquitination, $p = 2.38 \times 10^{-6}$; protein polyubiquitination, $p = 7.94 \times 10^{-6}$; and regulation of protein ubiquitination, $p = 2.72 \times 10^{-4}$), neuronal morphology (GO terms: neuron projection morphogenesis, $p = 1.01 \times 10^{-3}$; cell projection morphogenesis, $p = 1.69 \times 10^{-3}$; and cell part morphogenesis, $p = 1.99 \times 10^{-3}$), and histone modification (GO terms: regulation of histone modification, $p = 4.26 \times 10^{-5}$; and histone modification, $p = 1.29 \times 10^{-3}$) (Figures 2C and S3). KEGG annotation and enrichment revealed that the host genes of DEcircRNAs mainly participated in pathways of ubiquitin-mediated proteolysis ($p = 3.99 \times 10^{-4}$) (Figures 2D and S4).

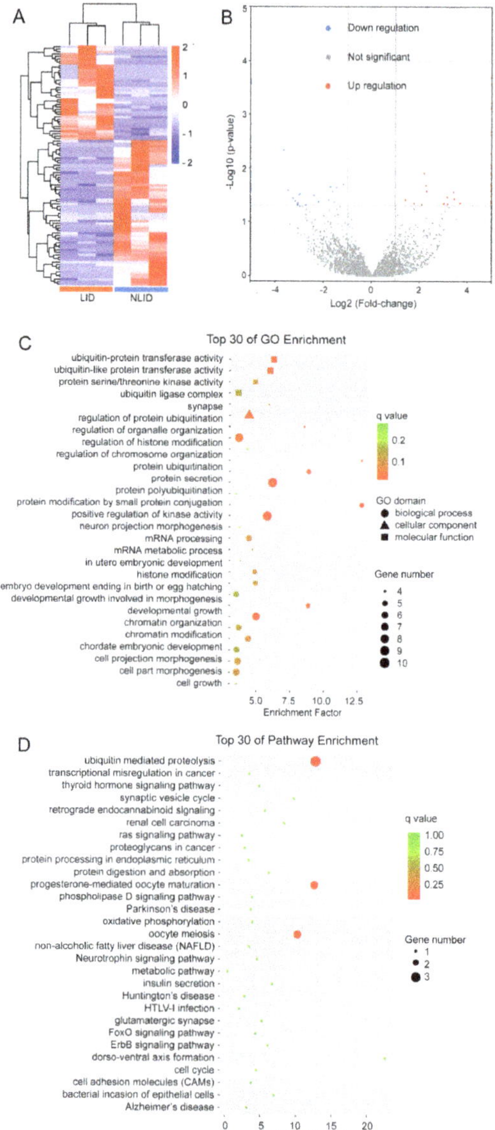

Figure 2. Expression of DEcircRNAs. (**A**) Hierarchical clustering of DEcircRNAs in LID (levodopa-induced dyskinesia) rats compared to non-LID (NLID) rats. The expression values are denoted using a color scale. The intensity increases from red (relatively higher expression) to blue (relatively lower expression). Different columns represent different samples (n = 6), and each row represents a single circRNA. (**B**) Volcano plot of circRNAs. A total of 39 up-regulated and 60 down-regulated circRNAs were screened out. Up-regulated and down-regulated circRNAs are denoted in red and blue, respectively. (**C**) Gene Ontology (GO) enrichment analysis. The abscissa is the rich factor, and the ordinate represents the GO terms. The size of the dot represents the number of genes annotated to the GO term. The color of the dot represents the q value. (**D**) Kyoto Encyclopedia of Genes and Genomes (KEGG) enrichment analysis. The size of the dot represents the number of genes annotated to the KEGG term. The color of the dot represents the q value.

3.4. Rno-Rsf1_0012 Expression Validated by qRT-PCR

Six most significant differential expressed up-regulated circRNAs (rno-Ick_0003, rno-N4bp1_0001, rno-Ell2_0005, rno-Rims2_0060, rno-Stk39_0001, and rno-Rsf1_0012) were selected for validation according to their functional annotations. Their expressions were analyzed by quantitative real-time (qRT-) PCR (Figure 3A). The PCR results showed that rno-Rsf1_0012, rno-N4bp1_0001, rno-Rims2_0060, and rno-Ell2_0005 were significantly higher in the LID group than in the NLID group, while there was no such difference in rno-Stk39_0001 and rno-Ick_0003. Among them, the difference of rno-Rsf1_0012 was the most significant, so subsequent studies focused on it. The mature sequence of rno-Rsf1_0012 was GTA AAA CCC AAA GGC AAA GTT CGA TGG ACT GGC TCT CGG ACA CGT GGC AGG TGG AAA TAC TCC AGC AAT GAT GAG AGC GAA GGG TCC GAG AGT GAC AAA TCC TCT GCC GCC TCG GAA GAG GAG GAA GGA AAG GAG AGT GAA GAA GCA GTC CTT CCA GAT GAC GAT GAA CCC TGC AAA AAG TGT GGC CTT CCG AAT CAC CCA GAA CTA. Fluorescence in situ hybridization (FISH) analysis showed that rno-Rsf1_0012 was mainly localized in the cytoplasm of neurons (Figure 3B).

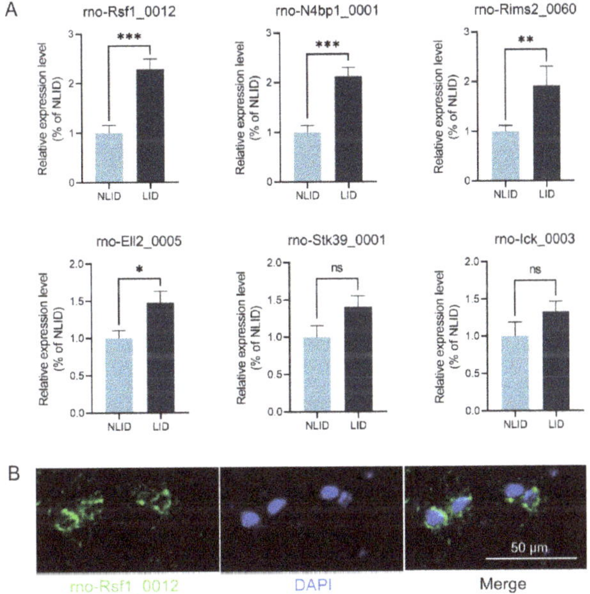

Figure 3. Expression profiles of rno-Rsf1_0012. (**A**) Validation of RNA sequencing results by qRT-PCR of six circRNAs. CircRNA rno-Rsf1_0012, rno-N4bp1_0001, rno-Rims2_0060, and rno-Ell2_0005 were significantly higher in the LID group ($n = 6$) than in the NLID group ($n = 6$). (**B**) Expression location of rno-Rsf1_0012. Immunofluorescence labeling of rno-Rsf1_0012 (green) and neuron markers (blue) in the striatum. * $p < 0.05$, ** $p < 0.01$, *** $p < 0.001$, ns non-significant.

3.5. Rno-Rsf1_0012 Regulates Expression of Target Genes via rno-miR-298-5p

CeRNA mechanisms include the RNA transcript competitively binding miRNA, resulting in diluting the concentrations of free miRNAs in cells, reducing the inhibition of miRNA on coding RNA, and increasing the expressions of target genes [29]. The ceRNA network analysis showed that rno-Rsf1_0012 regulated multiple target genes by sponging miRNAs, including rno-miR-298-5p, rno-miR-503-3p, and rno-miR-668 (Figure 4). Interaction of rno-Rsf1_0012 and rno-miR-298-5p was confirmed using the dual-luciferase reporter assay. Wild-type (WT) and mutant (MT) dual-luciferase reporter vectors of rno-Rsf1_0012 incorporating miRNA binding sites were constructed and co-transfected with miRNA mimics or

NC mimics into HEK 293 cells (Figure 5A). Compared with the NC mimics, rno-miR-298-5p significantly reduced the luciferase activity of the WT reporter, while rno-miR-298-5p did not affect the luciferase activity of the MT reporter (Figure 5B). The expression of two target genes (PCP4 and TBP) was confirmed using qRT-PCR. Compared with the NLID group, the expressions of PCP4 and TBP in the LID group were significantly increased (Figure 5C). These results indicated that rno-Rsf1_0012 may function as a molecular sponge of rno-miR-298-5p using its predicted binding sites.

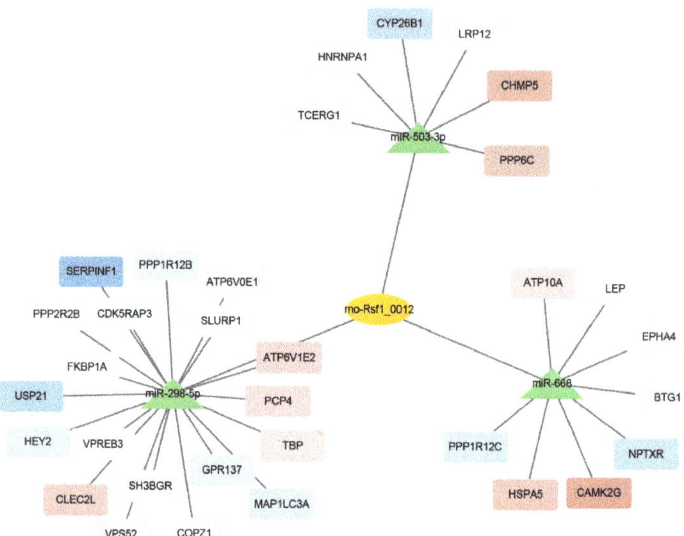

Figure 4. ceRNA network of rno-Rsf1_0012 with mRNAs. CircRNA rno-Rsf1_0012 and binding miRNAs are denoted by orange ellipse and green triangles, respectively. According to the trend of mRNA differential expression, the up-regulated and down-regulated mRNAs are denoted by red and blue rectangles, respectively.

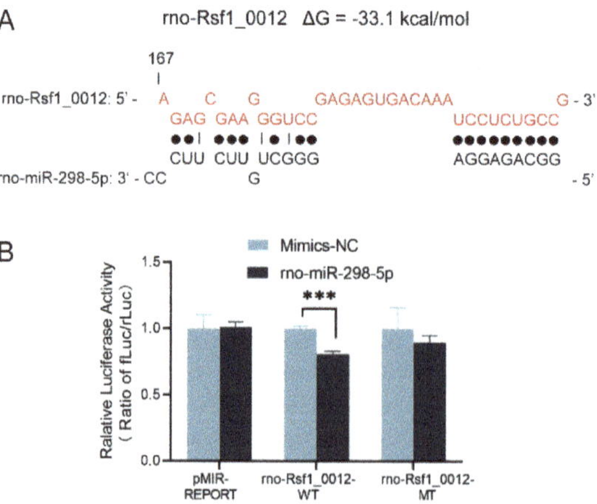

Figure 5. *Cont.*

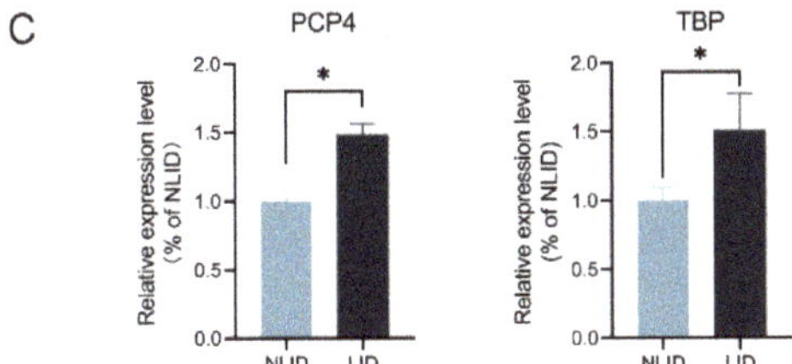

Figure 5. Dual-luciferase reporter assay and target genes expression. (**A**) Wild-type (WT) and mutated-type (MT) sequences of the putative binding sites between rno-Rsf1_0012 and rno-miR-298-5p. (**B**) Dual-luciferase reporter assays were performed to validate the association of rno-Rsf1_0012 and rno-miR-298-5p. Compared with the NC mimics, rno-miR-298-5p significantly reduced the luciferase activity of the WT reporter, while rno-miR-298-5p did not affect the luciferase activity of the MT reporter. (**C**) PCP4 and TBP in LID group were significantly increased in LID group ($n = 6$) compared with NLID group ($n = 6$). * $p < 0.05$, *** $p < 0.001$.

4. Discussion

Here we established a LID rat model using the well-recognized intraperitoneal injection of L-DOPA and benserazide after unilateral 6-OHDA lesioning of the nigrostriatal pathway. Because a subset of SD rats is LID resistant [30,31], we divided rats in our study into the LID and NLID groups according to whether the model was successful. High-throughput sequencing was then used to screen the DEcircRNAs in the striatum. Using bioinformatics approaches, we further narrowed the scope to focus on a subset of circRNAs that may have played a key regulatory role and found rno-Rsf1_0012 may be related to the occurrence and development of LID. This possibility was further verified by PCR results. According to the mechanism of ceRNA, we speculated that rno-Rsf1_0012 might function as miRNA sponges to abrogate the inhibitory impact of rno-miR-298-5p on target genes. The dual-luciferase reporter genes experiment confirmed this idea.

Although no animal model can fully replicate human PD, the unilateral striatum of SD rats damaged by 6-OHDA is one of the most well-studied PD models. The substantia nigra dopaminergic neuron degeneration and loss, glial cell proliferation, substantia nigra and striatum TH activity, and dopamine decrease are similar to that of human PD. Apomorphine-induced rotation behavior can be used to quantify PD behavior and facilitate evaluation by researchers. However, it must be noted that this model belongs to an acute injury model and cannot simulate the characteristics of the chronic progressive course of PD in humans. Human LID can be divided into different subtypes with different clinical manifestations and mechanisms, including peak-dose dyskinesia, diphasic dyskinesia, and "off" dystonia [32]. The widely used LID rat model adopted in this study can simulate the peak-dose dyskinesia well [33,34]. A certain percentage of SD rats showed resistance to L-DOPA during the preparation of the LID rat model [35]. Therefore, studying the differences between LID susceptible and LID resistant SD rats may be important in clarifying the molecular mechanism of LID. Starting from this phenomenon, Manfredsson et al. characterized the key regulatory role of Nurr1 in LID [36]. In the present study, the incidence and severity of AIMs in rats increased with an increase in levodopa dose, which was consistent with the phenomenon reported in LID patients [37].

The details of LID pathogenic mechanisms are not completely understood, similar to PD mechanisms. According to the classic neural circuit model, contrary to the formation of PD, after long-term administration of levodopa, the direct pathway is hyperactive, and the indirect pathway is inhibited. However, this general model cannot explain all the clinical phenomena and experimental results of LID. LID is associated with molecular changes mediated by dopamine D1 receptors in the striatum, including phosphorylation of ERK, MSK1, and histone H3 at the level of the medium spiny neuron of the striatal nigral pathway [26]. Activation of D1/DARPP32 by levodopa induces the translocation of phosphorylated ERK in the nucleus and subsequent activation of MSK1, which plays a key role in regulating synaptic plasticity and transcriptional activity [38].

Functional analysis showed that many host genes of DEcircRNAs such as N4bp1, Ubr5, Klhl2, Rc3h2, Ankib1, Ccnc, Birc6, Trip12, Cdc23, and Kdm2a were involved in ubiquitination. Previous studies have shown that the process of neuronal death in PD is related to ubiquitination, but there is no evidence that ubiquitination is involved in the pathological process of LID [39].

Rno-Rsf1_0012 is located on chr1 and is mainly expressed in brain tissue. There are no reports on the expression difference and functional verification of rno-Rsf1_0012 in PD or LID. Rno-Rsf1_0012 was mainly localized in the cytoplasm of neurons, which indicated that it might act in a ceRNA manner.

CircSNCA can increase SNCA expression by down-regulating miR-7 and inducing apoptosis in PD [40]. a-Synuclein is prone to aggregate protein which forms toxic aggregates and is a major component of Lewy bodies–hallmarks of PD [41]. Kumar et al. reported that circzip-2 may sponge miR-60-3p in the C. elegans model of PD [42]. However, the role of ceRNA in LID has been rarely reported. The interaction between rno-Rsf1_0012 and rno-mir-298-5p in this study was confirmed by dual-luciferase reporter gene assays, but still needs to be verified by functional experiments in vivo.

The target genes of rno-miR-298-5p, including PCP4 and TBP, are associated with some movement abnormality disorders such as Huntington's disease and spinocerebellar ataxia. PCP4 encodes a neuron-specific calmodulin-binding protein and may play a role in the pathophysiology of Huntington's disease and Alzheimer's disease [43]. It is highly and specifically expressed in Purkinje cells altered by spinocerebellar ataxia type 2 progression [44,45]. Polyglutamine expansion in the TBP can cause spinocerebellar ataxia type 17 [46,47], which prompted us to further investigate the relationship between these target genes and LID in future studies.

5. Conclusions

In conclusion, the present study reveals that a set of circRNAs are differentially expressed between LID and NLID rats. Among them, rno-Rsf1_0012 is increased in LID rats and can regulate the expression of target genes by binding rno-miR-298-5p. Rno-Rsf1_0012 may play a vital role in LID occurrence, but its specific mechanism needs to be verified by subsequent function studies.

Supplementary Materials: The following supporting information can be downloaded at: https://www.mdpi.com/article/10.3390/brainsci12091206/s1, Figure S1: Genome distribution map of four samples; Figure S2: The classification of predicted circRNAs; Figure S3: Go functional classification of the host genes of DEcircRNAs; Figure S4: KEGG classification of the host genes of DEcircRNAs.

Author Contributions: Data curation, Q.W., X.Z. and Y.-P.S.; formal analysis, Q.W. and Z.-B.L.; funding acquisition, F.-G.M., C.-L.H., Y.-L.X. and G.-E.C.; investigation, Q.W. and C.L.; methodology, Q.W. and C.-L.H.; project administration, F.-G.M. and C.-L.H.; resources, T.-T.D. and X.Z.; supervision, J.-G.Z. and F.-G.M.; validation, T.-T.D., Y.-L.X. and Y.-P.S.; visualization, Q.W.; writing—original draft preparation, C.-L.H. and Q.W.; writing—review and editing, C.-L.H., T.-T.D., C.-L.H. and Q.W. were major contributors in writing the manuscript. All authors have read and agreed to the published version of the manuscript.

Funding: This research was funded by the Beijing Natural Science Foundation Program and Scientific Research Key Program of the Beijing Municipal Commission of Education (Grant Number: KZ201910025036), the National Natural Science Foundation of China (Grant Number: 81971070, grant number: 81901314), and The Taishan Scholar Project of Shandong Province of China (grant number: tsqn202103200).

Institutional Review Board Statement: Animal experiments were conducted according to the Chinese Animal Welfare Act and Guidance for Animal Experimentation of Capital Medical University. The study protocol was approved by the Ethics Committee of the Beijing Neurosurgical Institute, Capital Medical University (Protocol No.: AEEI-2018-200).

Informed Consent Statement: Not applicable.

Data Availability Statement: The datasets generated and/or analyzed during the current study are not publicly available due we have unpublished studies from this data but are available from the corresponding author on reasonable request.

Conflicts of Interest: The authors declare no conflict of interest.

References

1. McGregor, M.M.; Nelson, A.B. Circuit Mechanisms of Parkinson's Disease. *Neuron* **2019**, *101*, 1042–1056. [CrossRef] [PubMed]
2. Dexter, D.T.; Jenner, P. Parkinson disease: From pathology to molecular disease mechanisms. *Free Radic. Biol. Med.* **2013**, *62*, 132–144. [CrossRef] [PubMed]
3. Ahlskog, J.E.; Muenter, M.D. Frequency of levodopa-related dyskinesias and motor fluctuations as estimated from the cumulative literature. *Mov. Disord.* **2001**, *16*, 448–458. [CrossRef]
4. Scott, N.W.; Macleod, A.D.; Counsell, C.E. Motor complications in an incident Parkinson's disease cohort. *Eur. J. Neurol.* **2016**, *23*, 304–312. [CrossRef] [PubMed]
5. Jenner, P. Preventing and controlling dyskinesia in Parkinson's disease—A view of current knowledge and future opportunities. *Mov. Disord.* **2008**, *23* (Suppl. 3), S585–S598. [CrossRef]
6. Thanvi, B.; Lo, N.; Robinson, T. Levodopa-induced dyskinesia in Parkinson's disease: Clinical features, pathogenesis, prevention and treatment. *Postgrad. Med. J.* **2007**, *83*, 384–388. [CrossRef]
7. Li, X.; Yang, L.; Chen, L.L. The Biogenesis, Functions, and Challenges of Circular RNAs. *Mol. Cell* **2018**, *71*, 428–442. [CrossRef]
8. Kristensen, L.S.; Andersen, M.S.; Stagsted, L.V.W.; Ebbesen, K.K.; Hansen, T.B.; Kjems, J. The biogenesis, biology and characterization of circular RNAs. *Nat. Rev. Genet.* **2019**, *20*, 675–691. [CrossRef]
9. Gasparini, S.; Licursi, V.; Presutti, C.; Mannironi, C. The Secret Garden of Neuronal circRNAs. *Cells* **2020**, *9*, 1815. [CrossRef]
10. Liu, J.; Zhao, K.; Huang, N.; Zhang, N. Circular RNAs and human glioma. *Cancer Biol. Med.* **2019**, *16*, 11–23. [CrossRef]
11. Huang, J.L.; Su, M.; Wu, D.P. Functional roles of circular RNAs in Alzheimer's disease. *Ageing Res. Rev.* **2020**, *60*, 101058. [CrossRef] [PubMed]
12. Hanan, M.; Simchovitz, A.; Yayon, N.; Vaknine, S.; Cohen-Fultheim, R.; Karmon, M.; Madrer, N.; Rohrlich, T.M.; Maman, M.; Bennett, E.R.; et al. A Parkinson's disease CircRNAs Resource reveals a link between circSLC8A1 and oxidative stress. *EMBO Mol. Med.* **2020**, *12*, e11942. [CrossRef]
13. Barros, A.S.; Crispim, R.Y.G.; Cavalcanti, J.U.; Souza, R.B.; Lemos, J.C.; Cristino Filho, G.; Bezerra, M.M.; Pinheiro, T.F.M.; de Vasconcelos, S.M.M.; Macêdo, D.S.; et al. Impact of the Chronic Omega-3 Fatty Acids Supplementation in Hemiparkinsonism Model Induced by 6-Hydroxydopamine in Rats. *Basic Clin. Pharmacol. Toxicol.* **2017**, *120*, 523–531. [CrossRef] [PubMed]
14. Steece-Collier, K.; Stancati, J.A.; Collier, N.J.; Sandoval, I.M.; Mercado, N.M.; Sortwell, C.E.; Collier, T.J.; Manfredsson, F.P. Genetic silencing of striatal CaV1.3 prevents and ameliorates levodopa dyskinesia. *Mov. Disord.* **2019**, *34*, 697–707. [CrossRef]
15. Lundblad, M.; Andersson, M.; Winkler, C.; Kirik, D.; Wierup, N.; Cenci, M.A. Pharmacological validation of behavioural measures of akinesia and dyskinesia in a rat model of Parkinson's disease. *Eur. J. Neurosci.* **2002**, *15*, 120–132. [CrossRef]
16. Li, H. Aligning sequence reads, clone sequences and assembly contigs with BWA-MEM. *arXiv preprint* **2013**, arXiv:1303.3997.
17. Gao, Y.; Wang, J.; Zhao, F. CIRI: An efficient and unbiased algorithm for de novo circular RNA identification. *Genome Biol.* **2015**, *16*, 4. [CrossRef] [PubMed]
18. Robinson, M.D.; McCarthy, D.J.; Smyth, G.K. edgeR: A Bioconductor package for differential expression analysis of digital gene expression data. *Bioinformatics* **2010**, *26*, 139–140. [CrossRef]
19. Kumar, S.; Reddy, P.H. Are circulating microRNAs peripheral biomarkers for Alzheimer's disease? *Biochim. Biophys. Acta* **2016**, *1862*, 1617–1627. [CrossRef]
20. Mao, X.; Cai, T.; Olyarchuk, J.G.; Wei, L. Automated genome annotation and pathway identification using the KEGG Orthology (KO) as a controlled vocabulary. *Bioinformatics* **2005**, *21*, 3787–3793. [CrossRef]
21. Miranda, K.C.; Huynh, T.; Tay, Y.; Ang, Y.S.; Tam, W.L.; Thomson, A.M.; Lim, B.; Rigoutsos, I. A pattern-based method for the identification of MicroRNA binding sites and their corresponding heteroduplexes. *Cell* **2006**, *126*, 1203–1217. [CrossRef] [PubMed]
22. Chen, Y.; Wang, X. miRDB: An online database for prediction of functional microRNA targets. *Nucleic Acids Res.* **2020**, *48*, D127–D131. [CrossRef] [PubMed]
23. Jenner, P. Molecular mechanisms of L-DOPA-induced dyskinesia. *Nat. Rev. Neurosci.* **2008**, *9*, 665–677. [CrossRef] [PubMed]
24. Berton, O.; Guigoni, C.; Li, Q.; Bioulac, B.H.; Aubert, I.; Gross, C.E.; Dileone, R.J.; Nestler, E.J.; Bezard, E. Striatal overexpression of DeltaJunD resets L-DOPA-induced dyskinesia in a primate model of Parkinson disease. *Biol. Psychiatry* **2009**, *66*, 554–561. [CrossRef]
25. Pavon, N.; Martin, A.B.; Mendialdua, A.; Moratalla, R. ERK phosphorylation and FosB expression are associated with L-DOPA-induced dyskinesia in hemiparkinsonian mice. *Biol. Psychiatry* **2006**, *59*, 64–74. [CrossRef]
26. Schuster, S.; Nadjar, A.; Guo, J.T.; Li, Q.; Ittrich, C.; Hengerer, B.; Bezard, E. The 3-hydroxy-3-methylglutaryl-CoA reductase inhibitor lovastatin reduces severity of L-DOPA-induced abnormal involuntary movements in experimental Parkinson's disease. *J. Neurosci.* **2008**, *28*, 4311–4316. [CrossRef]

27. Han, C.L.; Liu, Y.P.; Sui, Y.P.; Chen, N.; Du, T.T.; Jiang, Y.; Guo, C.J.; Wang, K.L.; Wang, Q.; Fan, S.Y.; et al. Integrated transcriptome expression profiling reveals a novel lncRNA associated with L-DOPA-induced dyskinesia in a rat model of Parkinson's disease. *Aging* **2020**, *12*, 718–739. [CrossRef]
28. Tay, Y.; Rinn, J.; Pandolfi, P.P. The multilayered complexity of ceRNA crosstalk and competition. *Nature* **2014**, *505*, 344–352. [CrossRef]
29. Konradi, C.; Westin, J.E.; Carta, M.; Eaton, M.E.; Kuter, K.; Dekundy, A.; Lundblad, M.; Cenci, M.A. Transcriptome analysis in a rat model of L-DOPA-induced dyskinesia. *Neurobiol. Dis.* **2004**, *17*, 219–236. [CrossRef]
30. Zhang, Y.; Meredith, G.E.; Mendoza-Elias, N.; Rademacher, D.J.; Tseng, K.Y.; Steece-Collier, K. Aberrant restoration of spines and their synapses in L-DOPA-induced dyskinesia: Involvement of corticostriatal but not thalamostriatal synapses. *J. Neurosci.* **2013**, *33*, 11655–11667. [CrossRef]
31. Espay, A.J.; Morgante, F.; Merola, A.; Fasano, A.; Marsili, L.; Fox, S.H.; Bezard, E.; Picconi, B.; Calabresi, P.; Lang, A.E. Levodopa-induced dyskinesia in Parkinson disease: Current and evolving concepts. *Ann. Neurol.* **2018**, *84*, 797–811. [CrossRef] [PubMed]
32. Cenci, M.A.; Lee, C.S.; Björklund, A. L-DOPA-induced dyskinesia in the rat is associated with striatal overexpression of prodynorphin- and glutamic acid decarboxylase mRNA. *Eur. J. Neurosci.* **1998**, *10*, 2694–2706. [CrossRef] [PubMed]
33. Lee, C.S.; Cenci, M.A.; Schulzer, M.; Björklund, A. Embryonic ventral mesencephalic grafts improve levodopa-induced dyskinesia in a rat model of Parkinson's disease. *Brain* **2000**, *123 Pt 7*, 1365–1379. [CrossRef]
34. Steece-Collier, K.; Collier, T.J.; Lipton, J.W.; Stancati, J.A.; Winn, M.E.; Cole-Strauss, A.; Sellnow, R.; Conti, M.M.; Mercado, N.M.; Nillni, E.A.; et al. Striatal Nurr1, but not FosB expression links a levodopa-induced dyskinesia phenotype to genotype in Fisher 344 vs. Lewis hemiparkinsonian rats. *Exp. Neurol.* **2020**, *330*, 113327. [CrossRef] [PubMed]
35. Sellnow, R.C.; Steece-Collier, K.; Altwal, F.; Sandoval, I.M.; Kordower, J.H.; Collier, T.J.; Sortwell, C.E.; West, A.R.; Manfredsson, F.P. Striatal Nurr1 Facilitates the Dyskinetic State and Exacerbates Levodopa-Induced Dyskinesia in a Rat Model of Parkinson's Disease. *J. Neurosci.* **2020**, *40*, 3675–3691. [CrossRef]
36. Lindgren, H.S.; Rylander, D.; Ohlin, K.E.; Lundblad, M.; Cenci, M.A. The "motor complication syndrome" in rats with 6-OHDA lesions treated chronically with L-DOPA: Relation to dose and route of administration. *Behav. Brain Res.* **2007**, *177*, 150–159. [CrossRef]
37. Santini, E.; Alcacer, C.; Cacciatore, S.; Heiman, M.; Herve, D.; Greengard, P.; Girault, J.A.; Valjent, E.; Fisone, G. L-DOPA activates ERK signaling and phosphorylates histone H3 in the striatonigral medium spiny neurons of hemiparkinsonian mice. *J. Neurochem.* **2009**, *108*, 621–633. [CrossRef]
38. Olanow, C.W. The pathogenesis of cell death in Parkinson's disease–2007. *Mov. Disord.* **2007**, *22* (Suppl. 17), S335–S342. [CrossRef]
39. Sang, Q.; Liu, X.; Wang, L.; Qi, L.; Sun, W.; Wang, W.; Sun, Y.; Zhang, H. CircSNCA downregulation by pramipexole treatment mediates cell apoptosis and autophagy in Parkinson's disease by targeting miR-7. *Aging* **2018**, *10*, 1281–1293. [CrossRef]
40. Kumar, L.; Shamsuzzama; Jadiya, P.; Haque, R.; Shukla, S.; Nazir, A. Functional Characterization of Novel Circular RNA Molecule, circzip-2 and Its Synthesizing Gene zip-2 in C. elegans Model of Parkinson's Disease. *Mol. Neurobiol.* **2018**, *55*, 6914–6926. [CrossRef]
41. Surguchov, A. Intracellular Dynamics of Synucleins: "Here, There and Everywhere". *Int. Rev. Cell Mol. Biol.* **2015**, *320*, 103–169. [CrossRef] [PubMed]
42. Utal, A.K.; Stopka, A.L.; Roy, M.; Coleman, P.D. PEP-19 immunohistochemistry defines the basal ganglia and associated structures in the adult human brain, and is dramatically reduced in Huntington's disease. *Neuroscience* **1998**, *86*, 1055–1063. [CrossRef]
43. Scoles, D.R.; Meera, P.; Schneider, M.D.; Paul, S.; Dansithong, W.; Figueroa, K.P.; Hung, G.; Rigo, F.; Bennett, C.F.; Otis, T.S.; et al. Antisense oligonucleotide therapy for spinocerebellar ataxia type 2. *Nature* **2017**, *544*, 362–366. [CrossRef]
44. Dansithong, W.; Paul, S.; Figueroa, K.P.; Rinehart, M.D.; Wiest, S.; Pflieger, L.T.; Scoles, D.R.; Pulst, S.M. Ataxin-2 regulates RGS8 translation in a new BAC-SCA2 transgenic mouse model. *PLoS Genet.* **2015**, *11*, e1005182. [CrossRef]
45. Liu, Q.; Pan, Y.; Li, X.J.; Li, S. Molecular Mechanisms and Therapeutics for SCA17. *Neurotherapeutics* **2019**, *16*, 1097–1105. [CrossRef] [PubMed]
46. Coutelier, M.; Coarelli, G.; Monin, M.L.; Konop, J.; Davoine, C.S.; Tesson, C.; Valter, R.; Anheim, M.; Behin, A.; Castelnovo, G.; et al. A panel study on patients with dominant cerebellar ataxia highlights the frequency of channelopathies. *Brain* **2017**, *140*, 1579–1594. [CrossRef]
47. Stevanin, G.; Fujigasaki, H.; Lebre, A.S.; Camuzat, A.; Jeannequin, C.; Dode, C.; Takahashi, J.; San, C.; Bellance, R.; Brice, A.; et al. Huntington's disease-like phenotype due to trinucleotide repeat expansions in the TBP and JPH3 genes. *Brain* **2003**, *126*, 1599–1603. [CrossRef]

Communication

Stochasticity, Entropy and Neurodegeneration

Peter K. Panegyres [1,2]

1 Neurodegenerative Disorders Research Pty Ltd., Perth 6005, Australia; research@ndr.org.au
2 School of Medicine, The University of Western Australia, Perth 6009, Australia

Abstract: We previously suggested that stochastic processes are fundamental in the development of sporadic adult onset neurodegenerative disorders. In this study, we develop a theoretical framework to explain stochastic processes at the protein, DNA and RNA levels. We propose that probability determines random sequencing changes, some of which favor neurodegeneration in particular anatomical spaces, and that more than one protein may be affected simultaneously. The stochastic protein changes happen in three-dimensional space and can be considered to be vectors in a space-time continuum, their trajectories and kinetics modified by physiological variables in the manifold of intra- and extra-cellular space. The molecular velocity of these degenerative proteins must obey the second law of thermodynamics, in which entropy is the driver of the inexorable progression of neurodegeneration in the context of the N-body problem of interacting proteins, time-space manifold of protein-protein interactions in phase space, and compounded by the intrinsic disorder of protein-protein networks. This model helps to elucidate the existence of multiple misfolded proteinopathies in adult sporadic neurodegenerative disorders.

Keywords: stochasticity; space-time; entropy; neurodegeneration

Citation: Panegyres, P.K. Stochasticity, Entropy and Neurodegeneration. *Brain Sci.* **2022**, *12*, 226. https://doi.org/10.3390/brainsci12020226

Academic Editors: Christina Piperi, Chiara Villa, Yam Nath Paudel and Andrew Clarkson

Received: 13 January 2022
Accepted: 2 February 2022
Published: 7 February 2022

Publisher's Note: MDPI stays neutral with regard to jurisdictional claims in published maps and institutional affiliations.

Copyright: © 2022 by the author. Licensee MDPI, Basel, Switzerland. This article is an open access article distributed under the terms and conditions of the Creative Commons Attribution (CC BY) license (https://creativecommons.org/licenses/by/4.0/).

1. Introduction

In our previous studies, we postulated that stochastic processes were important in the pathogenesis of sporadic adult onset neurodegenerative disorders [1]. To summarize, random sequence or other changes in proteins, generated at a DNA, RNA or peptide level, provide the kernel that generates, cultivates, and eventually propagates the neurodegenerative process in an appropriate intra- and extracellular milieu. Such considerations might also be relevant to other conditions such as cancer. In this study, we wish to enunciate this emphasizing a theoretical framework that enhances our understanding of the four dimensional nature of the pathophysiological mechanisms and the important role of entropy as an operator of the neurodegenerative process.

To the best of our knowledge, there are no communications that attempt to provide a conceptual mathematical basis for the origin, propagation, and progression of neurodegeneration [2,3]. A single publication deals with computer simulation of stochastic models of polypeptide amyloid-beta in Alzheimer's disease [4]. Stochastic mathematical concepts have been outlined to enhance the design of clinical trials in Alzheimer's disease [5], and to model biological networks [6].

2. Materials and Methods

This communication is theoretically based on our previously published investigations [1,7].

3. Results and Discussion

It is the author's suggestion that the probability that neurodegeneration occurs in a particular anatomical space, $P_{AS}(neurodeg)$, is a function, f, of P_{DNA}, P_{RNA}, and $P_{PROTEIN}$, where:

P_{DNA} expresses the probability that a DNA sequence will modify to advance neurodegeneration,

P_{RNA} expresses the probability that an RNA sequence will change to enable neurodegeneration,

$P_{PROTEIN}$ connotes the probability that a protein sequence, say the Aβ peptide important in Alzheimer's disease, will alter its composition to favor neurodegeneration.

It is possible that more than one of these probabilities may simultaneously be non-zero in the same neuron or other cell types, such as glial cells. It is also possible that each of these probabilities may take on different values in different anatomical brain regions, and that different proteins may be affected simultaneously.

Thus:

$$P_{AS}(neurodeg) = f(P_{DNA}, P_{RNA}, P_{PROTEIN})$$

Such a concept helps to comprehend the regional nature of young onset dementia; for example, frontal variants of Alzheimer's disease versus temporal linguistic forms.

Determining f is the ultimate goal of this stochastic approach. To make any progress, our study needs to mathematically analyze the motion of protein sequences. A protein sequence (PS) will change by stochastic forces and enable neurodegeneration:

$$PS_{PHYSIOL} \rightarrow PS_{NEURODEG}$$

This process occurs in three-dimensional intracellular and extracellular space $[R^3]$.

We will begin by representing a PS as a position vector, $\vec{PS}$, in R^3. The location of the origin and orientation of the axes is arbitrary for this illustration (Figure 1).

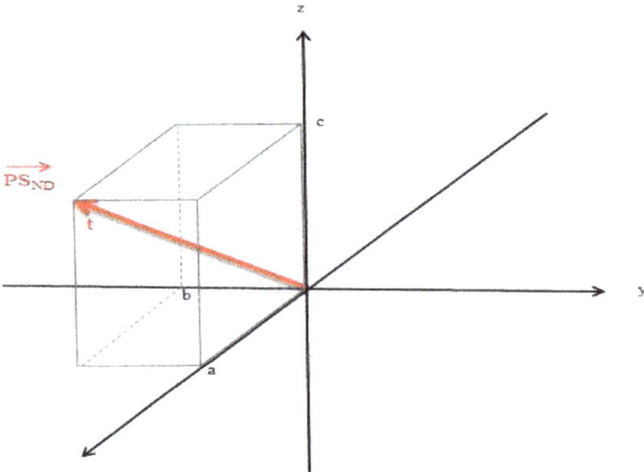

Figure 1. A protein sequence (PS) involved in neurodegeneration represented in three-dimensional space and time.

Figure 1 shows the vector broken down into its components, a, b, and c. This position vector is a function of time (t).

The pathway of single aberrant molecule PS_{ND} will drive the production of other PS_{ND} by aberrant biochemical feedback mechanisms at a DNA, RNA or protein level resulting in an overabundance of PS_{ND}:

$$PS_{ND} \rightarrow PS_{ND1} + PS_{ND2} + PS_{ND3} + \cdots + PS_{NDx}$$

The production of proteins is tightly regulated to maintain the constancy of the milieu interieur. If we take the example of prion protein production $PrP^C \rightarrow PrP^{SC}$, PrP^C is synthesized and post-translationally modified in the endoplasmic reticulum (ER), transported to the cell membrane after modification in the Golgi body. In the ER, the protein undergoes cleavage of the N– and C–terminal signal peptides, followed by the addition of N–linked glycan molecules at two sites, as well as a glycosylphosphatidylinositol

(GPI) anchor, a single disulfide bond is then formed. PrP^C travels to the cell membrane, some PrP^C is internalized into endosomes, most is recycled to the cell membrane; some PrP^C may be released into the extracellular space by cleavage within the GPI anchor. $PrP^C \rightarrow PrP^{SC}$ may occur in cell membranes, endosomes or lysosomes [8]. As this system is tightly regulated, we posit that a single aberrant PrP^{SC}, generated from alpha-helix rich PrP^C, to the β-sheet, is sufficient to disrupt the physiological function of PrP^C, resulting in PrP^{SC} overproduction → aggregation → insolubility → disruption of the control of PrP ssynthesis → overproduction → neurodegeneration, with PrP^{SC} binding PrP^C resulting in further intracellular disruption, and their abnormal conformations → self-propagation → resulting in disease [9]. Similar considerations apply to the proteins involved in Alzheimer's disease, Parkinson's disease, and frontotemporal dementia. Protective mechanisms may prevent this happening to every aberrant molecule.

The passage of our single PS_{ND} molecule will be affected by other PS_{ND} in three-dimensional intracellular and extracellular space, each with its own trajectory (Figure 2).

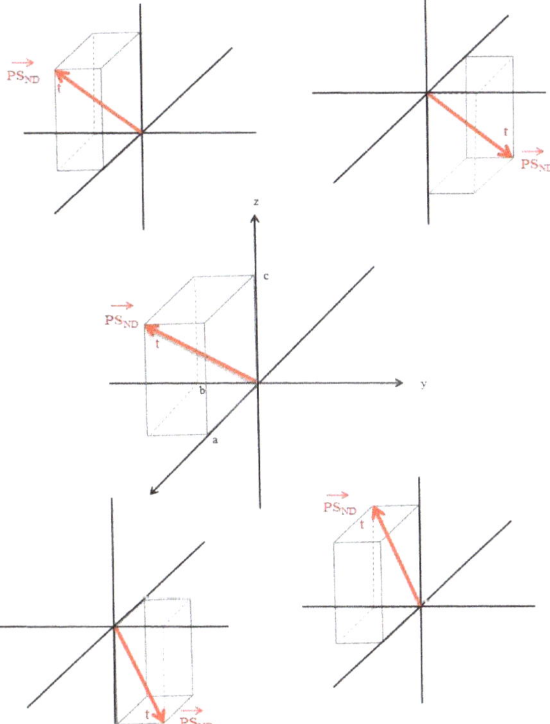

Figure 2. The trajectories of multiple protein sequences (PS) in space and time in neurodegeneration. PS_{ND} = protein sequence, neurodegeneration.

These three-dimension considerations allow us to factor time, anatomical location, biophysical microenvironmental effects—such as pH—atypical protein folding, degradation mechanisms, over-production and cellular mechanisms, such as immunological reactions, and microglial cellular responses that influence normal protein traffic. Furthermore, as the pathological molecules aggregate both intracellularly and extracellularly the probability increases that collisions between the molecules will occur, compromising PS_{ND} movement and flow, disrupting the physiological kinetics of protein intracellular/extracellular motion and advancing neurodegeneration. In the manifold of intracellular and extracellular space, as the physiological manifold disrupts the dimensions influencing PS_{ND} motion, the process moves ever forward towards neurodegeneration.

These considerations of the movement of molecules in R^3 intracellular space raises questions as to the kinetics and molecular velocity distribution of PS_{ND} in cellular and extracellular spaces and is of greater complexity than the concepts developed for kinetic theory of gases, but nevertheless must involve the second law of thermodynamics in which entropy is an intrinsic component of a thermodynamic system—like a cell. Boltzmann developed the H-theorem, in which molecular velocity distribution acts like thermodynamic entropy [10]. Probability and mechanics are important in PS_{ND} action in the manifold of intracellular space and time as developed above.

When systems reach the equilibrium of a normal cell, the interactions between proteins in the cell approach the classical N-body problem, an attempt to predict the movement of one molecule in relation to another. In physiological systems, there are many degrees of freedom and the interactions that determine equilibrium is expressed as:

$$H = \sum H_j$$

H = Hamiltonian = total energy of a system.
H_j = molecular kinetic energies.

This formula applies to states such as gases and solids and has some validity for the complexity of biological systems. The N-body complexity of interacting PS_{ND} may be approached by perturbation theory, which then becomes an extremely difficult problem for the brain cell and its intracellular proteins in dynamic physiological systems and even greater in the pathological state when equilibrium is disrupted. The N–body problem was developed as a means of predicting the individual motions of a group of interacting celestial bodies, where forces such as gravity are operative. In protein and cellular assemblies in structural biology, the Coulomb potential has the same form as the gravitational potential, and the charges may be positive or negative, resulting in repulsive as well as attractive forces which influence molecular interaction and movement in a subcellular and extracellular domain [11]. Perturbation theory, by which a solution to complex problem is approximated by solving a simpler but related problem, helps in this analysis [12].

The degrees of freedom for protein interaction in pathological states are enormous, with $N \rightarrow \infty$. Further complexity is added by the time-space manifold of protein interaction, its dynamic nature and the large value of t, measured in years. Quantum field theory may help to comprehend the interacting fields of proteins, space-time, kinesis, and neurodegeneration.

The Liouville equation, which describes the evolution of density p of the system in phase space, may be written as:

$$i\partial p / \partial t = \hat{L} p$$

where $\hat{L}$ is known as the Liouville operator and i is the unit imaginary number defined by $i^2 = -1$.

Phase space = a space in which all possible states of a system are represented, with each possible state representing one unique point in the phase space (Figure 3, Phase spaces). The Liouville equation is a partial differential equation for the phase space probability distribution function.

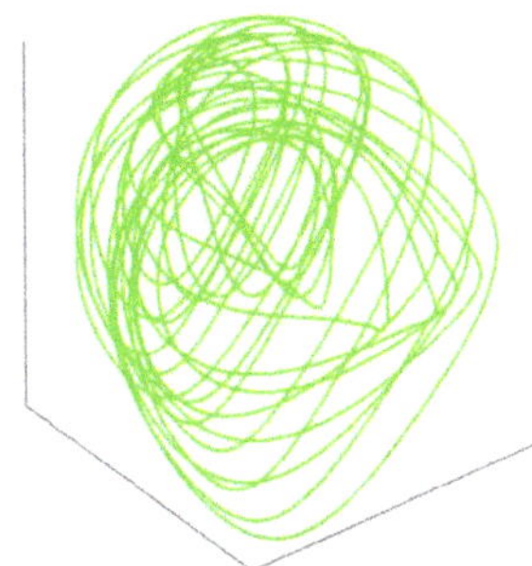

Figure 3. Idealized phase spaces representing all possible states in a neurodegenerative system, each phase state represented by one point of space-time, protein interaction, time, and movement (figure reproduced with permission from ref [13]. Copyright 2011 Elsevier).

Such spaces are of relevance to neurodegenerative processes where $N \to \infty$. In the context of interacting proteins in space and time, which result in a progressive neurodegenerative process, the evolution of the pathology is a result of the dynamics of correlations, influenced by the Liouville operator $\hat{L}$, which describes the time evolution of the phase space distribution function; such wave vectors allow the description of the evolution of molecular neurodegenerative pathophysiology as correlations derived from molecular interactions, i.e., an ensemble theory representing the probability distribution for the state of the system, and represented by the phase distribution function:

$$\rho$$

In large systems such as neurons, in which the molecular pathogenesis evolves, all properties (anatomical location, pH, t, phagosome function, etc.) exist in the limit $N \to \infty$ and result in the inexorable march of the neurodegenerative protein driven process and creates the mechanism of irreversibility. After time, this cascade leads to increasing degrees of freedom, and highly multiple and incoherent correlations, i.e., $PS_{ND} \to$ irreversibility as a result of continuous wave vectors as $V \to \infty$; the neurodegeneration progresses as a consequence of causality conditions as applied to N-body problems. The irreversibility of neurodegenerative molecular protein processes seems inevitable, determined by the nature of its physio-chemical properties, its non-equilibrium nature as the neuropathology progresses, and the role of entropy. The summation of all these factors causes the exponential transformation of the neurodegenerative process resulting in neuronal death, progressive neurological disability, and death of the subject as the pathology spreads through the intercellular space and into other brain cells.

It is posited that the theoretical framework developed so far on the nature of the interactions that influence the neurodegenerative process entropy is its driver. Living matter avoids the inert state of matter by drawing negative entropy from its environment (i.e., energy from food), being a neurone, the brain or the whole human [14]. Entropy is measurable and may be expressed by:

ENTROPY = K log D

K = Boltzmann constant (=3.2983×10^{24} cal/°C)

D = A quantitative measure of the atomistic disorder of the body.

This molecular disorder is at the basis of neurodegeneration as the march of PS_{ND} causes cellular and molecular disarray, increasing randomness of cellular and molecular machinery → breakdown of the neurone → loss of neurones as the molecular pathology spreads through the cell, extracellular space and the central nervous system:

→ Loss of brain structure (e.g., atrophy of the hippocampus in Alzheimer's disease)
→ Generalized cerebral atrophy
→ Dementia
→ Inanition
→ Death and decay (thermodynamic equilibrium).

The human organism transforming from a low entropy state → high entropy state—in both intra- and extra-cellular space.

The natural history of neurodegenerative disorders such as Alzheimer's disease, Parkinson's disease, frontotemporal dementia, Prion diseases, and motor neuron disease are all progressive. As the neurodegenerative process begins, it has an inexorable march toward a state of high entropy equilibrium. Many attempts to hold this process with medications or monoclonal antibodies have not been successful, suggesting that the instigation of the neurodegenerative process is irreversible, propelled by the progression to increased entropy as dictated by the second law of thermodynamics.

Support for the concepts presented here comes from recent clinic-pathological studies which show that patients with a clinical diagnosis of, say, Alzheimer's disease might have, on neuropathological examination, multiple other protein deposits such as α-synuclein and transactive response DNA binding protein 43 [15,16]. These later proteins are associated with Parkinson's disease, multiple system atrophy, frontotemporal dementia, and motor neurone disease. The presence of these other proteins is associated with a more severe clinical presentation and poorer prognosis. To explain, the stochastic forces determine multiple pathological proteins to be deposited in the brain; space-time interactions within cellular and extracellular compartments with molecular collision and entanglement with each other (and subcellular organelles) hastening the pathological process; with this disorder, entropy, increases exponentially, resulting in a more rapid clinical deterioration and poorer survival (Figure 4).

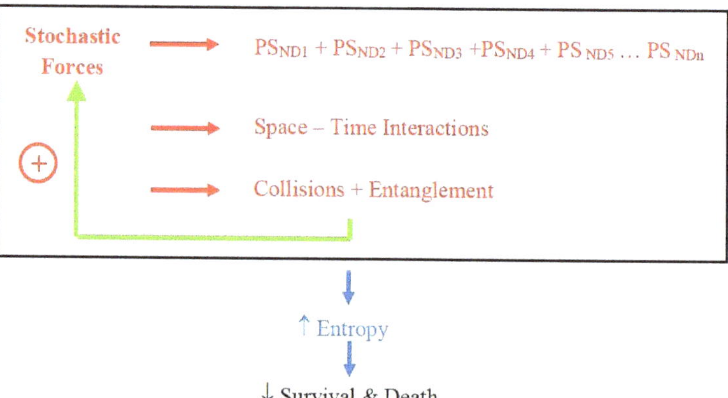

Figure 4. Stochastic mechanisms determine neurodegenerative protein sequences that interact in space-time, collide and become entangled with other proteins and structures → entropy increases and neurodegeneration ensues.

These considerations strongly suggest that prevention should be the optimum management strategy before the precise mechanisms enunciated in this paper are elucidated, including prevention of head injuries (even in children), a healthy diet, avoiding smoking and alcohol, good cardiovascular health—especially blood pressure management, regular physical exercise, cognitive and social engagement, and healthy sleep. Our findings submit that future research directed to understanding the stochastic basis (i.e., how random sequence changes at a peptide, RNA or DNA level → neurodegeneration) of molecular processes, the intra/extracellular kinetics of PS_{ND} in three-dimensional space and countering entropy might yield new treatments, as our current treatments are symptomatic and do not change the natural history.

To this complexity must be added considerations as to the functional organization of the proteome and the intrinsic disorder of proteins. In the brain certain functions are integrated by networks [17] and that brain anatomical networks—the connectome—has topological properties [18]. Graph theory has helped to elucidate the operation of these networks in health and disease [19].

Protein-protein interaction networks control the functions of these large-array networks at a molecular level and are universally fundamental in all neuronal operations including synaptic function, neurogenesis, cell-cell interactions, autophagy and neuronal death [20]. Disturbed protein-protein interaction networks are elemental in neurodegeneration [21,22] This proteome can be measured using mass spectroscopy of biological fluids including cerebral spinal fluid and brain tissue [23]. As a consequence of these, disrupted protein-protein networks protein agglomeration leads to neuronal and glial cells distribution, neurones die, the neuropathology advances through the connectome, further neuronal loss occurs, the clinical state of the patient declines, the neuropathology proceeds through the interactome, and the neurodegeneration promotes advancing dementia, or muscular weakness as in the case of motor neuron disease, causing death.

Furthermore, proteins have been shown to have intrinsic disorder which enables protein-protein networks to have "hub" connectivity, such that these proteins have the ability to bind and interact with a vast number of targets [24]. An analysis of experimental modeling in neurones of the ApoE ε4 allele, a risk factor for sporadic Alzheimer's disease, shows a molecular impression of disrupted gene-protein-protein interaction networks [25]. We propose that in sporadic neurodegenerative disorders—and if we take Alzheimer's disease as an example—the disturbed Aβ network interacts with the Tau network in three-dimensional space where they collide in space-time after a stochastic perturbation. This is conceptualized in Figure 5 where the disordered protein-protein interactive networks massively interact and contact other protein networks thought important in Alzheimer pathology—with other yet-to-be-discovered proteins contributing, possibly from the dark proteome [26] (these are shown unlabeled in Figure 5), i.e., the disrupted protein networks and protein network-protein network collisions compound the atomistic disorder promoting neurodegeneration (Figure 5).

Neurodegenerative disorders are characterized by neuronal loss and protein congregation. The pathological proteins project in the brain in a progressive manner, by seeding mechanisms and intercellular multiplication [27]. The popular processes believed to be operative in all sporadic neurodegenerative processes include neuroinflammation with microglial cells, the brain's endogenous macrophages playing an important role which are activated by toxins, pathogens, peripheral inflammation, age, and chronic stress. Autophagy in which some proteins are broken down: there is macroautophagy—the principal pathway for removing damaged cellular organelles or proteins that are no longer needed; microautophagy—in which lysosomes evaginate the cytoplasm; chaperone—mediated autophagy is a proteolytic pathway that eradicates cytosolic proteins under certain conditions, molecular chaperones such as heat shock proteins stimulate this process. Macroautophagy has specific processes such as mitophagy (removal of mitochondria), lipophagy. and ribophagy. Oxidative stress is a process in which certain molecular species develop free radicals by autooxidation. Aging, toxins, and mitochondrial dysfunction generate reactive oxygen species (ROS), which results in cellular damage [28].

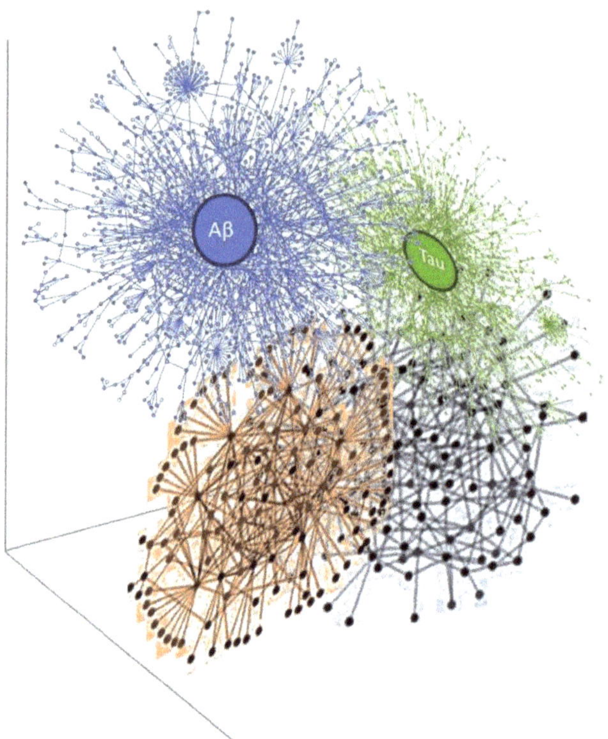

Figure 5. The interaction of protein networks and collisions in Alzheimer's disease, as an example of neurodegeneration, with engagement of known proteins Aβ and tau, and yet-to-be-discovered proteins—unlabeled.

We posit that stochastic mechanisms provide the aberrant protein sequences that stimulate neuroinflammation, distort autophagy, and promote damaging oxidative stress. Furthermore, these abnormal protein sequences perturb the balance between phase separation and irreversible aggregation of proteins. Liquid-liquid phase separation (LLPS) of proteins and nucleic acids drives the formation of membraneless organelles, including neuronal stress granules in the cytoplasm and nucleoli, and paraspeckles in the nucleus [29,30]. These molecular condensates control subcellular chemical reactions and genetic flow of information from the nucleus. We posit that stochastically generated protein sequences disrupt these membraneless organelles and that randomly produced protein sequences, with prion-like realms and post-translational exchanges, distort the phase behavior of protein networks, resulting in disease [31].

With these insights, our theory provides a mechanistic insight into the generation of neurodegenerative disorders, through which randomly generated sequences provoke neuroinflammation, autophagy, oxidative stress, and disrupt phase separation of proteins and nucleic acid causing the irreversible aggregation of proteins—with entropy providing the irreversible force resulting in inexorable neurodegeneration (Figure 6).

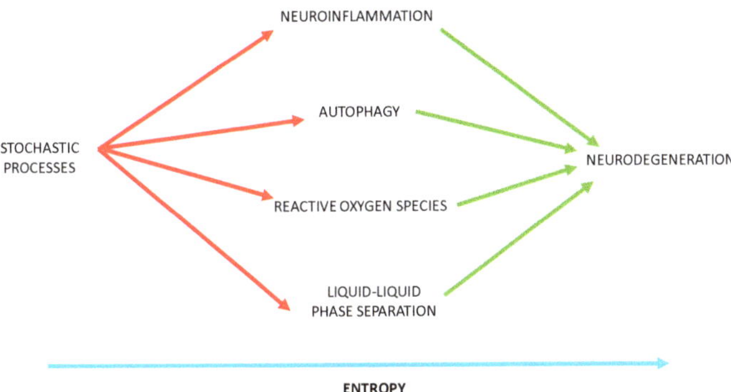

Figure 6. Stochastic processes and the mechanisms of neurodegeneration.

The limitations of this model are to hypothesize how stochastic variation in molecular sequences arises and what promotes their formation other than probability. This question raises the possibility of experimental verification of this model. Furthermore, in what cells—neurones, glia, and microglia—do these stochastic changes originate, and why do stochastic alterations favor certain anatomical locations; e.g., mesial temporal structures in Alzheimer's disease and frontal lobes in frontotemporal dementia? The techniques of single cell proteomics offer an opportunity to examine the proteome of neurons and glial cells in tissue from brain bank collections to elucidate the cellular and anatomical diversity of stochastic protein variation in single cells from patients with Alzheimer's disease, Parkinson's disease, and other neurodegenerative conditions.

Additionally, DNA, RNA, and proteins could be extracted from brain tissue such that the total spectrum of sequence variation could be determined, and these sequences injected into experimental animals to assay relative pathogenicity—those sequences showing the greatest likelihood of promoting disease, allowing the development of molecular inhibitors

Organisms survive on the negative entropy extracted from food. Clinical trials using energy producing molecules such as glucose or treholose, given by infusions, might stay the progression of neurodegeneration.

Our findings support recent studies that suggest Alzheimer's disease clinical symptomatology begins when tau aggregates exist in multiple brain regions; that is, inhibiting the initial stochastic sequence(s) that enhance neurodegeneration before propagation is likely to be an effective treatment and not preventing spread [32].

4. Conclusions

We submit that random changes in the sequences of DNA, RNA, or proteins is the fundamental pathological lesion in sporadic neurodegenerative disorders, and occurs in space-time vectors, with abnormal protein configurations propagating in intracellular and extracellular space leading to disrupted protein networks and cell death with entropy being the operator of relentless clinical deterioration and death. The merits of this concept are to understand that the majority of adult neurodegenerative disorders are sporadic and not related to gene mutations [33]. Our inquiries lead to novel therapeutic approaches into neurodegeneration such as space-time vectors of interacting protein networks and halting entropy's relentless march. Stochasticity is considered a fundamental biological process important in our evolution and driven by entropy [34].

Finally, we propose that probability determines random sequence changes at a DNA, RNA, or protein level that result in pathogenic protein sequences, which misfold and cause neurodegeneration. This probability operates in anatomical space and may be represented by vectors in a phase space, influenced by time. These processes are determined by the second law of thermodynamics in which entropy is the driver, resulting in progressive neurodegeneration compounded by the atomistic disorder of protein network interactions.

Our clinical studies of young onset dementia provide clinical evidence for these stochastic processes [7].

Funding: This research was supported by Neurodegenerative Disorders Research Pty Ltd. No external funding was received.

Institutional Review Board Statement: Not applicable.

Informed Consent Statement: Not applicable.

Data Availability Statement: All data generated or analyzed during this study are included in this published article.

Acknowledgments: I thank Allan Karasavas for his helpful input on the mathematical equations and Allana Gurney for assistance with the illustrations.

Conflicts of Interest: The author declares no conflict of interest.

References

1. Panegyres, P.K. Stochastic considerations into the origins of sporadic adult onset neurodegenerative disorders. *J. Alzheimer's Dis. Parkinsonism* **2019**, *9*, 473.
2. Ravits, J. Focality, stochasticity and neuroanatomic propagation in ALS pathogenesis. *Exp. Neurol.* **2014**, *262*, 121–126. [CrossRef] [PubMed]
3. Colby, D.W.; Cassady, J.P.; Lin, G.C.; Ingram, V.M.; Wittrup, K.D. Stochastic kinetics of intracellular huntingtin aggregate formation. *Nat. Chem. Biol.* **2006**, *2*, 319–323. [CrossRef] [PubMed]
4. Proctor, C.J.; Pienaar, I.S.; Joanna, L.; Elson, J.L.; Kirkwood, T.B.L. Aggregation, impaired degradation and immunization targeting of amyloid-beta dimers in Alzheimer's disease: A stochastic modelling approach. *Mol. Neurodegen.* **2012**, *7*, 32. [CrossRef]
5. Hadjichrysanthou, C.; Ower, A.K.; de Wolf, F.; Anderson, R.M. The development of a stochastic mathematical model of Alzheimer's disease to help improve the design of clinical trials of potential treatments. *PLoS ONE* **2018**, *13*, e0190615. [CrossRef]
6. Altarawni, I.; Samarasinghe, S.; Kulasiri, D. An improved stochastic modelling framework for biological networks. In Proceedings of the MODSIM2019, 23rd International Congress on Modelling and Simulation, Canberra, Australia, 1–6 December 2019; Modelling and Simulation Society of Australia and New Zealand: Canberra, Australia, 2019.
7. Panegyres, P.K. The clinical spectrum of young onset dementia points to its stochastic origins. *J. Alzheimer's Dis. Rep.* **2021**, *5*, 663–679. [CrossRef]
8. Das, A.S.; Zou, W.Q. Prions: Beyond a single protein. *Clin. Microbiol. Rev.* **2016**, *29*, 633–658. [CrossRef]
9. Prusiner, S.B. Biology and genetics of prions causing neurodegeneration. *Annu. Rev. Genet.* **2013**, *47*, 601–623. [CrossRef]
10. Prigogine, I. *Introduction to Non-Equilibrium Thermodynamics*; Wiley-Interscience: New York, NY, USA, 1962.
11. Krumscheid, S. *Benchmark of Fast Coulomb Solvers for Open and Periodic Boundary Conditions*; FZJ-JSC-IB-2010-01; Forschungszentrum Jülich GmbH: Jülich, Germany, 2010.
12. Broer, H.W. Normal forms in perturbation theory. In *Encyclopaedia of Complexity & System Science*; Meyer, R., Ed.; Springer: New York, NY, USA, 2009; pp. 6310–6329. [CrossRef]
13. Paul, S.; Wahi, P.; Verma, M. Bifurcations and Chaos in Large Prandtl-Number Rayleigh-Bénard Convection. *Int. J. Non-Linear Mech.* **2011**, *46*, 772–781. [CrossRef]
14. Schrödinger, E. *What Is Life?* Cambridge University Press: Cambridge, UK, 1943.
15. Karanth, S.; Nelson, P.T.; Katsumata, Y.; Kryscio, R.J.; Schmitt, F.A.; Fardo, D.W.; Cykowski, M.D.; Jicha, G.A.; Van Eldik, L.J.; Abner, E.L. Prevalence and clinical phenotype of quadruple misfolded proteins in older adults. *JAMA Neurol.* **2020**, *77*, 1299–1307. [CrossRef]
16. Robinson, J.L.; Lee, E.B.; Xie, S.X.; Rennert, L.; Suh, E.; Bredenberg, C.; Caswell, C.; Van Deerlin, V.M.; Yan, N.; Yousef, A.; et al. Neurodegenerative disease concomitant proteinopathies are prevalent, age-related and APOE4-associated. *Brain* **2018**, *141*, 2181–2193. [CrossRef] [PubMed]
17. McIntosh, A.R. Towards a network theory of cognition. *Neural Netw.* **2000**, *13*, 861–870. [CrossRef]
18. Papo, D.; Buldú, J.M.; Boccaletti, S.; Bullmore, E.T. Complex network theory and the brain. *Philos. Trans. R. Soc.* **2014**, *369*, 20130520. [CrossRef]
19. Vecchio, F.; Miraglia, F.; Rossini, P.M. Connectome: Graph theory application in functional brain network architecture. *Clin. Neurophys. Prac.* **2017**, *2*, 206–213. [CrossRef] [PubMed]
20. Kitchen, R.; Rozowsky, J.S.; Gerstein, M.B.; Nairn, A.C. Decoding neuroproteomics: Integrating the genome, translatome and functional anatomy. *Nat. Neurosci.* **2014**, *17*, 1491–1499. [CrossRef] [PubMed]
21. Bai, B.; Wang, X.; Li, Y.; Chen, P.C.; Yu, K.; Dey, K.K.; Jarbro, J.M.; Han, X.; Lutz, B.M.; Rao, S.; et al. Deep multilayer brain proteomics identifies molecular networks in Alzheimer's disease progression. *Neuron* **2020**, *105*, 975–991. [CrossRef] [PubMed]

22. Hosp, F.; Gutiérrez-Ángel, S.; Schaefer, M.H.; Cox, J.; Meissner, F.; Hipp, M.S.; Hartl, F.U.; Klein, R.; Dudanova, I.; Man, M. Spatiotemporal proteomic profiling of Huntington's disease inclusions reveals widespread loss of protein function. *Cell Rep.* **2017**, *21*, 2291–2303. [CrossRef]
23. Basu, A.; Ash, P.E.; Wolozin, B.; Emili, A. Protein interaction network biology in neuroscience. *Proteomics* **2021**, *21*, e1900311. [CrossRef]
24. Dunker, A.K.; Cortese, M.S.; Romero, P.; Iakoucheva, L.M.; Uversky, V.N. Flexible nets: The roles of intrinsic disorder in protein interaction networks. *FEBS J.* **2005**, *272*, 5129–5148. [CrossRef]
25. Kim, H.; Yoo, J.; Shin, J.; Chang, Y.; Jung, J.; Jo, D.-G.; Kim, J.; Jang, W.; Lengner, C.J.; Kim, B.-S.; et al. Modelling APOE ε3/4 allele-associated sporadic Alzheimer's disease in an induced neuron. *Brain* **2017**, *140*, 2193–2209. [CrossRef]
26. Kunkle, B.W.; Grenier-Boley, B.; Sims, R.; Bis, J.C.; Damotte, V.; Naj, A.C.; Boland, A.; Vronskaya, M.; van der Lee, S.J.; Amlie-Wolf, A.; et al. Genetic meta-analysis of diagnosed Alzheimer's disease identifies new risk loci and implicates Aβ, tau, immunity and lipid processing. *Nat. Genet.* **2019**, *51*, 414–430. [CrossRef] [PubMed]
27. Kovacs, G.G. Molecular pathology of neurodegenerative diseases: Principles and practice. *J. Clin. Pathol.* **2019**, *72*, 725–735. [CrossRef] [PubMed]
28. Zaman, V.; Shields, D.C.; Shams, R.; Drasites, K.P.; Matzelle, D.; Hague, A.; Banik, N.L. Cellular and molecular pathophysiology in the progression of Parkinson's disease. *Metab. Brain Dis.* **2021**, *36*, 815–827. [CrossRef] [PubMed]
29. Shorter, J. Phase separation of RNA-binding proteins in physiology and disease: An introduction to the JBC Reviews thematic series. *J. Biol. Chem.* **2019**, *294*, 7113–7114. [CrossRef] [PubMed]
30. Jeon, P.; Lee, J.A. Dr Jekyll and Mr Hyde? Physiology and pathology of neuronal stress granules. *Front. Cell Dev. Biol.* **2021**, *9*, 609698. [CrossRef]
31. Webber, C.J.; Lei, S.E.; Wolozin, B. The pathophysiology of neurodegenerative disease: Disturbing the balance between phase separation and irreversible aggregation. *Prog. Mol. Biol. Transl. Sci.* **2020**, *174*, 187–223.
32. Meisl, G.; Hidari, E.; Allinson, K.; Rittman, T.; Devos, S.L.; Sanchez, J.; Xu, C.K.; Duff, K.E.; Rowe, J.B.; Hyman, B.T.; et al. In vivo rate-determining steps of tau seed accumulation in Alzheimer's disease. *Sci. Adv.* **2021**, *7*, eabh1448. [CrossRef]
33. Jarmolowicz, A.I.; Chen, H.Y.; Panegyres, P.K. Patterns of inheritance in early onset dementia: Alzheimer's disease and frontotemporal dementia. *Am. J. Alzheimer's Dis. Other Dement.* **2015**, *30*, 299–306. [CrossRef]
34. Maynard, S.; Fang, E.F.; Scheibye-Knudsen, M.; Croteau, D.L.; Bohr, V.A. DNA damage, DNA repair, aging, and neurodegeneration. *Cold Spring Harb. Perspect. Med.* **2015**, *5*, a025130. [CrossRef]

MDPI
St. Alban-Anlage 66
4052 Basel
Switzerland
Tel. +41 61 683 77 34
Fax +41 61 302 89 18
www.mdpi.com

Brain Sciences Editorial Office
E-mail: brainsci@mdpi.com
www.mdpi.com/journal/brainsci